Lighting for Driving

Roads, Vehicles, Signs, and Signals

Lighting for Driving

Roads, Vehicles, Signs, and Signals

Peter R. Boyce

CRC Press
Taylor & Francis Group
Boca Raton   London   New York

CRC Press is an imprint of the
Taylor & Francis Group, an **informa** business

CRC Press
Taylor & Francis Group
6000 Broken Sound Parkway NW, Suite 300
Boca Raton, FL 33487-2742

© 2009 by Taylor & Francis Group, LLC
CRC Press is an imprint of Taylor & Francis Group, an Informa business

No claim to original U.S. Government works

International Standard Book Number-13: 978-0-8493-8529-2 (Hardcover)

Library of Congress Cataloging-in-Publication Data

Boyce, P. R.
   Lighting for driving : roads, vehicles, signs, and signals / Peter R. Boyce
     p. cm.
   Includes bibliographical references and index.
   ISBN-13: 978-0-8493-8529-2 (alk. paper)
   ISBN-10: 0-8493-8529-6 (alk. paper)
   1. Roads--Lighting. I. Title.

TE228.B66 2008
628.9'5--dc22
                                                    2008026909

Visit the Taylor & Francis Web site at
http://www.taylorandfrancis.com

and the CRC Press Web site at
http://www.crcpress.com

Dedication

In memory of
Robert James Boyce, motor mechanic, 1907–1989

Contents

Preface

I have two reasons for writing this book, one personal and one professional. The personal reason can be dealt with quickly. When I retired from the Lighting Research Center at Rensselaer Polytechnic Institute and returned to the United Kingdom, my daughter said to my wife, "If Dad doesn't have something to do, he will be intolerable." This book is one result of having something to do.

As for the professional reason, this is simply my belief that a review of lighting for driving is necessary and timely. It is necessary to ensure that lighting makes its full contribution to road safety at night. This may not occur with present practices because of the failure to integrate road lighting, vehicle lighting, signs, and signals into a coherent system. For many years, the practitioners of road lighting and vehicle lighting have studiously ignored each other. Further, much of the road lighting literature and much of the vehicle lighting literature are dominated by discussions of technological and financial issues rather than effectiveness. The main concern seems to be to ensure that the lighting meets existing standards rather than asking whether or not existing standards are meaningful. This does not mean that the question of the effectiveness of road and vehicle lighting is totally ignored. Rather, the effectiveness of road lighting, vehicle lighting, and signage is dealt with, at one level, through the efforts of human factors experts and, at another, by epidemiologists. This book attempts to integrate these diverse strands of evidence. The philosophy behind this integration is that the primary role of all forms of lighting for driving is to provide information to the driver about the road ahead and the presence and intentions of other people on and near the road.

As for timeliness, the fact is a wave of new technology is about to hit vehicle lighting and, to a lesser extent, road lighting. The new technology will allow much more variability in light spectrum and light distribution. This book provides some guidance as to how this flexibility might be used to improve lighting for driving. At the same time, road lighting is coming under increasing pressure to justify its existence. An understanding of what road lighting can contribute to road safety at night is necessary to ensure road lighting is used where and when it is of value.

All books have limitations. This book has at least two. The first is that of geography. The standards and technology relevant to lighting for driving that are discussed are those of the United States and the United Kingdom. This is because I have lived and driven extensively in both countries and am familiar with their road systems. But the literature relevant to the effectiveness of road lighting, vehicle lighting, signs, and signals comes from a wider range of countries, notably the United States, the United Kingdom, France, Germany, The Netherlands, Denmark, Sweden, Norway, Finland, Japan, Australia, and New Zealand. These countries have a number of things in common. They have a high number of vehicles relative to the population, extensive road networks, and strict regulation of people, vehicles, and other equipment allowed on the road. I have assumed that results obtained in any one of these countries will be applicable to all countries with similar characteristics. The second limitation relates to the coverage of vehicle lighting. Here, attention is focused on the most common vehicle types: cars, vans, trucks, and motorcycles, and their lighting systems. Instrument panel lighting in all vehicles and special forms of interior lighting, as in buses and ambulances, are not covered.

This book could not have been completed without the help of a number of people. Of particular value have been the contributions of John Bullough, Steve Fotios, Naomi Miller, Sabine Raphael, Peter Raynham, Mick Stevens, and Yutao Zhou. I thank them all for their willingness to assist with this endeavor. Last but not least, I want to record my gratitude to my wife, Susan, for her constant support and for the fact that she has tolerated having her dining table cluttered up with piles of paper for many months, with very little complaint.

Peter R. Boyce
Canterbury, United Kingdom

Acknowledgements

The cooperation of the following individuals and publishers in granting permission for reproduction of copyright material is gratefully acknowledged:

Fabian Stahl for Figure 3.6

Lei Deng for Figures 2.5 and 2.6

McGraw Hill Inc. for Figures 3.2, 3.4, 3.5, and 3.16

Mick Stevens for Figures 4.2, 5.2, 7.4, 10.7, and 13.2

Naomi Miller for Figure 9.7

Society of Automotive Engineers for Figure 6.8

The Chartered Institution of Building Services Engineers for Figures 6.4 and 6.17

The Illuminating Engineering Society of North America for Figures 2.4, 4.1 and 4.4

Translation

George Bernard Shaw described America and England as two nations divided by a common language. This is certainly the case for the words used to describe roads and vehicles. In this book I have used terms from both countries according to my everyday speech. To avoid confusion, the list below offers a translation.

English	American
Amber traffic signal	Yellow traffic signal
Autumn	Fall
Bonnet	Hood
Boot	Trunk
Car park	Parking lot
Hire car	Rental car
Multi-storey car park	Parking garage
Crossroads	Intersection
Dipped beam	Low beam
Junction	Intersection
Level crossing	Grade crossing
Lorry	Heavy truck
Main beam	High beam
Motorway	Freeway
MPV	Minivan
Pavement	Sidewalk
Pedestrian crossing	Crosswalk
Railway	Railroad
Silencer	Muffler
Road surface	Pavement
Road works	Worksite
Roundabout	Rotary or traffic circle
Waistcoat	Vest
Windscreen	Windshield

As well as words, there are also differences in the measures used for different quantities in different countries. For example, both the United States and the United Kingdom use miles per hour for speed, but most of the rest of the world uses kilometres per hour. Conversion factors for some commonly used measures are given below.

1 mile per hour (mph) = 1.6093 kilometres per hour (km/h)
1 kilometre per hour (km/h) = 0.6214 miles per hour (mph)
1 footcandle (Fc) = 10.76 lumens/metre^2 (lx)
1 lumen/metre^2 (lx) = 0.0929 footcandles (Fc)
1 mile = 1.6093 kilometres

1 kilometre = 0.6214 miles
1 metre (m) = 3.281 feet (ft)
1 foot (ft) = 0.305 metres (m)
1 degree = 60 minutes of arc
1 minute of arc = 0.0166 degrees

1 Driving and Accidents

1.1 INTRODUCTION

Lighting as an aid to driving is all around us in the form of road lighting, vehicle lighting, road signs, and traffic signals. But all is not well with lighting as an aid to driving. The design principles of road lighting have changed little since the 1930s, yet vehicle lighting has changed out of all recognition. The contribution of vehicle lighting to visibility is rarely considered by the designers of road lighting, and the contribution of road lighting to visibility is rarely considered by the designers of vehicle lighting. The driver's task has become more difficult as competition for attention has increased. Traffic densities are higher, traffic speeds are faster, sources of information relevant to the driver are more frequent, and sources of distraction are seldom absent. As if this were not enough, there is pressure on road lighting from people concerned with the collateral damage it causes by consuming electricity and by generating light pollution. Finally, the nexus of sensors, high levels of computer power in small packages, and wireless communication offers unheard-of flexibility for road lighting, vehicle lighting, signs, and signals. These possibilities, together with the limitations of current practice, make a review of the contribution of lighting to the safety of drivers and others on and near the road, timely. Such is the purpose of this book.

1.2 DRIVING AS A VISUAL TASK

Driving is a visual task. If you doubt this and are foolish enough to try it, you could attempt to drive with your eyes shut and measure how far you can go without running off the road or into something or somebody. It will not be far. This means that vision is vital to driving, but like almost all so-called visual tasks, the complete task involves a lot more than seeing. Most visual tasks have three components: visual, cognitive, and motor. The visual component refers to the process of extracting information relevant to the performance of the task using the sense of sight. The cognitive component is the process by which sensory stimuli are interpreted and the appropriate action determined. The motor component is the process by which the stimuli are manipulated to extract information and/or the actions decided upon are carried out. Of course, these three components interact to produce a complex pattern between stimulus and response that will be different for different tasks and for different drivers with different levels of experience (CIE 1992a).

As examples of the visual tasks associated with driving, consider your likely responses to the onset of stop lamps on a vehicle ahead and detecting a movement at the edge of the road ahead. Your response to the former will depend on your distance

1

from the vehicle ahead. If you are close behind, the decision is simple, you have to brake immediately. If the distance is large, no response may be necessary immediately other than to pay attention to the vehicle later. The response to detecting a movement at the edge of the road ahead will depend on the nature of the movement. If the movement is recognizable as a pedestrian, the next step is to determine the direction of movement. If the direction is into the road, there are a number of options ranging from nothing through swerving away to an emergency stop, depending on the distance to the pedestrian. If the movement is away from the road, no response is necessary. If the movement detected is a ball bouncing into the road, braking is an appropriate response because experience will tell you that a bouncing ball is likely to be followed by a child, even though the child is not yet visible.

These simple examples are enough to indicate that while vision is a necessary condition for being able to drive, alone it is not sufficient. The quality of driving is also influenced by cognitive skills such as learning, remembering, and decision making, largely derived from experience, and personality variables, such as the threshold for boredom and levels of risk aversion. Further, when driving we receive information through sensors other than vision. When in motion, we receive auditory information from the noise of the vehicle itself and from the environment through which it moves, as well as information about the forces acting on the body obtained from the kinesthetic and vestibular mechanisms. Sometimes, these other senses are the first indication of how the driver should use the sense of vision, e.g., hearing the siren of an emergency vehicle will usually initiate a visual search to locate the vehicle, but most of the time, the senses are mutually supportive. For example, accelerating into a corner will produce changes in the flow of visual information indicating increasing speed and a change of heading (see Figure 3.16); changes in the auditory stimuli produced by the engine indicating an increasing speed; and changes in the kinesthetic and vestibular mechanisms indicating a centrifugal force.

Despite this mutual support between the senses, there can be little doubt that vision is the primary sensory input for driving. In most countries, there are no restrictions on deaf people with regard to driving but there are restrictions on people with limited vision (see Section 12.5). Put succinctly, driving requires the driver to control the velocity, acceleration, and direction of the vehicle by seeing the line of the road ahead and the relative position and movement of other vehicles, people, animals, and objects on and around the road; detecting and understanding information presented on the vehicle's instruments and through signs and signals on the road, as well as keeping a look-out for the unexpected, such as a spilt load from a truck. Further, this has to be done dynamically at a speed that is much faster than that for which the visual system evolved, that is, the maximum speed of a human running (approximately 36 km/h or 22 mph). Such is the nature of driving as a visual task.

1.3 THE ROLE OF LIGHTING IN DRIVING

The role of lighting in driving is to enable the transfer of information, either directly or indirectly. Direct transfer of information occurs when the light source itself conveys the information, as in a traffic signal or a flashing turn lamp on a vehicle.

Indirect transfer of information occurs when the light is used to illuminate a surface that is then searched by the driver for the information it contains. Such surfaces may contain written information, e.g., road signs, or they may be empty but need to be searched for content, e.g., a road surface. For both direct and indirect information to be transferred, the information has to be visible, so the first question that needs to be considered is what makes things visible?

The answer to this question can be divided into three parts: the stimulus the object presents to the visual system, the quality of the retinal image, and the operating state of the visual system. There are three variables that are always relevant to the stimulus the object presents to the visual system: visual size, luminance contrast, and colour difference.

1.3.1 VISUAL SIZE

There are several different ways to express the size of a stimulus presented to the visual system but all of them are angular measures. The visual size of a stimulus for detection is usually given by the solid angle the stimulus subtends at the eye. The solid angle is the quotient of the areal extent of the object and the square of the distance from which it is viewed and is measured in steradians (sr). The larger the solid angle, the easier the stimulus is to detect.

The visual size for resolution is usually given as the angle the critical dimension of the stimulus subtends at the eye. What the critical dimension is depends on the stimulus. For two points, such as headlamps far away, the critical dimension is the separation between them. For two parallel lines such as are used in road markings, it is the separation between the two lines. For a letter C on a road sign, it is the size of the gap that differentiates the letter C from the letter O. The larger is the visual size of detail in a stimulus, the easier it is to resolve that detail.

For complex stimuli, the measure used to express the size of detail is the spatial frequency distribution. Spatial frequency is the reciprocal of the angular subtense of a critical detail, expressed as cycles per degree. Complex stimuli have many spatial frequencies and hence a spatial frequency distribution. The match between the luminance contrast at each spatial frequency of the stimulus and the contrast sensitivity function of the visual system determines if the stimulus will be seen and what detail will be resolved (see Section 3.4.3). Lighting can do little to change the visual size of two-dimensional objects such as lettering on a road sign, but shadows can be used to enhance the effective visual size of three-dimensional objects such as vehicles on the road.

1.3.2 LUMINANCE CONTRAST

The luminance contrast of a stimulus expresses its luminance relative to its immediate background. The higher the luminance contrast, the easier it is to detect the stimulus. There are three different forms of luminance contrast commonly used for uniform luminance targets seen against a uniform luminance background. There is no agreement on how to measure luminance contrast for complex objects when contrast can occur within the target (Peli 1990).

For uniform targets seen against a uniform background, luminance contrast is defined as

$$C = |L_t - L_b| / L_b$$

where C is the luminance contrast, L_t is the luminance of the target (cd/m^2), and L_b is the luminance of the background (cd/m^2). This formula gives luminance contrasts that range from zero to unity for targets that are darker than the background and from zero to infinity for targets that are brighter than the background.

Another form of luminance contrast for a uniform targets seen against a uniform background is defined as

$$C = L_t / L_b$$

where C is the luminance contrast, L_t is the luminance of the target (cd/m^2), and L_b is the luminance of the background (cd/m^2). This formula gives luminance contrasts that can vary from zero when the target has zero luminance, to infinity when the background has zero luminance. It is often used for self-luminous targets.

For targets that have a periodic luminance pattern, e.g., a multiple chevron road sign indicating a sharp bend, the luminance contrast is given by

$$C = (L_{max} - L_{min}) / (L_{max} + L_{min})$$

where C is the luminance contrast, L_{max} is the maximum luminance (cd/m^2), and L_{min} is the minimum luminance (cd/m^2). This formula gives luminance contrasts that range from zero to unity, regardless of the relative luminances of the target and background. It is sometimes called the luminance modulation.

Lighting can change the luminance contrast of a stimulus by producing disability glare in the eye (see Section 6.6).

1.3.3 COLOUR DIFFERENCE

Luminance quantifies the amount of light emitted from a stimulus but ignores the combination of wavelengths making up that light. It is the wavelengths emitted from the stimulus that influence its colour. It is possible to have a stimulus with zero luminance contrast that can still be detected because it differs from its background in colour (Eklund 1999). There is no widely accepted measure of colour difference, although various suggestions have been made (Tansley and Boynton 1978) and a number of metrics could be constructed from the location of the object and the immediate background on the colour planes of one of the CIE colour spaces (see Section 2.4). Colour difference only becomes important for visibility when luminance contrast is low, typically below 0.3. Lighting using light sources with different spectral content may alter the colour difference between the stimulus and its background.

Visual size, luminance contrast, and colour difference can be used to characterize a stimulus but that stimulus cannot be said to have been presented to the visual system until it reaches the retina of the eye. This raises the next factor involved in determining the visibility of an object, the retinal image quality.

1.3.4 RETINAL IMAGE QUALITY

As with all image processing systems, the visual system works best when it is presented with a sharp image. The sharpness of the image of the stimulus can be quantified by its spatial frequency distribution: a sharp image will have high spatial frequency components present, a blurred image will not. The sharpness of the retinal image is determined by the extent to which the atmosphere and materials through which light passes scatters that light and the ability of the visual system to focus the image on the retina. The greater the amount of scatter and the more limited the ability to focus, the poorer will be the retinal image quality. For driving, scattering can be caused by water droplets in the atmosphere and by dirt or mist on the windscreen. Scattering of light also occurs within the eye, increasingly so as we age (see Section 12.4.1). As for the ability to bring the object to focus on the retina, the range of distances over which we can do this becomes more restricted as we age. Lighting can do little to alter any of these factors, although it has been shown that light sources that are rich in the short wavelengths produce smaller pupil sizes for the same luminance than light sources that are deficient in the short wavelengths (Berman et al. 1992). A smaller pupil size produces a better quality retinal image because it implies a greater depth of field and less spherical and chromatic aberration.

The final factor determining whether a given object will be visible is the operating state of the visual system. This is determined by the retinal illuminance.

1.3.5 RETINAL ILLUMINANCE

The illuminance on the retina determines the state of adaptation of the visual system and therefore alters the capabilities of the visual system (see Section 3.3). The retinal illuminance is given by the equation

$$E_r = e_t . \tau . (\cos \theta / k^2)$$

where E_r is the retinal illuminance (lx), τ is the ocular transmittance, θ is the angular displacement of the surface from the line of sight (degrees), k is a constant equal to 15, and e_t is the amount of light entering the eye (trolands). The amount of light entering the eye, measured in trolands, is given by the equation

$$e_t = L . \rho$$

where L is the surface luminance (cd/m²) and ρ is the pupil area (mm²).

The amount of light entering the eye, e_t is often referred to as retinal illumination but it does not take the transmittance of the optic media into account and therefore does not truly represent the luminous flux density on the retina. The amount of light entering the eye is mainly determined by the luminances in the field of view. For drivers in daytime, these luminances will be determined by the extent to which the sky and sun can be seen and the reflectances of the surfaces in the field of view and the illuminances on them. For drivers at night, the relevant luminances are those of the reflecting surfaces illuminated by road lighting or vehicle forward lighting, such as the road, and of self-luminous sources, such as the headlamps of approaching vehicles.

To summarize, it is the interaction between the object to be seen, the background against which it is seen, the lighting of both object and background, the atmosphere and materials through which light passes, and the capabilities of the optics of the eye that determine the retinal image of the stimulus and the operating state of the visual system. These two factors together largely determine the visibility of the object. The visibility of the object influences the driver's response. But, as should be obvious from what was said earlier, on the road, visibility and response are not linked by rods of iron, but rather by bands of spaghetti. The driver's response is also influenced by the decision-making process, which is itself dependent on the driver's knowledge, experience, attention, fatigue, and biochemistry. Visibility is necessary for a response to occur but that response is dependent on many other factors. A timely and correct response to every relevant stimulus is what is needed in order to drive safely. Timely and correct responses also make a valuable contribution to ensuring that traffic flow is smooth rather than turbulent and that the experience of driving is one of comfort rather than stress.

1.4 THE EFFECTIVENESS OF LIGHTING

Given that the fundamental purpose of lighting on vehicles and roads is to enhance the safety of road users by increasing the visibility of the road ahead and objects on and around it, one way to examine the effectiveness of such lighting is to consider the number of accidents that occur by night and day. Different countries have different ways of recording accidents. Not all of them identify the lighting conditions at the time of the accident but some do. One that does is the compilation of crash data derived from the Fatality Analysis Reporting System (FARS) and the National Automatic Sampling System General Estimate System (GES) of the National Highway Traffic Safety Administration (NHTSA 2006a) in the US. Table 1.1 shows the number of road accidents involving fatalities, personal injuries, and property damage alone, occurring in 2005, under different lighting conditions but in good weather conditions, in the 50 states of the US, the District of Columbia, and Puerto Rico (NHTSA 2006b). At first glance, the data in Table 1.1 suggest that current standards of road and vehicle lighting in the US are more than adequate. Fewer fatal accidents occur after dark than in daylight and more fatal accidents occur when vehicle lighting alone is being used (dark) than when road lighting is present as well (dark but lit). However, two others features of Table 1.1 suggest that such a conclusion is premature. First, there are more accidents of all three types during daytime than after dark, yet the visibility provided by daylight is certainly greater than anything provided by vehicle lighting, alone or with road lighting. Second, the numbers of personal injury and property damage only accidents are greater for the dark but lit condition than for the dark condition, which implies that road lighting makes such accidents more likely.

TABLE 1.1

Number of Road Accidents Involving Fatalities, Personal Injury, and Property Damage Only Occurring in Good Weather Conditions, in the Fifty States of the US as well as the District of Columbia and Puerto Rico, under Different Lighting Conditions, in 2005

Nature of Accident	Daylight	Dark but Lit	Dark	Dawn or Dusk
Fatality	17,332	5,455	10,224	1,381
Personal injury	1,118,000	243,000	161,000	54,000
Property damage only	2,577,000	505,000	417,000	130,000

From NHTSA (2006b).

The reason for these unexpected observations is that the number of accidents represents only one part of the quantity needed to evaluate the benefits of vehicle and road lighting. The other part is the level of exposure. The quantity needed to evaluate the benefits of any traffic safety measure is the accident risk, which is the ratio of the outcome of exposure, i.e., an accident, to the level of exposure itself (Yannis et al. 2005). This offers two possible approaches to quantifying the effectiveness of road or vehicle lighting. The simpler is to measure the accident risks, after dark, before and after a change in road or vehicle lighting. The difference in accident risks will establish if the modified lighting is an improvement. This approach has a limitation. It does not identify how large an improvement could be achieved. The second approach addresses this limitation by looking at the change in accident risks for night and day, separately, following a change in the road or vehicle lighting. The accident risk in daytime provides a measure of what might be achieved by lighting. Lamm et al. (1985) report a study using a slightly more elaborate version of the second approach to examine the effectiveness of the lighting of motorways in Germany. Specifically, the research was carried out over a period of nine years between 1972 and 1981 on a stretch of motorway divided into three sections, A, B, and C, of lengths 1.9, 3.7, and 2.3 km (1.2, 2.3, and 1.5 miles), respectively. For the first year all three sections were unlit. For the next 5 years, sections A and B were lit all night but section C remained unlit. For the following three years, the lighting in sections A and B was usually turned off after 10 p.m. while section C remained unlit. Table 1.2 shows the accident risk by day and night in each section, measured as accidents per million vehicle kilometres of travel (Schreuder 1998). The accident risks shown in Table 1.2 reveal a confusing picture. When all three sections of motorway are unlit, the accident risk is higher at night than in the day, everywhere. When road lighting was introduced into sections A and B, there was the hoped-for reduction in accident risk at night. However, there was also a dramatic reduction in accident risk by day in sections A and B, a change that is difficult to relate to the presence of road lighting. When the road lighting on sections A and B is extinguished at 10 p.m., there is a decrease in accident risk for section A by night, but an increase in the corresponding measure for section B.

TABLE 1.2

Accident Risk, Measured as Accidents per Million Vehicle Kilometres, for Three Sections of a Motorway, by Day and Night, Lit and Unlit

Lighting Condition	Accident Risk for Section A, by Day	Accident Risk for Section A, by Night	Accident Risk for Section B, by Day	Accident Risk for Section B, by Night	Accident Risk for Section C, by Day	Accident Risk for Section C, by Night
All sections unlit	2.03	4.28	1.80	3.33	1.41	1.85
Sections A and B lit all night; Section C unlit	0.88	1.97	1.08	1.66	1.13	2.08
Sections A and B lit until 10 p.m. Section C unlit	0.62	1.76	0.90	2.18	0.93	1.89

From Schreuder (1998).

What these changes suggest is that the crude measure of exposure, vehicle kilometres of travel, is not an adequate representation of the difficulties facing a driver on these sections of road and that, over the years, other changes have taken place that affect the probability of an accident, such as changes in road layout, changes in driver demographics, changes in traffic volumes, and so on. The sad fact is that accidents are influenced by many factors other than lighting. The longer the collection of data goes on, the more likely it is that some of these other factors will change, thereby increasing the level of noise in the data.

An alternative to the prolonged before-and-after study of the type undertaken by Lamm et al. (1985) is the relatively rapid collection of accident data from a large number of similar sites with different types and levels of lighting. This approach reduces the likelihood of changes occurring at a site but inevitably ignores the differences between sites, including different levels of exposure. This was the approach used for a study of the effect of road lighting on traffic safety undertaken in the UK (Scott 1980). In this study, photometric measurements were taken of the lighting conditions at up to eighty-nine different sites using a mobile laboratory (Green and Hargroves 1979). The sites were all at least 1 km long with homogeneous lighting conditions and both the lighting and the road features had been unchanged for at least three years. The sites were all two-way urban roads with a 48 km/h (30 mph) speed limit. The photometric measurements were made with the road dry and the accidents considered were only those that occurred when the roads were dry. Multiple regression analysis was used to determine the importance of various characteristics of the lighting on the night/day accident ratio. The average road surface luminance was found to be the best predictor of the effect of the lighting on the night/day accident ratio. Figure 1.1 shows the night/day accident ratios for the sites plotted against the average road surface luminance. The best-fitting exponential curve through the data is shown, the night/day accident ratios

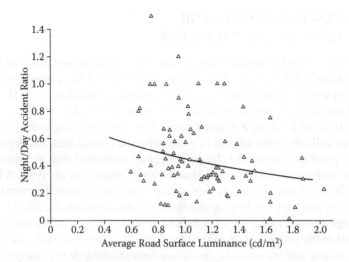

FIGURE 1.1 Night/day accident ratios plotted against average road surface luminance (cd/m²). The curve is the best fitting exponential through the data, after weighting each ratio for the number of accidents to which it relates (after Hargroves and Scott 1979).

being weighted to give greater importance to those sites where accidents occurred most frequently. The equation for the curve is

$$N_R = 0.66 \, e^{-0.42L}$$

where N_R is the night/day accident ratio and L is the average road surface luminance (cd/m²).

It is clear from Figure 1.1 that increasing the average road surface luminance does contribute something to a reduction in accidents at night, but the wide scatter in the individual night/day accident ratios indicates that there are many factors other than the road surface luminance provided by the lighting that matter.

Another method for overcoming the high level of noise evident in Figure 1.1 is meta-analysis. In meta-analysis, the results from a number of independent studies of the same question are combined to increase the statistical power of the analysis. Meta-analysis is useful where the independent studies are of good quality but of small size because combining them then increases the statistical power of the analysis. It can be misleading when the underlying independent studies are of dubious quality. Elvik (1995) carried out a meta-analysis of the safety benefits of introducing road lighting to previously unlit roads. Based on results from 37 studies from different countries in which the effects had been measured in terms of the change in either number of nighttime accidents or number of nighttime accidents per million vehicle kilometres of travel, it was estimated that introducing road lighting should lead to a 65 percent reduction in nighttime fatal accidents, a 30 percent reduction in nighttime injury accidents, and a 15 percent reduction in nighttime property damage accidents.

1.5 THE BENEFITS OF LIGHT FOR DIFFERENT TYPES OF ACCIDENT

The above discussion has served two purposes. It has demonstrated that road lighting does have some value as an accident countermeasure and it has exposed the limitations of simple evaluations based on accident numbers or accident risks. The results discussed above are bedeviled by noise, this being due to the multiple causes of accidents. If we wish to have a clearer picture of the role of lighting in traffic safety, a method that will reduce the amount of noise in the data is required. An elegant solution to this problem is to use the change in lighting associated with the introduction of daylight saving time (Tanner and Harris 1956; Ferguson et al. 1995; Whittaker 1996). In the usual daylight saving time system, the clock is moved forward by one hour in spring and back one hour in autumn. On both occasions, the effect is to suddenly change a period of driving from light to dark or vice versa. If it is assumed that activity and traffic patterns are governed by clock time, then it is likely that levels of exposure, fatigue and intoxication, and driver demographics do not change substantially shortly before and shortly after the daylight saving time changeover, so any difference in accidents can plausibly be ascribed to the change in lighting conditions. Sullivan and Flannagan (2002) used data from the years 1987 to 1997 in the FARS database to determine the total number of fatal collisions involving pedestrians in 46 of the 50 states in the US, for the hour close to the dark limit of civil twilight that showed the greatest change in light level at the daylight saving time change (Arizona, Hawaii, and Indiana were excluded because they do not have daylight saving time, and Alaska was excluded because its solar cycle is markedly different from the other states included). The dark limit of civil twilight is defined as occurring when the centre of the sun is 6 degrees below the horizon. The effect of the daylight saving time change on a spring morning is to move the lighting conditions from twilight to night and then back through twilight to day, as day length increases. Figure 1.2a shows the total number of fatal pedestrian accidents occurring at twilight, for the morning transition, in the 9 weeks before and after the spring daylight saving change. It can be seen that in the weeks before the change, there is a steady decrease in the number of fatal accidents, but at the daylight saving change, there is a rapid return to a high level of accidents, a level that then reduces with the increasing day length. Figure 1.2b shows analogous data for the spring evening twilight, for the 9 weeks before and after the daylight saving change. For the evening, the effect of the daylight saving change is to change driving conditions from night to day. The dramatic decrease in the number of fatal pedestrian accidents with this transition is obvious.

This approach has recently been adopted to examine the effect of the change from light to dark on a number of accident types using two databases (Sullivan and Flannagan, 2007). The first is the FARS database (NHTSA 2006a). The second is the North Carolina Department of Transportation crash dataset (NCDOT). For each dataset, accidents that occurred in the one-hour time window that changed from dark to light or from light to dark in the evening when the spring or autumn daylight savings time change occurred were totaled over several years. The FARS dataset was used to examine fatal accidents of different types over 18 years (1987–2004). The NCDOT dataset was used to examine different types of fatal, personal injury, and

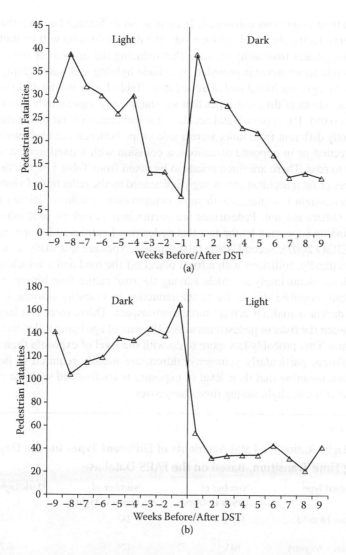

FIGURE 1.2 Cumulative number of pedestrian fatalities in forty-six states of the United States, over the years 1987 to 1997, during twilight, for the 9 weeks before and after the spring daylight saving time (DST) change (a) for morning, (b) for evening (after Sullivan and Flannagan 2002).

property damage only accidents over 9 years (1991–1999). For both databases, the time window for accidents starts at the dark limit of civil twilight based on Standard Time and extends forward by one hour. In spring, this window changes from dark to light following the daylight saving time change. In autumn, this evening window changes from light to dark following the daylight saving time change. Accidents occurring during the evenings of the five weeks either side of the daylight saving time changes were compiled and the ratio of accidents of each type occurring in dark and light conditions calculated. If there is no difference between the number of accidents during dark and light periods, the dark/light ratio will be unity. Dark/light ratios greater than unity indicate that reducing the amount of light available from daylight to whatever is provided by vehicle lighting and road lighting, if present, leads to a greater likelihood of an accident. Table 1.3 shows the dark/light ratios for fatal accidents of different types that are statistically significantly different from unity ($p < 0.05$). The types of fatal accident that had dark/light ratios not statistically significantly different from unity were a side swipe between vehicles moving in the same direction or in opposite directions, a collision with a fixed item, and a collision rear to rear. There are three points to be noted from Table 1.3. The first is that some types of fatal accident are strongly influenced by the reduction of visibility that occurs as daylight is replaced with some combination of vehicle lighting and road lighting. Others are not. Pedestrians are particularly at risk of fatal injuries after dark, a finding supported by the fact that pedestrian fatalities in Europe increase in winter (ERSO 2007). Second, some fatal accident types are less likely to occur after dark, specifically, collisions with a fixed object off the road and a vehicle overturning. Such accidents imply a vehicle leaving the road rather than hitting something on the road. Possibly, this is due to the reduction in visibility associated with the onset of darkness making drivers more circumspect. Third, there is a large difference between the risks of pedestrians under 18 years of age being killed and for adult pedestrians. This probably has more to do with the level of exposure than anything else. Children, particularly younger children, are usually required to be indoors before dark meaning that their level of exposure is confounded with the change in light level at the daylight saving time changeover.

TABLE 1.3

Dark/Light Ratios for Fatal Accidents of Different Types for the Daylight Saving Time Transition, Based on the FARS Database

Accident Type	Number of Accidents In Dark	Number of Accidents in Light	Dark/Light Ratio
Pedestrians 18 to 65 years	1635	243	6.73
Pedestrians > 65 years	845	126	6.71
Animals	61	11	5.55
Rear end collision	440	198	2.22
Head-on collision	1058	748	1.41
Collision with vehicle parked on road	82	58	1.41
Pedestrians <18 years	349	252	1.38
Angle collision	1507	1239	1.22
Miscellaneous	522	460	1.13
Collision with fixed object off road	955	1088	0.88
Overturn	492	691	0.71

From Sullivan and Flannagan (2007).

Unlike the FARS database, the NCDOT database is dominated by non-fatal accidents. In the daylight saving time sample drawn from the NCDOT database, fatal accidents constituted only 0.5 percent of the total. Sixty percent of the accidents were property damage only, the rest being accidents involving personal injuries. Table 1.4 shows the dark/light ratios for what are essentially non-fatal accidents that are statistically significantly different from unity ($p < 0.05$). The non-fatal accident types that had dark/light ratios not statistically significantly different from unity were colliding with elderly pedestrians, a rear-end collision when turning, a collision at an angle, a collision while turning right, a side swipe, a collision with an object in the road, a collision with a fixed object, running off the road to right or left, and a collision while backing.

There are a number of differences between Tables 1.3 and 1.4. Some of these differences are due to the different accident classification systems used in the two databases, but where the same accident type is considered in both databases, there is some consistency. Adult but not elderly pedestrians are at greater risk of both fatal and non-fatal accidents after dark. Both fatal and non-fatal accidents involving animals are more likely after dark. Both fatal and non-fatal rear-end and head-on collisions are more likely after dark. Both fatal and non-fatal accidents involving collision with a parked vehicle are more likely after dark. Both fatal and non-fatal accidents involving a vehicle overturning are less likely after dark.

Of course, there are also some discrepancies. The dark/light ratio for non-fatal accidents involving pedestrians under the age of 18 years is less than unity, while the

TABLE 1.4

Dark/Light Ratios for Non-Fatal Accidents of Different Types for the Daylight Saving Time Transition, Based on the NCDOT Database

Accident Type	Number of Accidents in Dark	Number of Accidents in Light	Dark / Light Ratio
Animals	4,656	560	8.31
Pedestrians 18 to 65 years	292	115	2.54
Ran off road — straight ahead	205	96	2.14
Rear end collision — slow	5,466	3,708	1.47
Left turn	2,265	1,819	1.25
Collision with parked vehicle	894	747	1.20
Head on collision	205	162	1.18
Right turn cross traffic	362	310	1.17
Left turn cross traffic	1,340	1,167	1.15
Pedestrians < 18 years	80	117	0.68
Overturn	52	98	0.53

From Sullivan and Flannagan (2007).

dark/light ratio for fatal accidents is greater than unity. This discrepancy is also probably due to the confounding of children's level of exposure with light level. Another anomaly involves the dark/light ratio for accidents involving animals. The dark/light ratio for non-fatal accidents involving animals is higher than that for fatal accidents. This discrepancy is probably a matter of absolute numbers. The number of accidents associated with animals that prove fatal to humans is small but the number involving personal injury or property damage is large. Small numbers of accidents make the estimation of dark/light ratios uncertain.

Another interesting feature revealed by a comparison of Tables 1.3 and 1.4 is that for the same accident type, the dark/light ratio for fatal accidents is usually larger than for nonfatal accidents. This may be plausibly explained by the fact that fatal accidents often involve higher speeds than nonfatal accidents. Higher speeds allow less time to respond before collision, a time limit that is shortened further by low visibilities. This suggests that better road or vehicle lighting may be of greater importance for fatal accidents than non-fatal accidents because they offer the possibility of increasing the time available for a response.

The data contained in Tables 1.3 and 1.4 are useful for three reasons. First, they indicate that some types of accident are more sensitive to the reduction in visibility that follows the end of the day than others. If it were possible to identify where the accident types most sensitive to poor visibility were likely to happen, it would be possible to use light as an accident countermeasure more effectively. Figure 1.3 shows a slip road from a major road leading to a T-junction notorious for people running off the road straight ahead. The data in Table 1.4 suggest that the installation of a road light on the T-junction would reduce the number of such occurrences after dark. Second, the data in Tables 1.3 and 1.4 indicate that whatever the standards are for vehicle lighting and road lighting in the US, they are capable of improvement. Ideally, vehicle and road lighting should reduce the dark/light ratio to unity. Third, the dark/light ratios can be used to assess the effectiveness of proposed lighting changes. For example, Sullivan and Flannagan (2007) used dark/light ratios for fatal and non-fatal accidents to evaluate the likely effectiveness of several innovative forms of vehicle forward lighting (see Section 6.7.2). For road lighting, the dark/light ratios combined with the frequency and cost of each accident type can be used to provide a monetary value for the benefits of road lighting to set against its undoubted cost.

Of course, the dark/light ratios derived by the daylight saving time changeover method are not without limitations. They are derived from the data of one country. Different dark/light ratios are likely to be found in other countries where different driving habits prevail. What constitutes dark will vary from site to site depending on whether road lighting is installed, so they tell us more about the absence of daylight than the value of different types of vehicle and road lighting. But the main limit is that they are based on drivers who are traveling around dusk. This may exaggerate the role of animals in accidents, because some large animals, such as deer, are crepuscular and so are most active around dusk. It may also show bias because the characteristics of drivers change through the night. Depending on the time of year and the latitude of the country, dusk can range from late afternoon to late evening, clock time. People driving at dusk are much less likely to be intoxicated than those driving late at night (NHTSA 2006b) but they are also more likely

FIGURE 1.3 A slip road off a strategic route with a T-junction at the top that is unlit at night. Drivers regularly crash into the trees on the other side of the T-junction, despite, or maybe because of, the excess of road signs. The dark/light ratios for non-fatal crashes (Table 1.4) suggest that lighting the junction would solve the problem.

to be exposed to higher-density traffic, so in what direction the bias would occur is not at all clear.

Such limitations suggest that the dark/light ratios given in Tables 1.3 and 1.4 should be considered as indicative rather than definitive. Fortunately, the pattern of dark/light ratio values conforms to common sense. The accident types with the highest dark/light ratios are those involving unlighted objects, such as pedestrians and animals, or where objects, which may be lighted or unlighted, appear unexpectedly in the road, or where the road suddenly changes direction. Unlighted objects, such as pedestrians, will have a low visibility after dark compared to lit objects, such as vehicles. Unexpected objects and unexpected road configurations require a response within a limited time. Improving visibility through better road marking, better road lighting, or better vehicle lighting allows more time to make a response. Thus, there can be little doubt that lighting has a role to play in reducing accidents, but before getting too carried away, it would be as well to remember the saying "When all you have is a hammer, everything looks like a nail." Visibility can be improved by means other than lighting. For example, a pedestrian wearing light clothing will be more visible than one wearing dark clothing, and one wearing a fluorescent jacket will be even more so. Although this book is concerned with lighting, care will be taken not to ignore alternative means to achieve the desired end, an enhancement of traffic safety.

1.6 SUMMARY

Driving is a visual task but driving involves a lot more than seeing. Most visual tasks have three components: visual, cognitive, and motor. The visual component is the process of extracting information relevant to the performance of the task using the sense of sight. The cognitive component is the process by which sensory stimuli are interpreted and the appropriate action determined. The motor component is the process by which the stimuli are manipulated to extract information and/or the actions decided upon are carried out. All three components are involved in the process of driving but lighting affects only the visual component.

The role of lighting in driving is to enable the transfer of information from the environment to the human visual system. Lighting achieves this role by making objects visible. Whether or not an object is visible will depend on the stimulus the object presents to the visual system, the quality of the retinal image, and the operating state of the visual system. There are three measures that determine the stimulus: visual size, luminance contrast, and colour difference. The quality of the retinal image is determined by the extent to which light is scattered during its passage to the retina and the ability of the visual system to focus the image on the retina. As for the operating state of the visual system, that is determined by the retinal illuminance. It is the interaction between the object to be seen, the background against which it is seen, the lighting of both object and background, the atmosphere and materials through which light passes, and the capabilities of the optics of the eye that determine the retinal image of the object and the operating state of that system. It is these two factors together that mainly determine the visibility of the object. The visibility of the object influences the driver's response.

Given that the ultimate purpose of lighting on vehicles and roads is to enhance the safety of road users by increasing the visibility of the road ahead and objects on and around it, one way to examine the effectiveness of such lighting is to consider the number of accidents that occur by night and day. This has proved to be more difficult than would be expected. Simply counting the number of accidents occurring by night and day is not enough because it ignores the other side of the equation, the exposure. The quantity needed to evaluate the benefits of any traffic safety measure is the accident risk, which is the ratio of the outcome of exposure, i.e., an accident, to the exposure itself. Even when this is done, there remains a lot of noise in the data. The basic problem is that road accidents have many interacting causes and visibility influences only some of them. Nonetheless, a meta-analysis of multiple studies of the effect of road lighting on accidents has led to the conclusion that introducing road lighting to previously unlit roads should lead to a 65 percent reduction in nighttime fatal accidents, a 30 percent reduction in nighttime injury accidents, and a 15 percent reduction in nighttime property damage accidents.

Of course, these are overall figures and offer little guidance as to where introducing road lighting might be most effectively employed. An alternative approach based on the sudden change in light level at the same clock time that occurs at the daylight saving time change has been used to examine the consequences of reduced visibility. The results indicate that some types of accident are more sensitive to the reduction in visibility that follows the end of the day than others. For example, adult pedestrians are almost seven times more likely to be killed after dark than during daytime, but fatalities associated with overturning the vehicle are less likely after dark. Further, the pattern of sensitivity to reduced visibility conforms to common sense. The accident types with the highest sensitivity to reduced visibility are those involving unlighted objects, such as pedestrians and animals, or where objects appear unexpectedly in the road, or where the road suddenly changes direction. Unlighted objects will have a low visibility after dark compared to lighted objects. Unexpected objects and unexpected road configurations require a response within a limited time. Improving visibility through better road lighting and vehicle lighting allows more time to make a response. There can be little doubt that lighting has a role to play in improving traffic safety through greater visibility.

2 Light

2.1 INTRODUCTION

This book is concerned with how lighting can be used to enhance the safety of drivers and others on and near the roads. To understand the benefits and limitations of all forms of lighting, it is first necessary to understand what light is, how its characteristics can be quantified and how it is produced and controlled. These topics are the subject of this chapter.

2.2 LIGHT AND RADIATION

To the physicist, light is simply part of the electromagnetic spectrum that stretches from cosmic rays with wavelengths of the order of femtometres to radio waves with wavelengths of the order of kilometres (Figure 2.1). What distinguishes the wavelength region between 380 and 780 nm from the rest of the electromagnetic spectrum is the response of the human visual system. Photoreceptors in the human eye absorb energy in this wavelength range and thereby initiate the process of seeing. Other creatures are sensitive to different parts of the electromagnetic spectrum but light is defined by the visual response of humans.

The response of the human visual system is not the same at all wavelengths in the visible range. This makes it impossible to adopt the radiometric quantities conventionally used to measure the characteristics of the electromagnetic spectrum for quantifying light. Rather, a special set of quantities has to be derived from the radiometric quantities by weighting them by the spectral sensitivity of the human visual system.

Unfortunately, a unique spectral sensitivity curve applicable to all people in all conditions does not and cannot exist. This is because the human retina has two classes of visual photoreceptors, one class operating when light is plentiful, the cone photoreceptors, and the other operating when light is very limited, the rod photoreceptors (see section 3.2.2). When only the cone photoreceptors are active, the visual system is said to be in the photopic state and when only the rod photoreceptors are active, it is in the scotopic state. These two photoreceptor types have very different spectral sensitivities. What these spectral sensitivities are has been the subject of international agreement. The body that organizes these agreements is the Commission Internationale de l'Eclairage (CIE). In 1924, the CIE adopted the CIE Standard Photopic Observer, based on the work of Gibson and Tyndell (1923), who took data from several experiments and proposed a smooth and symmetric spectral sensitivity curve (Viikari et al. 2005). The experiments from which the data were taken used small test fields, usually less than two degrees in diameter, and the amount of light was sufficient to put the visual system into the photopic state.

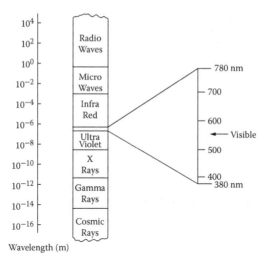

FIGURE 2.1 A schematic diagram of the electromagnetic spectrum showing the location of the visible spectrum. The divisions between the different types of electromagnetic radiation are indicative only.

Later work by Judd (1951) showed that the CIE Standard Photopic Observer was too insensitive at short wavelengths, a result that eventually led the CIE to formally recognize a modified photopic spectral sensitivity curve (CIE 1990a) with greater sensitivity than the CIE Standard Photopic Observer at wavelengths below 460 nm. This CIE Modified Photopic Observer was presented as a supplement to the CIE Standard Photopic Observer not a replacement for it. As a result, the CIE Standard Photopic Observer has continued to be widely used by the lighting industry. This is acceptable because the modified sensitivity at wavelengths below 460 nm has been shown to make little difference to the photometric properties of nominally white light sources that emit radiation over a wide range of wavelengths. It is only for light sources that emit large amounts of radiation below 460 nm that changing from the CIE Standard Photopic Observer to the CIE Modified Photopic Observer can be expected to make a significant difference to measured photometric properties (CIE 1978). Some coloured signals, coloured signs, and narrow-band light sources, such as blue light-emitting diodes, fall into this category.

In 1951, the CIE adopted the CIE Standard Scotopic Observer, based on measurements by Wald (1945) and Crawford (1949) using an area covering the central 20 degrees of the visual field and at a light level low enough to ensure the visual system was in the scotopic state. While this is scientifically interesting because it represents the spectral response of the rod photoreceptors, until recently the CIE Standard Scotopic Observer has rarely been used by the lighting industry because the provision of almost any lighting installation worthy of the name will take the human visual system out of the scotopic state. However, in the last decade, interest in mesopic vision, where both rod and cone photoreceptors are active, has increased the use of the CIE Standard Scotopic Observer (see Section 2.4.3.3.).

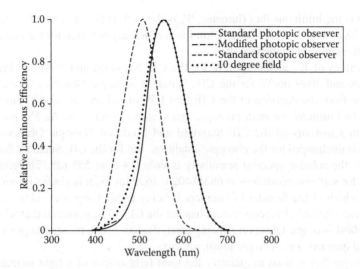

FIGURE 2.2 The relative luminous efficiency functions for the CIE Standard Photopic Observer, the CIE Modified Photopic Observer, the CIE Standard Scotopic Observer, and the relative luminous efficiency function for a 10-degree field of view in photopic conditions.

The CIE Standard and Modified Photopic Observers and the CIE Standard Scotopic Observer are shown in Figure 2.2, the Standard and Modified Photopic Observers having maximum sensitivities at 555 nm and the Standard Scotopic Observer having a maximum sensitivity at 507 nm (CIE 1983, 1990a). These relative spectral sensitivity curves are formally known as the 1924 CIE spectral luminous efficiency function for photopic vision, the CIE 1988 modified 2-degree spectral luminous efficiency function for photopic vision, and the 1951 CIE spectral luminous efficiency function for scotopic vision. More commonly, they are known as the CIE V (λ), CIE $V_M(\lambda)$, and the CIE V'(λ) curves, respectively. These curves are the basis of the conversion from radiometric quantities to photometric quantities, the quantities used to characterize light.

2.3 THE MEASUREMENT OF LIGHT—PHOTOMETRY

2.3.1 DEFINITIONS

The most fundamental measure of the electromagnetic radiation emitted by a source is its radiant flux. This is a measure of the rate of flow of energy emitted and is measured in watts. The most fundamental quantity used to measure light is luminous flux. Luminous flux is radiant flux multiplied, wavelength by wavelength, by the relative spectral sensitivity of the human visual system, over the wavelength range 380 nm to 780 nm. This process can be represented by the equation

$$\Phi = K_m \Sigma\, \Psi_\lambda\, V_\lambda\, \Delta\lambda$$

where Φ is the luminous flux (lumens), Ψ_λ is the radiant flux in a small wavelength interval $\Delta\lambda$ (watts), V_λ is the relative luminous efficiency function for the conditions, and K_m is a constant (lumens/watt).

The values of K_m are 683 lm/W for the CIE Standard and Modified Photopic Observers and 1699 lm/W for the CIE Standard Scotopic Observer. These numbers arise from the decision of the CIE that 1 W of radiant flux at 555 nm should produce 683 lumens, for both photopic and scotopic conditions. As 555 nm is the maximum sensitivity of the CIE Standard and Modified Photopic Observers, the constant is unchanged for the photopic condition. But for the CIE Standard Scotopic Observer, the relative spectral sensitivity is only 0.402 at 555 nm. Therefore, the constant for scotopic conditions is 683/0.402 = 1699 lm/W. It is always important to identify which of the Standard Observers is being used in any particular measurement or calculation. This requirement has led the CIE to recommend that whenever the Standard Scotopic Observer is being used, the word scotopic should precede the measured quantity, i.e., scotopic luminous flux.

Luminous flux is used to quantify the total light output of a light source in all directions. While this is important, for lighting practice it is also important to be able to quantify the luminous flux emitted in a given direction. The measure that quantifies this concept is luminous intensity. Luminous intensity is the luminous flux emitted/unit solid angle, in a specified direction. The unit of measurement is the candela, which is equivalent to one lumen/steradian. Luminous intensity is used to quantify the distribution of light from luminaires such as road lighting lanterns and vehicle headlamps.

Both luminous flux and luminous intensity have area measures associated with them. The luminous flux falling on unit area of a surface is called the illuminance. The unit of measurement of illuminance is the lumen/meter^2 or lux. The luminous intensity emitted per unit projected area of a source in a given direction is the luminance. The unit of measurement of luminance is the candela/meter^2. Table 2.1 summarizes these photometric quantities.

As might be expected, there is a relationship between the amount of light incident on a surface and the amount of light reflected from the same surface. The simplest form of the relationship is quantified by the luminance coefficient. The luminance coefficient is the ratio of the luminance of the surface to the illuminance incident on the surface and has units of candela/lumen. The luminance coefficient of a given surface is dependent on the nature of the surface and the geometry between the lighting, surface, and observer.

There are two other quantities commonly used to express the relationship between the luminance of a surface and the illuminance incident on it. For a perfectly diffusely reflecting surface, the relationship is give by the equation

$$L = E.\rho/\pi$$

where L is the luminance (cd/m^2), E is the illuminance (lm/m^2), and ρ is the reflectance.

TABLE 2.1

The Photometric Quantities

Measure	Definition	Units
Luminous flux	That quantity of radiant flux which expresses its capacity to produce visual sensation	Lumens (lm)
Luminous intensity	The luminous flux emitted in a very narrow cone containing the given direction divided by the solid angle of the cone, i.e., luminous flux/unit solid angle	Candela (cd)
Illuminance*	The luminous flux/unit area at a point on a surface	Lumen/meter²
Luminance	The luminous flux emitted in a given direction divided by the product of the projected area of the source element perpendicular to the direction and the solid angle containing that direction, i.e., luminous intensity/unit area	Candela/meter²
Luminance coefficient	The ratio of the luminance of a surface to the illuminance incident on it	Candela/lumen
Reflectance	The ratio of the luminous flux reflected from a surface to the luminous flux incident on it	
For a diffuse surface	Luminance = (Illuminance × Reflectance)/π	
Luminance factor	The ratio of the luminance of a reflecting surface viewed from a given direction to that of a perfect white uniform diffusing surface identically illuminated	
For a non-diffuse surface, for a specific direction and lighting geometry	Luminance = (Illuminance × Luminance Factor)/π	

* In the US, where imperial units are still used, the unit for illuminance is lumens/ft² or footcandle.

Reflectance is defined as the ratio of reflected luminous flux to incident luminous flux. For a non-diffusely reflecting surface, i.e., a surface with some specularity, the same equation between luminance and illuminance applies, but reflectance is replaced with luminance factor. Luminance factor is defined as the ratio of the luminance of the surface viewed from a specific position and lit in a specified way to the luminance of a diffusely reflecting white surface viewed from the same direction and lit in the same way.

Both illuminance and luminance are widely used in lighting practice to quantify the end result of installing a lighting system and the stimulus to the visual system. Being able to define these quantities is useful but, in addition, it is always helpful to have an idea of what are representative magnitudes for these quantities in different situations. Table 2.2 shows some illuminances and luminances typical of commonly occurring interior and exterior lighting situations, all measured using the CIE Standard Photopic Observer.

2.3.2 SOME LIMITATIONS

Although the photopic photometric quantities defined above can be calculated or measured precisely, it is important to appreciate that they only represent the visual effect of light in a particular state. Specifically, they represent the brightness response of the central two degrees of the retina in high light level conditions with a minimum contribution from the colour vision channels (see Section 3.2.5). Changing field size or light level or using coloured stimuli can change the spectral sensitivity of the visual system.

The effect of field size was recognized by the CIE in 1964 when a provisional relative spectral sensitivity curve for the central 10 degrees of the visual field in photopic conditions was approved (CIE 1986; see Figure 2.2). This curve shows greater sensitivity to short wavelength light than the CIE Standard Photopic Observer because the visual field extends beyond the macula, an area covering the central five degrees of the retina and containing a pigment that attenuates short wavelength light, and into the area where short-wavelength cone photoreceptors are found.

As for the effect of changing light level, in addition to the obvious change from cone-based photopic vision to rod-based scotopic vision there is an intermediate zone where both cones and rods are active, called mesopic vision (see Section 3.3.4). For the central 2 degrees of the retina, the fovea, the CIE Standard Photopic Observer still applies in the mesopic state because there are only medium and long wavelength cones present in the fovea, which is what the CIE Standard Photopic Observer is based on. In the rest of the visual field the spectral sensitivity is in a state of continual change as the balance between rod and cone photoreceptors changes with light level, until either rods dominate, as in scotopic vision, or cones dominate, as in photopic vision. Mesopic vision is important for driving because road lighting usually provides visual conditions that are in the mesopic range. Nonetheless, all the photometric quantities that are used to characterize road lighting and vehicle lighting are

TABLE 2.2

Typical Illuminance and Luminance Values for Various Lighting Situations

Situation	Typical Surface	Illuminance (lm/m²)	Luminance (cd/m²)
Daylight from a clear sky in summer in a temperate zone	Grass	150,000	2,900
Daylight from an overcast sky in summer in a temperate zone	Grass	16,000	300
Electric lighting for textile inspection	Light grey cloth	1,500	140
Electric lighting for office work	White paper	500	120
Electric lighting for heavy engineering work	Steel	300	20
Motorway lighting	Asphalt road	70	1.5
Residential road lighting	Asphalt road	10	0.2
Moonlight	Asphalt road	0.1	0.002

based on the CIE Standard Photopic Observer. This practice can lead to situations where the photometric measurements bear little relation to the visual effect of the light source for off-axis vision. The absence of a CIE mesopic photometry system is not for want of trying (CIE 1989). Indeed, several different systems have been suggested, most based on brightness perception and some weighted combination of photopic and scotopic measurements (Palmer 1968; Ikeda and Shimozono 1981; Trezona 1991; Sagawa and Takeichi 1992). Others have abandoned the perception of brightness as the basis for measuring spectral sensitivity and, using reaction time, have developed a unified system of photometry that covers photopic, mesopic, and scotopic light levels (Rea et al. 2004). Yet others have developed a system of mesopic photometry based on the performance of a number of tasks associated with driving (Eloholma and Halonen 2006; Goodman et al. 2007). The significance of these different approaches will be taken up in Section 4.6.

The presence of a coloured stimulus, such as a traffic signal, is important because such stimuli ensure that the colour vision channels as well as the achromatic channel of the human visual system contribute to the measured spectral sensitivity. At the low luminances typically found on the roads at night, the activity in the colour channels leads to spectral sensitivities that are multi-peaked, with a shift in the wavelength for maximum sensitivity, relative to the conventional CIE Standard Photopic Observer (Varady et al. 2007).

Two other systematic effects that lead to different relative spectral sensitivities from the values represented by the CIE Standard Photopic Observer occur with age and with defective colour vision. As discussed in Section 12.4, as the eye ages, the transmittance of the lens decreases, particularly at the short wavelength end of the visible spectrum. This will lead to a reduced sensitivity in this wavelength region for older people (Sagawa and Takahashi 2001). For people with defective colour vision, either there are missing photopigments or the photopigments are different from the normal. In either case, if the medium or long wavelength cone photoreceptors are involved, these individuals' relative spectral sensitivity is likely to depart from that of the CIE Standard Photopic Observer.

Although there are undoubted limitations, the CIE Standard Observers have definite value. They lead to a system of photometry that provides a globally agreed means for regulators to specify what is required, for the lighting industry to quantify the performance of its products, and for designers to quantify what their lighting systems deliver. Despite the utility of such measures, whenever considering the photometric quantities for a given lighting situation it is always important to ask whether the CIE Standard Observer being used in the calculation of the photometric quantities is appropriate to the situation. If it is not, then the apparent precision of the measurement may be misleading.

2.4 THE MEASUREMENT OF LIGHT—COLOURIMETRY

The photometric quantities described above do not take into account the wavelength combination of the light received at the eye. Thus it is possible for two luminous fields to have the same luminance but to be made up of totally different combinations of wavelengths. In this situation, and provided either photopic or mesopic conditions prevail, the two fields will look different in colour. Exactly what colour will be seen depends not only on the spectral distribution of the radiation incident on the retina but also on such factors as the luminance and colour of the surroundings and the state of adaptation of the observer (Purves and Beau Lotto 2003). Colour is a perception developed in the brain from past experience and the information contained in the retinal image. Light itself is not coloured. Nonetheless, to have a means of characterizing the colour perception associated with different light sources and other stimuli to the visual system, some way had to be found to provide quantitative measures of colour. The CIE colourimetry system provides such measures.

2.4.1 The CIE Colourimetry System

The basis of the CIE colourimetry system is colour matching, an activity in which an observer is asked to determine whether two fields are the same or different in colour. From extensive colour matching measurements, the CIE Colour Matching Functions have been determined. These functions are essentially the relative spectral sensitivity curves of human observers with normal colour vision and can be considered as another form of standard observer. There are three colour matching functions, as might be expected from the fact that humans with normal colour vision can match any colour of light with a combination of not more than three wavelengths of light from the long, medium, and short wavelength regions of the visible electromagnetic spectrum.

The colour of a light source can be represented mathematically by multiplying the spectral power distribution of the light source, wavelength by wavelength, by each of the three colour matching functions $x(\lambda)$, $y(\lambda)$, and $z(\lambda)$, the outcome being the amounts of three imaginary primary colours X, Y, and Z required to match the light source colour. In the form of equations, X, Y, and Z are given by

$$X = h\Sigma\ S(\lambda).x(\lambda).\Delta\lambda$$

$$Y = h\Sigma\ S(\lambda).y(\lambda).\Delta\lambda$$

$$Z = h\Sigma\ S(\lambda).z(\lambda).\Delta\lambda$$

where $S(\lambda)$ is the spectral radiant flux of the light source (W/nm), $x(\lambda)$, $y(\lambda)$, $z(\lambda)$ are the spectral tristimulus values from the appropriate colour matching function, $\Delta\lambda$ is the wavelength interval (nm), and h is an arbitrary constant.

If only relative values of the X, Y, and Z are required, an appropriate value of h is one that makes Y equal to 100. If absolute values of the X, Y, and Z are required it is convenient to take h equal to 683 since then the value of Y is the luminous flux in lumens. If the colour being calculated is for light reflected from a surface or transmitted through a material, the spectral reflectance or spectral transmittance is included as a multiplier in the above equations. For a reflecting surface, an appropriate value of h is one that makes Y equal to 100 for a reference white because then the actual value of Y is the percentage reflectance of the surface.

Having obtained the X, Y, and Z values, the next step is to express their individual values as proportions of their sum, i.e.,

$$x = X / (X + Y + Z) \quad y = Y / (X + Y + Z) \quad z = Z / (X + Y + Z)$$

The values x, y, and z are known as the CIE chromaticity coordinates. As x + y + z = 1, only two of the coordinates are required to define the chromaticity of a colour. By convention, the x and y coordinates are used. Given that a colour can be represented by two coordinates, then all colours can be represented on a two dimensional surface. Figure 2.3 shows the CIE 1931 chromaticity diagram, the two axes being the x and y chromaticity coordinates. It is possible to identify a number of interesting features on the CIE 1931 chromaticity diagram. The outer curved boundary is called the spectrum locus. All pure colours, i.e., those that consist of a single wavelength, lie on this curve. The straight line joining the ends of the spectrum locus is the purple boundary and is the locus of the most saturated purples obtainable. At the centre of the diagram is a point called the equal energy point (x = 0.33, y = 0.33). This is the point where a colourless surface will be located. Close to the equal energy point is a curve called the Planckian locus. This curve passes through the chromaticity coordinates of objects that operate as a black body, i.e., the spectral power distribution of the light source is determined solely by its temperature.

The CIE 1931 chromaticity diagram can be considered as a map of the relative location of colours. The saturation of a colour increases as the chromaticity coordinates get closer to the spectrum locus and further from the equal energy point. The hue of the colour is determined by the direction in which the chromaticity coordinates move from the equal energy point. Strictly, any discussion as to how a specific combination of wavelengths will appear, based on the chromaticity diagram, is nonsense. The only thing that a set of chromaticity coordinates tells us about a

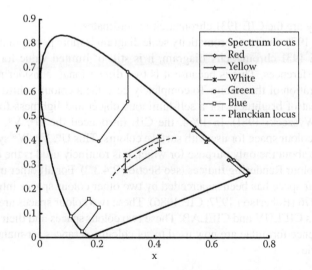

FIGURE 2.3 The CIE 1931 chromaticity diagram showing the spectrum locus and purple boundary, the Planckian locus from 2,500 K to 30,000 K, and the areas defining the chromaticity coordinates suitable for red, yellow, white, green, and blue signal colours (after CIE 1994a).

colour is that colours with the same chromaticity coordinates will match. They tell us nothing about the appearance of the matched colours. But this is an argument for colour vision zealots. The fact is, a red surface lit by a nominally white light source will always plot in the bottom right corner of the diagram and a green in the upper-most part of the diagram. Thus, although the CIE 1931 chromaticity diagram is not theoretically pure, it is useful for indicating approximately how a colour will appear, a value recognized by the CIE when it specified chromaticity coordinate limits for traffic signals and traffic sign surfaces so that they will be recognized as red, yellow, white, green, or blue (CIE 1994a).

Given that different colours plot at different positions on the CIE 1931 chromatic-ity diagram, it would seem reasonable to expect that the distance between two sets of chromaticity coordinates would be correlated with how different the two colours represented by the chromaticity coordinates appear. While this is approximately true, the correlation is very low. This is because the CIE 1931 chromaticity diagram is perceptually non-uniform. Green colours cover a large area while red colours are compressed in the bottom right corner (Figure 2.3). This perceptual non-uniformity makes any attempt to quantify large colour differences using the CIE 1931 chroma-ticity diagram futile. In an attempt to improve this situation, in 1976 the CIE recom-mended the use of the CIE 1976 uniform chromaticity scale (UCS) diagram. This diagram is simply a linear transformation of the CIE 1931 chromaticity diagram. The axes for the CIE 1976 UCS diagram are

$$u' = 4x / (-2x + 12y + 3)$$

$$v' = 9y / (-2x + 12y + 3)$$

where x and y are the CIE 1931 chromaticity coordinates.

While the 1976 uniform chromaticity scale diagram is more perceptually uniform than the CIE 1931 chromaticity diagram, it is still of limited value for determin-ing colour differences. This is because it is two-dimensional, considering only the hue and saturation of the colour. To completely describe a colour a third dimension is needed, that of brightness for a self-luminous object and lightness for a reflect-ing object (Wyszecki 1981). In 1964, the CIE introduced the U*, V*, W* three-dimensional colour space for use with surface colours. This U*, V*, W* system is lit-tle used now; about the only purpose for which it is routinely used is the calculation of the CIE Colour Rendering Indices (see Section 2.4.3.2). For all other uses the U*, V*, W* colour space has been superceded by two other colour spaces introduced by the CIE in 1976 (Robertson 1977; CIE 1986). These two colour spaces are known by the initialisms CIELUV and CIELAB. These two colour spaces and their associated colour difference formulae are now used to set colour tolerances for manufacture in many industries.

2.4.2 Colour Order Systems

While the CIE colourimetry system is valuable for quantifying colours, it does lack a physical presence. This need is met by a variety of colour ordering systems. A colour ordering system is a physical, three-dimensional representation of colour space. In a sense it is an atlas of colours and like an atlas, the separation between adjacent colours is intended to be uniform in all directions. There are several different colour ordering systems used in different parts of the world (Billmeyer 1987). One of the most widely used is the Munsell system, the organization of which is shown in Figure 2.4. The azimuthal hue dimension consists of 100 steps arranged around a circle, with five principal hues (red, yellow, green, blue, and purple) and five intermediate hues (yellow-red, green-yellow, blue-green, purple-blue, and red-purple). The vertical value scale contains ten steps from black to white. The horizontal chroma scale contains up to twenty steps from gray to highly saturated. Each of the three scales is designed to provide equal steps of perception for an observer with normal colour vision looking at the samples lit by daylight, with a gray or white surround. The position of any colour in the Munsell system is identified by an alphanumeric reference made up of three terms, hue, value, and chroma, e.g., a strong red is given the alphanumeric 7.5R/4/12. Achromatic surfaces, i.e., colours that lie along the vertical value axis and hence have no hue or chroma, are coded as neutral 1, neutral 2, etc., depending on their reflectance. The utility of a colour ordering system is that it makes colours manifest and hence makes it easy to communicate about colour in a more precise way than words permit.

The existence of several different colour atlas systems used in different parts of the world, as well as the quantitative CIE colourimetry system, would seem to be a recipe for confusion. Fortunately, this is usually avoided by the fact that conversions are available between many of the colour ordering systems and the CIE colourimetry system. For example, the German DIN system provides both Munsell and CIE equivalents of its components (Richter and Witt 1986). The name categories of

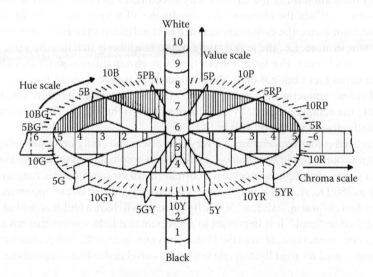

FIGURE 2.4   The organization of the Munsell colour order system. The hue letters are R = red, YR = yellow/red, Y = yellow, GY = green/yellow, G = green, BG = blue/green, B = blue, PB = purple/blue, P = purple, RP = red/purple (from IESNA 2000).

the Inter-Society Colour Council–National Bureau of Standards Method (Kelly and Judd 1965) are given in terms of the Munsell system (National Bureau of Standards 1976). Conversions between the Munsell system and the CIE colourimetry system are given in the American Society for Testing and Materials Test Method D1535 (ASTM 1996), based on Nickerson (1957).

2.4.3 Application Metrics

While the CIE colourimetry system is the most complete and most widely accepted means of quantifying colour, it is undeniably complex. Therefore, the lighting industry has used the CIE colourimetry system to derive two single-number metrics to characterize the colour properties of light sources: correlated colour temperature (CCT) and the CIE General Colour Rendering Index (CRI). Correlated colour temperature is a metric quantifying the colour appearance of the light emitted by a light source. The CIE General Colour Rendering Index is a metric quantifying the effect a light source has on the appearance of surface colours.

2.4.3.1 Correlated Colour Temperature

In principle, the colour of the light emitted by a light source can be characterized by its chromaticity coordinates. This approach is used for signal lights (Figure 2.3), but for applications where colour is less critical and nominally white the correlated colour temperature is used. The basis of this measure is the fact that the spectral emission of a black body is defined by Planck's Radiation Law and hence is a function of temperature only. Figure 2.3 shows the Planckian locus for a limited temperature range. The locus is the curved line in the middle of the CIE 1931 chromaticity diagram joining the chromaticity coordinates of black bodies at different temperatures. When the chromaticity coordinates of a light source lie directly on the Planckian locus, the colour appearance of that light source is expressed by the colour temperature, i.e., the temperature of the black body that has the same chromaticity coordinates. For light sources that have chromaticity coordinates close to the Planckian locus but not on it, their colour appearance is quantified as the correlated colour temperature, i.e., the temperature of the iso-temperature line that is closest to the actual chromaticity coordinates of the light source. The temperatures are usually given in degrees Kelvin (K).

Correlated colour temperature is a very convenient and easily understandable metric of light source colour appearance, applicable to nominally white light sources. As a rough guide, such light sources have correlated colour temperatures ranging from 2700 K to 7500 K. A 2700 K light source will have a yellowish colour appearance and be described as "warm," while a 7500 K light source will have a bluish appearance and be described as "cool." It is important to appreciate that light sources that have chromaticity coordinates distant from the Planckian locus, such as the low-pressure sodium light source used for road lighting, do not have a correlated colour temperature.

2.4.3.2 CIE Colour Rendering Index

The effect a given light source will have on the appearance of a surface colour can be estimated by calculating the chromaticity coordinates of the reflected radiation.

This is reasonable if a specific set of surface colours is of interest, as is the case for road sign colours, but for most lighting applications, where many different but unspecified colours are used, more general advice is desirable. This is where the CIE colour rendering index (CRI) comes in. The CIE colour rendering index measures how well a given light source renders a set of standard test colours relative to their rendering under a reference light source of the same correlated colour temperature as the light source of interest (CIE 1995a). The reference light source used is an incandescent light source for light sources with a correlated colour temperature below 5000 K and some form of daylight for light sources with correlated colour temperature above 5000 K. The actual calculation involves obtaining the positions of a surface colour in the CIE 1964 U^*, V^*, W^* colour space under the reference light source and under the light source of interest and expressing the difference between the two positions on a scale that gives perfect agreement between the two positions a value of 100. The CIE has fourteen standard test colours. The first eight form a set of pastel colours arranged around the hue circle. Test colours nine to fourteen represent colours of special significance, such as skin tones and vegetation. The result of the calculation for any single colour is called the CIE special colour rendering index, for that colour. The average of the special colour rendering indices for the first eight test colours is called the CIE general colour rendering index. It is this latter index that is usually presented in light source manufacturers' catalogues.

The CIE general colour rendering index has its limitations. First, it should be appreciated that just because two light sources have the same general colour rendering index, it does not mean that they render colours the same way. The general CRI is an average and there are many combinations of special CRI values that give the same average. Second, different light sources are being compared with different reference light sources. This makes the meaning of comparisons between different light sources uncertain, yet comparing light sources is what the general CRI is most often used for. Third, there is considerable argument about the method used to correct for chromatic adaptation. These limitations should be borne in mind when evaluating the CIE general colour rendering indices for different light sources.

2.4.3.3 Scotopic/Photopic Ratio

One other measure of light source colour characteristics that has been gaining interest in recent years is the scotopic/photopic ratio (Berman 1992). This is calculated by taking the spectral power distribution of the light source and weighting it by the CIE Standard Scotopic and Photopic Observers. The resulting scotopic lumens and photopic lumens are divided by the power of the light source in watts to get the scotopic and photopic luminous efficacies, which are then used to form the scotopic/photopic ratio. The value of the scotopic/photopic ratio is that it expresses the relative effectiveness of a light source in stimulating the rod and cone photoreceptors in the human visual system. A light source with a higher scotopic/photopic ratio will stimulate the rods more than a light source with a lower scotopic/photopic ratio when both produce the same photopic luminous flux. Table 2.3 gives scotopic/photopic ratios for a number of commonly used light sources.

2.5 LIGHT SOURCES

In the early days of motoring, light sources powered by gas were used for road and vehicle lighting. Today, all the light sources used for both road and vehicle lighting are powered by electricity. The lighting industry makes several thousand different types of electric light sources. They can be conveniently divided into three classes: incandescent, discharge, and solid state.

2.5.1 The Incandescent Light Source

The incandescent was, until recently, the main light source used in traffic signals, vehicle signal lamps, and vehicle interior lighting. The incandescent light source produces light by heating a thin tungsten filament to incandescence in an inert gas atmosphere. The spectral emission of the incandescent light source is a continuum over the visible spectrum (Figure 2.5), although the exact spectrum is determined by the temperature of the filament. This is easily seen when an incandescent light source is dimmed. Reducing the voltage reduces the current through the filament and hence the temperature of the filament. The result is that the colour appearance of the light emitted by the light source becomes more yellow and then red until, at very low voltages, no light can be seen at all, although the light source may still be emitting infrared radiation. The design of an incandescent light source is a matter of balancing luminous efficacy against life. A high light output and hence a higher luminous efficacy can be achieved by heating the filament to just below its melting point but then the life is short. When a long light source life is desirable, as in traffic signals, the filament is heated to a lower than usual temperature by reducing the applied voltage. Incandescent light sources are sensitive to voltage fluctuations and to vibration, unless specifically constructed to handle mechanical shock.

TABLE 2.3
Scotopic/Photopic Ratios for a Number of Widely Used Electric Light Sources

Light Source	Photopic Luminous Efficacy (lm/W)	Scotopic Luminous Efficacy (Scotopic lm/W)	Scotopic/Photopic Ratio
Incandescent	14.7	20.3	1.38
Fluorescent (3500 K)	84.9	115.9	1.36
Mercury vapour	52.3	66.8	1.28
Metal halide	107.4	181.7	1.69
High-pressure sodium	126.9	80.5	0.63
Low-pressure sodium	180.0	40.8	0.23

From He et al. (1997).

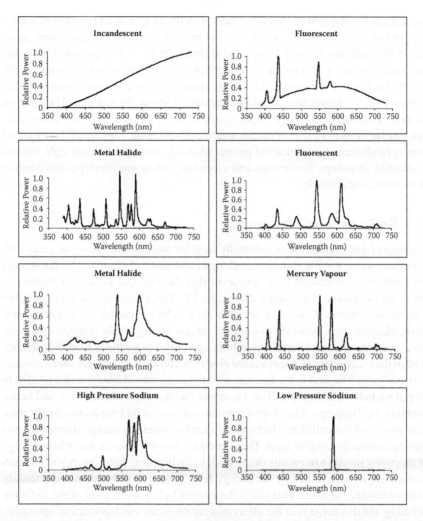

FIGURE 2.5 Relative spectral power distributions for incandescent, two forms of metal halide, high-pressure sodium, two forms of fluorescent, mercury vapour, and low-pressure sodium light sources. All the spectral power distributions are normalized to unity for the wavelength with the maximum output. The two forms of metal halide differ in the chemicals used in the arc tube. The two forms of fluorescent differ in the mixture of phosphors used.

2.5.2 THE TUNGSTEN-HALOGEN LIGHT SOURCE

Tungsten-halogen light sources are widely used in vehicle headlamps and some vehicle signal lamps. The tungsten-halogen light source is essentially an incandescent light source with a halogen in the gas filling. The inclusion of the halogen allows the filament to be run at a higher temperature without excessive envelope blackening because, although the tungsten is evapourated off the filament faster at the higher temperature, the halogen chemically reacts with the evapourated tungsten to form a tungsten-halogen compound, which will not precipitate onto the light source envelope. If the tungsten-halogen compound diffuses back to the filament, the higher temperature causes it to separate into tungsten and halogen, depositing some of the tungsten back onto the filament. This cycle ensures the light output is maintained at a higher level for longer than would be the case without the halogen. The spectral emission of the tungsten-halogen light source is a continuum across the visible spectrum similar in form to that of the incandescent light source, as would be expected given its fundamental incandescent nature. Similarly, tungsten-halogen light sources are sensitive to voltage fluctuations and vibration, unless specifically constructed to handle mechanical shock.

2.5.3 The Fluorescent Light Source

Fluorescent light sources are occasionally used for road lighting in urban areas and, in compact form, for illuminating road signs and traffic bollards. The fluorescent light source is a discharge light source in that the physical means for producing light is the excitation of a gaseous discharge. The fluorescent light source, in either its linear or compact form, consists of a glass tube containing a mercury atmosphere. Heating the electrodes produces a stream of negatively charged electrons. These electrons are accelerated through the mercury gas by the potential difference between the electrodes. The accelerated electrons collide with the gas atoms producing two effects. The first is the ionization of the atom into electrons and a positively charged particle called an ion. This increases the electron concentration and hence maintains the discharge. The second possible outcome is that the atom absorbs most of the energy of the colliding electron and thereby raises the energy state of its own captive electrons to a higher level. These energy levels are discrete and when the captive electrons shortly afterwards decay back to their resting level, energy is radiated at a wavelength determined by the energy level structure of the atom. For mercury at a low pressure, which is what fills a fluorescent light source, most of the radiation emitted by the discharge is in the ultraviolet region of the electromagnetic spectrum. To produce radiation in the visible spectrum, the inner surface of the glass tube is coated with a phosphor. This absorbs the ultraviolet radiation from the discharge and emits radiation in the visible spectrum. This two-step process is evident in the spectral emission of the fluorescent light source (see Figure 2.5), which usually consists of a series of strong emission lines, from the discharge, superimposed on a continuous emission spectrum, from the phosphor. By changing the phosphor mix, different spectral emissions can be created so fluorescent light sources are available with a wide range of colour properties. The fluorescent light source is a discharge light source and therefore needs to have a control system to alter the electrical conditions from those necessary to start the discharge to those required to maintain it. This control gear, which is sometimes called a ballast, can be electromagnetic or electronic, the latter providing greater circuit luminous efficacy than the former. The light output of the fluorescent light source is a maximum at one ambient temperature. Higher or lower temperatures reduce light output.

2.5.4 THE MERCURY VAPOUR LIGHT SOURCE

The mercury vapour light source is now obsolete but it can still be found in some road lighting installations. The mercury vapour light source is similar to the fluorescent light source in that it is a discharge light source based on a mercury atmosphere in an arc tube. The difference is that the mercury vapour light source is a high pressure light source. The result is that the spectral emission of the gas discharge is moved into the visible region, although it still consists of a series of intense spectral lines (see Figure 2.5). Today, the mercury vapour light source is usually sold with a phosphor coating on the inside of the envelope, the phosphor coating being used to improve its colour properties.

2.5.5 THE METAL HALIDE LIGHT SOURCE

The metal halide light source is increasingly being used, in one form, for road lighting in urban areas and, in another form, for vehicle headlamps. The metal halide light source is a high-pressure gas discharge light source based on a mercury discharge, but it is different from the mercury vapour light source in that it has metal halides, such as scandium and sodium iodides, in the arc tube. When the arc tube reaches operating temperature, the metal halides are vapourized. At the core of the discharge the metal halides separate into the metals and halogen, the metals emitting radiation in the visible region. At the cooler edge of the arc tube, the metals and halogen recombine and then repeat the process. The result is a spectrum consisting of discrete spectral lines (see Figure 2.5). As may be imagined, the chemistry of the metal halide light source is very complex. The result is that early metal halide light sources gained a reputation for showing shifts in colour properties over life and even between different light sources from the same manufacturer when new. However, developments in arc tube materials and design have gone a long way to alleviating this problem (van Lierop et al. 2000).

The operational difference in the forms of metal halide light sources used for road lighting and vehicle headlamps lies in the run-up time. Run-up time is the time taken to reach full light output. Metal halide light sources used for road lighting have run-up times of about six minutes. Such times are acceptable for road lighting but completely unacceptable for vehicle lighting. The metal halide light sources used for vehicle headlamps have xenon added to the discharge and are commonly known as xenon or high-intensity discharge (HID) headlamps. The effect of the xenon is to make a significant amount of the light output immediately available on switch on and to make the run-up to full light output a matter of seconds rather than minutes.

2.5.6 THE LOW-PRESSURE SODIUM LIGHT SOURCE

The low-pressure sodium light source is widely used for road lighting in the UK and The Netherlands but only sparingly in other countries. The low-pressure sodium light source is a discharge light source based around sodium. Electrically, the low-pressure sodium light source operates in the same manner as the fluorescent light source but in this case a phosphor is unnecessary because the spectral emission from the sodium discharge is concentrated in two spectral lines, which are both close to 598 nm (see Figure 2.5). Because this wavelength is near to the maximum sensitivity of the human visual system at 555 nm, as shown by the CIE Standard Photopic Observer, the low-pressure sodium light source has the highest luminous efficacy of all the artificial light sources (up to 180 lm/W). Unfortunately, its colour properties are what might be expected from an almost monochromatic source, non-existent.

2.5.7 THE HIGH-PRESSURE SODIUM LIGHT SOURCE

The high-pressure sodium light source is the most common light source used for road lighting. Conceptually, the high-pressure sodium light source is the same as the low-pressure sodium light source but the much higher pressure has an effect on the spectral emission. The increased pressure in the discharge leads to self-absorption of radiation within the discharge and interactions between the closely packed atoms. The combined effect of these phenomena is to reduce the power at 598 nm and to spread the spectral emission over a much wider range of wavelengths (see Figure 2.5). The result is a combination of high luminous efficacy and modest colour properties.

2.5.8 INDUCTION LIGHT SOURCES

Induction light sources are occasionally used for road lighting where access is difficult, e.g., bridges. Induction light sources differ from the discharge light sources discussed above in that they use an electromagnetic field, at either radio or microwave frequencies, to create a discharge. Induction light sources are essentially a fluorescent light source without electrodes. The electromagnetic field excites the mercury in an envelope that then emits radiation mainly in the ultraviolet. This is then absorbed by a phosphor and re-radiated in the visible region. The luminous efficacy and colour properties of induction light sources are similar to those of fluorescent light sources, their main advantage being the much longer life produced by not having any electrodes to fail.

2.5.9 LIGHT EMITTING DIODES (LEDs)

The LED is a solid-state light source that has already replaced the incandescent light source in traffic signals, is in the process of replacing incandescent light sources in vehicle signal lamps, and will soon be available for vehicle headlamps. The LED is a semiconductor that emits light when a current is passed through it. The spectral emission of the LED depends on the materials used to form the semiconductor. For light, the most common LED material combinations are now aluminium indium gallium phosphide (AlInGaP) and indium gallium nitride (InGaN). These combinations are replacing older combinations of gallium arsenide phosphide (GaAsP), aluminium gallium phosphide (AlGaP), and aluminium gallium arsenide (AlGaAs). LEDs typically produce narrow-band radiation, the spectral emission being characterized by the wavelength at which the maximum emission occurs (peak wavelength) and the half bandwidth, this being half the difference in wavelengths at which the radi-ant flux is half the maximum radiant flux. The half-bandwidth for AlInGaP LEDs is about 17 nm while that for InGaN LEDs is 35 nm. The narrow spectrum of the light emitted by LEDs make them an attractive proposition for traffic signals and vehicle signal lamps where a specific colour is required. With an incandescent light source, a coloured filter is required but with an LED a filter may be unnecessary to produce the desired colour.

The light output of LEDs is determined by the current through the semiconductor and its temperature. LEDs made from different materials can vary widely in life, from a few thousand hours to a projected 100,000 hours, depending on the current used and the operating conditions (Narendran et al. 2001). The fact that LEDs are solid-state devices and hence are less likely than incandescent light sources to fail due to mechanical damage caused by vibration, have made them an attractive option for vehicle signal lights.

It might be thought that the fact that the LED is a narrow-band source of light would preclude its use where white light is required, but this is not so. White light can be produced from LEDs by combining the light outputs from red, green, and blue LEDs in the right proportions or by adding a phosphor to the epoxy that encapsulates an InGaN LED with a peak wavelength of 470 nm. Figure 2.6 shows the spectral emission of a phosphor-based white LED. LEDs have been used in a prototype post-top lantern for use in rural areas (Brandston et al. 2000), proposed for a standard road lighting luminaire for New York City, and used to form head-lamp systems.

The LEDs discussed above are all point sources of light and use inorganic com-pounds as their light-emitting elements. However, there are also organic LEDs or light-emitting polymers (Kwong et al. 2001). Light-emitting polymers are organic semiconducting materials that possess similar light-emitting properties to the inor-ganic LEDs but also have the mechanical properties of plastics. This makes them an interesting possibility for vehicle interior lighting.

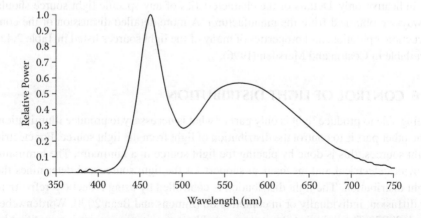

FIGURE 2.6 The relative spectral power distribution of a white light-emitting diode. This light-emitting diode is basically an indium gallium nitride LED with an integral phosphor.

2.5.10 THE ELECTROLUMINESCENT LIGHT SOURCE

Electroluminescent light sources are a solid-state light source consisting of a sandwich in the form of a flat area conductor, a layer of dielectric-phosphor mixture, and another area conductor that is transparent. When a high, alternating voltage is applied across the two area conductors, the phosphor is excited and light is emitted. The colour of the light emitted depends on the dielectric-phosphor combination used and the frequency of the applied voltage. Spectral emissions that are perceived as blue, yellow, green, and pink are available. Electroluminescent light sources have a lower luminous efficacy than incandescent light sources. However, the fact that they have a long life, low power requirements, and can be formed as either rigid ceramic or flexible plastic sheets or tapes have made them an attractive option for instrument panels and for backlighting liquid crystal displays.

2.5.11 LIGHT SOURCE CHARACTERISTICS

Electric light sources can be characterized on several different dimensions. They are:

- Luminous efficacy is the ratio of luminous flux produced to power supplied (lumens/watt). If the light source needs control gear to operate, the watts supplied should include the power demand of the control gear.
- Correlated colour temperature (see Section 2.4.3.1).
- CIE general colour rendering index (see Section 2.4.3.2).
- Light source life is the number of burning hours until either light source failure or a stated percentage reduction in light output occurs.
- Run-up time is the time from switch-on to full light output.

Table 2.4 summarizes these characteristics for many of the light sources used for road, vehicle, sign, and signal lighting. The values in Table 2.4 should be treated as indicative only. Details of the characteristics of any specific light source should always be obtained from the manufacturer. A more detailed discussion of the construction, operation, and properties of many of the light sources listed in Table 2.4 is available in Coaton and Marsden (1996).

2.6 CONTROL OF LIGHT DISTRIBUTION

Being able to produce light is only part of what is necessary to produce illumination. The other part is to control the distribution of light from the light source. For electric light sources, this is done by placing the light source in a luminaire. The luminaire provides electrical and mechanical support for the light source and determines the light distribution. The light distribution is controlled by using reflection, refraction, or diffusion, individually or in combination (Simons and Bean 2000; Wordenweber et al. 2007). The light distribution provided by a specific luminaire is quantified by the luminous intensity distribution. The light distribution allowed from vehicle lighting is strictly regulated but that from road lighting luminaires is not. All reputable

TABLE 2.4

A Summary of the Properties of Some Electric Light Sources Used for Road, Vehicle, Sign, and Signal Lighting

Light Source	Luminous Efficacy (lm/W)	CCT (K)	CRI	Lamp Life (hrs)	Run-up Time (mins)
Incandescent	8–14	2,500–2,700	100	1,000	Instant
Tungsten-halogen	15–25	2,700–3,200	100	1,500–4,000	Instant
Fluorescent	50–96	2,700–7,500	50–95	8,000–14,000	0.5
Compact fluorescent	30–70	2,700–4,100	80–85	8,000	0.25–1.5
Mercury vapour	33–57	3,200–3,900	40–50	8,000–10,000	4
Metal halide (HID)	60–98	3,000–6,000	60–90	2,000–10,000	0.5–8
High-pressure sodium	53–142	1,900–2,100	19–25	10,000–20,000	3–7
Low-pressure sodium	70–180	n.a.	n.a.	15,000–20,000	10–20
Induction	47–80	2,500–4,000	80	60,000	1
White LED	30–55	3,000–6,500	70–93	50,000	Instant
Red LED	30	n.a.	n.a.	50,000	Instant

CCT = correlated colour temperature; CRI = CIE general colour rendering index; n.a. = not applicable.

road lighting luminaire manufacturers provide luminous intensity distributions for their luminaires.

Luminaires used for road lighting have to be capable of withstanding the elements. This is done by the choice of materials and careful sealing between components. The extent of protection provided by different luminaires is given by the Ingress Protection (IP) system, the degree of protection being indicated by the letters IP followed by two numbers. The first number gives the degree of protection against the ingress of foreign bodies and dust. The second gives the protection against the ingress of moisture. Table 2.5 shows the degree of protection indicated by each number. Most road lighting luminaires are IP 65 and better. The IP system is not applied to vehicle lighting, the requirements for resisting the elements being included in the relevant regulations.

2.7 THE CONTROL OF LIGHTING

The control of lighting involves the control of both light output and light distribution. The control of light output is provided by switching or dimming. Switching systems can vary from the conventional manual switch in a vehicle through simple time switches for road lighting to slightly more sophisticated photoelectric control

TABLE 2.5

The IP Classification of Luminaires According to the Degree of Protection against the Ingress of Foreign Bodies, Dust, and Moisture

First Number	Degree of Protection	Second Number	Degree of Protection
0	Not protected	0	Not protected
1	Protected against solid objects greater than 50 mm	1	Protected against dripping water
2	Protected against solid objects greater than 12 mm	2	Protected against dripping water when tilted up to 15 degrees
3	Protected against solid objects greater than 2.5 mm	3	Protected against spraying water
4	Protected against solid objects greater than 1.0 mm	4	Protected against splashing
5	Dust-protected	5	Protected against water jets
6	Dust-tight	6	Protected against heavy seas
		7	Protected against the effects of immersion
		8	Protected against submersion to a specified depth

systems that sense the level of natural light available and turn on or turn off the vehicle lighting or road lighting, as appropriate. Recently, vehicle forward lighting systems have been developed that integrate information from sensors that detect the vehicle speed, the movement of the steering, and ambient light levels, to modify the light distribution of the headlights to better match the conditions and the manoeuvre being attempted (see Section 6.7.2). For example, some vehicles are now being fitted with headlamps that are aimed around a bend as the driver changes course.

A similar level of sophistication is also being applied to road lighting. Signaling down the power supply cables provides information to a local computer, which sends the information through a wireless network to a central monitoring station. This enables the authority responsible for the road lighting to monitor its state and, through two-way communication, to adjust the light output over a limited range. This possibility has given rise to the idea that road lighting illuminances might be adjusted according to traffic densities and other demand factors (see Section 14.4.2). Another control variable of interest is the light distribution of the luminaire. Changing the light distribution is difficult when a single light source is used in the luminaire. However, where multiple light sources are used, as is the case with LEDs, there exists the possibility of changing the light output of different LEDs and hence the light distribution. The point to take from this brief discussion is that the reduction in costs and the increase of capabilities of sensors, computer networks, and communication networks have made much more sophisticated control of light output and light distribution possible.

2.8 FLUORESCENCE AND RETROREFLECTION

Although they are neither light sources nor luminaires, the phenomena of fluorescence and retroreflection have important roles to play in making objects visible on and around the road. Both are part of the high visibility clothing widely used by road workers and police officers and by some pedestrians, horse-riders, and cyclists to increase their conspicuity.

Fluorescence occurs when electromagnetic radiation at one wavelength is absorbed by a chemical that then re-radiates some of it at a longer wavelength. The chemical incorporated into the material forming high-visibility clothing absorbs radiation in the ultraviolet and short wavelength visible region of the spectrum and then re-radiates some of it at longer wavelengths in the visible region. The material used in high-visibility clothing has a higher reflectance than most of the other substances against which it will be seen. This alone will give the clothing a high visibility for most backgrounds. However, the luminance caused by fluorescence is added to the luminance produced by reflection to produce an even higher luminance, so much so that the lightness constancy can be broken (see Section 3.5), a perception that will increase the conspicuity of the wearer. Fluorescence is effective as a means of increasing luminance during daytime as daylight is a light source rich in the relevant radiation. After dark, the benefits of fluorescence are much reduced because few light sources have much of the required radiation. After dark, retroreflection is a more effective means of increasing visibility (CIE 1987).

A retroreflector is a device that reflects light back from whence it came, over a wide range of angles of incidence. When the headlamps of a car illuminate a retroreflective surface, the reflected light is directed back towards the car and its driver. Retroreflectors come in a number of forms, the most common being the corner cube and the refractive/reflective combination. As its name suggests, the cube is an arrangement of three mutually perpendicular plane reflectors forming a corner, usually manufactured as a solid transparent cube. The corner cube relies on total internal reflection to redirect the light. The refractive/reflective combination has the reflective element coinciding with the image plane of the refractive element. A special case of this relevant to drivers is a transparent sphere that has a refractive index twice that of air. This will act as a retroreflector because the back surface of the sphere forms the image plane of the refracting front surface.

Retroreflectors are good for visibility when placed amongst diffusely reflecting surfaces because they concentrate the reflected light in the direction from which it came, thereby giving the retroreflector a much higher luminance than the surroundings and hence a much higher luminance contrast when viewed from that direction. Of course, this is only true for an observer sitting close, in angular terms, to a small source of high luminous intensity, such as a driver in a vehicle with the headlamps on, at night. Retroreflectors do nothing for the visibility of objects by people remote from the light source, such as pedestrians.

Retroreflectors can be small enough for large numbers to be incorporated into sheet material for use in road signs, into paint or plastic for road marking, and into tape for application to clothing. Retroreflectors can also be large enough for individual units to be embedded in the road with a slight projection above the road surface.

These are the road studs familiar to drivers in many countries (see Section 5.2). Road studs are visible to drivers from a long way away and are particularly valuable where fog is frequent. Road studs are not generally used in areas that regularly experience snow during winter as snowploughs tend to damage them.

2.9 SUMMARY

This chapter is concerned with what light is, how it can be measured, and how it is produced and controlled. Light is part of the electromagnetic spectrum between 380 and 780 nm. What differentiates this wavelength range from the rest of the electromagnetic spectrum is that the human visual system responds to it. The actual human spectral response has been standardized in an internationally agreed form represented by the CIE Standard Photopic and Scotopic Observers. Using the appropriate spectral sensitivity curve, the four basic photometric quantities can be derived—luminous flux, luminous intensity, illuminance, and luminance.

These measures are all concerned with the overall amount of light and not with its colour. The CIE colourimetry system has been developed to quantify colour. By using the CIE colourimetry system, the colour appearance and colour rendering properties of light sources can be characterized, using such measures as correlated colour temperature and CIE general colour rendering index.

It is important to appreciate that while there are numerous metrics used to characterize a lighting situation or a light source, these metrics are simultaneously precise and inaccurate. The precision arises because the metrics can often be measured or calculated exactly. If they are regarded as simple physical measures they can be considered accurate. But they are not simple physical measures. The whole reason for having photometric and colourimetric quantities is to quantify the visual effect of light. Because of the complexity and flexibility of the human visual system and the differences between individuals, any one standardized metric of visual effect is inevitably an approximation. This is particularly so for driving because all the photometric quantities used in the design and specification of vehicle lighting and road lighting use the CIE Standard Photopic Observer but when driving at night, large parts of the visual field will be operating in the mesopic state.

Having considered how light can be measured, the physical principles and properties of the electric light sources used for roads, vehicles, signs, and signals are considered. Electric light sources used for roads, vehicles, signs, and signals fall into three types: incandescent, discharge, and solid state. Light sources differ in their physical size, light output, luminous efficacy, colour properties, life, run-up time, and sensitivity to the environment in which they have to operate. Road lighting gives priority to light output, luminous efficacy, and life. Vehicle lighting gives priority to physical size, light output, colour properties, and run-up time. Traffic signals and the lighting of traffic signs give priority to physical size, colour properties, life, and run-up time.

Having chosen the light source, the next thing to consider is the characteristics of the luminaire in which it is contained. The luminaire provides electrical and mechanical support for the light source and controls the light distribution. The light distribution is controlled by using reflection, refraction, or diffusion, individually or in combination. Finally, the whole light output of the whole system is subject to control. Lighting control systems for road and vehicle lighting are becoming much more sophisticated through a combination of sensors, computers, and communication networks.

The final optical phenomena of relevance to driving are fluorescence and retroreflection. Both are used in the high-visibility clothing commonly worn by people working on the roads. Fluorescence occurs when electromagnetic radiation at one wavelength is absorbed by a chemical that then re-radiates some of it at a longer visible wavelength. Fluorescence is effective as a means of increasing luminance during daytime. After dark, the benefits of fluorescence are much reduced and retroreflection is a more effective means of increasing visibility. A retroreflector is a device that reflects light back from whence it came, over a wide range of angles of incidence. When the headlamps of a car illuminate a retroreflective surface, the reflected light is directed back towards the car and its driver, thereby increasing the luminance of the retroreflector relative to the more diffusely reflecting surfaces around it. Retroreflectors are used on road surfaces for edge and lane delineation, on road signs to make them readable at greater distances, and on vehicles and clothing to enhance conspicuity.

3 Sight

3.1 INTRODUCTION

Light is essential for sight. With light we can see, without light we cannot. But how different lighting conditions influence sight depends on the operation of the human visual system. The construction, capabilities, and constraints of the human visual system are the subjects of this chapter.

3.2 THE STRUCTURE OF THE VISUAL SYSTEM

The human visual system is an image processing system consisting of the eye and brain working together to construct a model of the external world (Wolfe et al. 2006; Wordenweber et al. 2007). The model developed resolves the ambiguities inherent in the retinal image on the basis of probabilities derived from experience. Despite this complexity, the obvious starting point for any discussion of the visual system is the eye.

3.2.1 THE STRUCTURE OF THE EYE

Figure 3.1 shows a section through the eye, the upper and lower halves being adjusted for focus at near and far distances, respectively. The eye is basically spherical with a diameter of about 24 mm. The sphere is formed from three concentric layers. The outermost layer, called the sclera, protects the contents of the eye and maintains its shape under pressure. Over most of the eye's surface, the sclera looks white, but at the front of the eye the sclera bulges up and becomes transparent. It is through this area, called the cornea, that light enters the eye. The next layer is the vascular tunic, or choroid. This layer contains a dense network of small blood vessels that provide oxygen and nutrients to the innermost layer, the retina. As the choroid approaches the front of the eye it separates from the sclera and forms the ciliary body. This element produces the watery fluid that lies between the cornea and the lens, called the aqueous humour. The aqueous humour provides oxygen and nutrients to the cornea and the lens, and takes away their waste products. Elsewhere in the eye this is done by blood but on the optical pathway through the eye, a transparent medium is necessary.

As the ciliary body extends further away from the sclera, it becomes the iris. The iris forms a circular opening, called the pupil, that admits light into the rear of the eye. Pupil size varies with the amount of light reaching the retina but it is also influenced by the distance of the object from the eye, the age of the observer, and by emotional factors such as fear, excitement, and anger (Duke-Elder 1944).

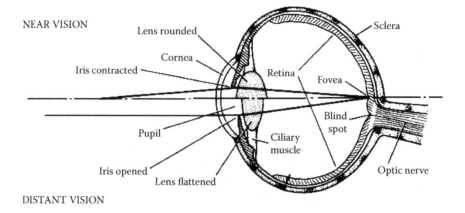

FIGURE 3.1   A section through the eye adjusted for near and distant vision..

After passing through the pupil, light reaches the crystalline lens. The lens is fixed in position, but varies its focal length by changing its shape. The change in shape is achieved by contracting or relaxing the ciliary muscles. For objects close to the eye, the lens is fattened. For objects far away, the lens is flattened.

The space between the lens and the retina is filled with another transparent material, the jelly-like vitreous humour. After passing though the vitreous humour, light reaches the retina where light is absorbed and converted to electrical signals. The retina is a complex structure (Figure 3.2). It can be considered as having three layers: a layer of photoreceptors, a layer of collector cells that provide links between multiple photoreceptors, and a layer of ganglion cells. The axons of the ganglion cells form the optic nerve, which produces the blind spot where it passes through the retina out of the eye. Light reaching the retina passes through the ganglion and collector cell layers before reaching the photoreceptors, where it is absorbed. Any light that gets past the photoreceptor layer is absorbed by the pigment epithelium.

3.2.2 THE RETINA

The retina is an extension of the brain. The human visual system has four photoreceptor types in the retina, each containing a different photopigment. These four types are conventionally grouped into two classes, rods and cones, the names coming from the appearance of the photoreceptors under a microscope. All rod photoreceptors contain the same photopigment and therefore have the same spectral sensitivity. The other three photoreceptor types are all cones. Each contains a different photopigment and hence has a different spectral sensitivity. The three cone types are called short (S), medium (M), and long (L) wavelength cones after the wavelength region where they have the greatest sensitivity.

Rods and cones are distributed differently across the retina (Figure 3.3). The cones are concentrated in one small area that lies on the visual axis of the eye, called the fovea, although there is a low density of cones across the rest of the retina. Rods are absent from the fovea and reach a maximum concentration about 20 degrees off the

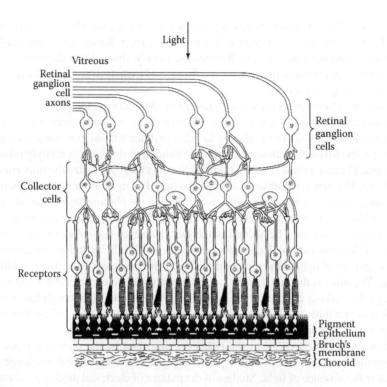

FIGURE 3.2 A section through the retina (from Sekular and Blake 1994).

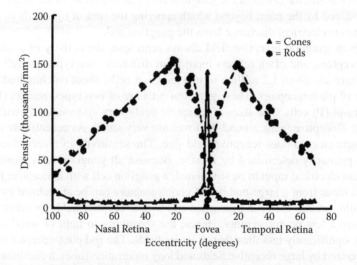

FIGURE 3.3 The distribution of rod and cone photoreceptors across the retina. Zero degrees indicates the position of the fovea.

visual axis. The three cone types are also not distributed equally across the retina. The L and M cones are concentrated in the fovea, their density declining gradually with increasing eccentricity. The S cones are largely absent from the fovea, reach a maximum concentration just outside the fovea, and then decline gradually in density with increasing eccentricity.

Over the whole retina there are many more rods than cones, approximately 120 million rods and 8 million cones. The fact that there are many more rod than cone photoreceptors should not be taken to indicate that human vision is dominated by the rods. It is the fovea that allows resolution of detail and the fovea is entirely inhabited by cones. There are three other anatomical features that emphasize the importance of the fovea. The first is the absence of blood vessels. For most of the retina, light passes through a network of blood vessels before reaching the photoreceptors but blood vessels avoid crossing over the central area of the retina, called the macula, at the centre of which is the fovea. The second is the fact that, even the collector and ganglion layers of the retina are pulled away over the fovea. This helps reduce the absorption and scattering of light in the region of the fovea and hence enhances the resolution of detail. The third is the fact that the outer limb of the cone photoreceptor can act as a waveguide, making cones more sensitive to light rays passing through the centre of the lens (Crawford 1972). This characteristic, known as the Stiles-Crawford effect, compensates to some extent for the poor quality of the eye's optics.

Photons of light arriving at the retina are absorbed by the photoreceptors. This absorption results in changes to the pattern of spontaneous electrical discharges produced in the absence of light. Studies of the pattern of electrical discharges from single ganglion cells have revealed two important aspects of the operation of the retina. The first is the existence of receptive fields. A receptive field is the area of the retina that determines the output from a single ganglion cell. The size of a receptive field is measured by exploring the retina with a very small spot of light while measuring the electrical discharges from the ganglion cell. The boundary of the receptive field is determined by the point beyond which applying the spot of light fails to alter the spontaneous electrical discharge from the ganglion cell.

A given ganglion receptive field always represents the activity of a number of photoreceptors, and often reflects input from different cone types as well as from rods. There are about 1.2 million retinal ganglion cells, about one hundredth of the number of photoreceptors. These ganglion cells are of two types, midget (M) cells and parasol (P) cells. The sizes of receptive fields vary systematically with retinal location. Receptive fields around the fovea are very small. As eccentricity from the fovea increases, so does receptive field size. The sensitivity of a receptive field to light is primarily determined by its size. Because all ganglion cells require some minimum electrical input to be stimulated, a ganglion cell with a receptive field that receives input from a large number of photoreceptors can be stimulated by a lower retinal illuminance than can a ganglion cell with a receptive field that receives input from only a few photoreceptors. Hence, the sensitivity to light of small receptive fields is significantly less than that of larger fields. The rod photoreceptor system is characterized by large receptive fields and long integration times, a combination that makes the rod photoreceptor system significantly more sensitive to light than the cone photoreceptor system.

The second important aspect of the operation of the retina revealed by studies of electric discharges is that within each ganglion cell receptive field there is a structure. Specifically, retinal ganglion receptive fields consist of a central circular area and a surrounding annular area. These two areas have opposing effects on the ganglion cell's electrical discharge. Either light falling on the central area increases the rate of electrical discharge while light falling on the annular surround decreases it, or, in other receptive fields, the reverse occurs. These types of receptive fields are known as on-center/off-surround and off-center/on-surround fields, respectively. If either of these two types of retinal ganglion receptive fields is illuminated uniformly, the two types of effect on electrical discharge cancel each other, a process called lateral inhibition. However, if the illumination is not uniform across the two parts of the receptive field, a net effect on the ganglion cell discharge is evident. This means that the image processing of the retina is devoted to detecting differences in luminance, not absolute luminance, a process that emphasizes the contrasts in the scene and discounts absolute light level.

There are an approximately equal number of on-center/off-surround and off-center/on-surround receptive fields in the retina. The electrical signals from the two types of receptive field do not cancel each other. Rather, the signals from the two types of receptive field are kept separate, indicating that they serve different aspects of vision. Further, different ganglion cells respond most strongly to different spatial frequencies.

There are a number of important differences between the M cells and P cells. First, the axons of the M cells are thicker than the axons of the P cells, indicating that signals are transmitted more rapidly from the M cells than from the P cells. Second there are many more P cells than M cells and they are distributed differently across the retina. The P cells dominate in the fovea and parafovea and the M cells dominate in the periphery. Third, for a given eccentricity, the P cells have smaller receptive fields than the M cells. Fourth, the M cells and P cells are sensitive to different aspects of the retinal image. The M cells are more sensitive to rapidly varying stimuli and to small differences in illumination but are insensitive to differences in colour. The P cells are more sensitive to small areas of light and to colour.

Overall, this brief description should have demonstrated that the retina is really the first stage of an image processing system. The retina extracts information on boundaries in the retinal image and then seeks out specific aspects of the stimulus within the boundaries, such as colour. These aspects are then transmitted up the optic nerve, formed from the axons of the retinal ganglion cells, along different channels. For a more detailed description of the operation of the retina, see Wolfe et al. (2006).

3.2.3 THE CENTRAL VISUAL PATHWAYS

Figure 3.4 shows the pathways over which signals from the retina are transmitted to the visual cortex. The optic nerves leaving the two eyes are brought together at the optic chiasm, rearranged and then extended to the lateral geniculate nuclei (LGN). Somewhere between leaving the eyes and arriving at the lateral geniculate nuclei, some optic nerve fibres are diverted to the superior colliculus, which is located at the top of the brain stem and is responsible for controlling eye movements. After the

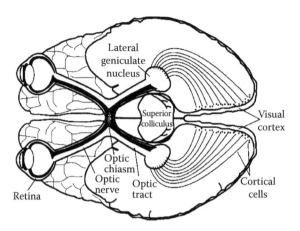

FIGURE 3.4   A schematic diagram of the pathways from the eyes to the visual cortex (from Sekular and Blake 1994).

lateral geniculate nuclei, the two optic nerves spread out to supply information to various parts of the visual cortex, the part of the brain where vision occurs.

At the optic chiasm, the optic nerve from each eye is split and then parts of the optic nerves from the same side of the two eyes are combined. This arrangement ensures that the signals from the same side of the two eyes are received together on the same side of the visual cortex.

The signals from the same side of the two eyes pass from the optic chiasm to the LGN. Anatomically, an LGN shows six distinct layers. Two of these layers receive signals from the M cells of the retina while the other four layers receive signals from the P cells. Each layer is arranged so that the location of the M cells and P cells on the retina is preserved. As in the retina, electrophysiological recording of discharges from individual LGN cells have shown the existence of receptive fields, with either an on-center/off-surround or an off-center/on-surround structure. The division of function found in the retina is also present in the LGN. The M cell layers respond to large, fast-moving objects but do not respond to colour differences, while the P cell layers process detailed information about stationary targets, including colour. Indeed, some P cell receptive fields show strong responses when the center is illuminated by one colour and the surround by another. The specific colour combinations are red and green or yellow and blue, this being the basis of human colour vision (see Section 3.2.5).

From the above description it might seem that the LGN are just relay stations between the retinas of the two eyes and the visual cortex. However, they are more than this. There are more connections from other parts of the brain to the LGN than there are connections from the LGN to the visual cortex. Thus, the LGN is where information from other parts of the brain modifies the transfer of information to the visual cortex. For example, the LGN receive signals from the reticular activating system, a part of the brain stem that determines the general level of arousal. It has been shown that low levels of arousal lead to an attenuated visual signal being transmitted from the LGN to the visual cortex (Livingston and Hubel 1981).

3.2.4 THE VISUAL CORTEX

The visual cortex is located at the back of the cerebral hemispheres. The primary visual cortex consists of another layered array, containing about 200 million cells. At one level, the primary visual cortex is similar to the organization of the retina and the lateral geniculate nuclei. For example, the M cell and P cell channels remain separated; signals from the M cell layers of the LGN are received in one layer of the visual cortex, while signals from the P cell layers go elsewhere. Further, each cortical cell reacts only to signals from a limited area of the retina, and the arrangement of the cortical cells replicates the arrangement of ganglion cells on the retina. But two features separate the primary visual cortex from the LGN and the retina. The first is cortical magnification. What this means is that the number of cells allocated to each part of the retina enhances the importance of the fovea. About 80 percent of the cortical cells are devoted to the central 10 degrees of the visual field (Drasdo 1977), in the middle of which is the fovea. The second is the shape of the receptive fields. On-center/off-surround and off-center/on-surround receptive fields are found, but now they are elongated in shape rather than circular. These receptive fields show sensitivity to the orientation of a line, edge, or grating and, for the last of these three, different cells are sensitive to the different spatial frequencies of the grating. Other cells respond to moving edges and lines, but only movement in one direction. Yet other cells respond more to signals from the left eye and others to the right eye but the majority respond to signals from both eyes. It is in the primary visual cortex that the information from the two eyes is brought together. All this cellular diversity occurs at the entry level of the visual cortex. There is a much more complex structure beyond this in the higher areas of the visual cortex (Wolfe et al. 2006). Investigation of these areas has shown that different parts of the visual cortex are dedicated to specific discriminations. For example, areas have been identified in the visual cortex of monkeys that are concerned with analysing colour, motion, and even faces viewed from particular angles (Desimone 1991). Visually, monkeys are very like humans.

3.2.5 COLOUR VISION

The differences between rod and cone photoreceptors have been considered above, but the value of having cones with three different photopigments has not. The value of having three different cone photoreceptor types is that it allows colour vision to occur. The number of photoreceptor types used to form a colour system is a matter of compromise. A single photoreceptor type containing a single photopigment is unable to discriminate differences in wavelength from differences in irradiance and so does not support colour vision, e.g., rod photoreceptors. A system with many different photoreceptors, each containing a different photopigment, would be able to make many discriminations between wavelengths but at the cost of taking up more of the neural capacity of the visual system.

The ability to discriminate the wavelength content of incident light makes a dramatic difference to the information that can be extracted from a scene. Creatures with only one type of photopigment, i.e., creatures without colour vision, can only discriminate shades of gray, from black to white. Approximately 100 such discriminations can be made. Having two photopigment types increases the number of different combinations of irradiance and spectral content that can be discriminated to about 10,000. Having three types of photopigment increases the number of discriminations to approximately 1,000,000 (Neitz et al. 2001). Thus, colour vision is a valuable part of the visual system, and not a luxury that adds little to utility.

Figure 3.5 shows how the outputs from the three cone photoreceptor types are believed to be arranged into one non-opponent achromatic system and two opponent chromatic systems. The achromatic channel receives inputs from the M and L cones only. The opponent blue-yellow channel produces the difference between the outputs of the S cones and the sum of the M and L cones. The other opponent channel, the red-green channel, produces the difference between the outputs of the M cones and the sum of the L and S cones. This opponent structure for colour vision influences the perception of colours. This was shown in an experiment by Boynton and Gordon (1965). They presented monochromatic lights of different wavelengths and asked subjects to describe the appearance of each stimulus using only the colour names, red, green, yellow, and blue. Either one or two names could be used for each colour presented. The interesting result was that people very rarely described a colour as red-green or yellow-blue, while yellow-red and green-blue were frequently used. Even four-month-old infants divide incident light into four categories, corresponding to what adults call red, yellow, green, or blue (Bornstein et al. 1976).

Physiologically, the electrical outputs from the three different cone types are organized into opponent and non-opponent classes in the retina. Outputs in the non-opponent class always give an increase in activity with increasing retinal irradiance,

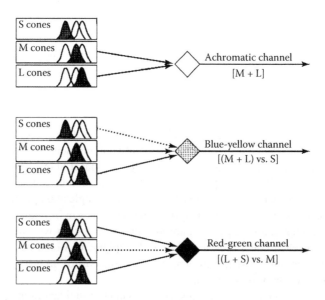

FIGURE 3.5   The organization of the human colour system showing how the three cone photoreceptor types are believed to feed into one achromatic, non-opponent channel and two chromatic, opponent channels (from Sekular and Blake 1994).

although the magnitude of that increase will vary with the wavelength of the incident light. Outputs in the opponent class can show either an increase or a decrease in activity depending on the wavelengths of the incident light. Cells of the non-opponent type are believed to constitute the achromatic channel shown in Figure 3.5, while cells of the opponent type form the opponent channels. The achromatic information is transmitted to the visual cortex by the M cell channel, while the chromatic information proceeds via the P cell channel. Much more detail on the structure and capabilities of the human colour vision can be obtained from Kaiser and Boynton (1996).

By now it should be clear that the visual system is not a camera. The retinal image is not simply transferred, lock, stock, and barrel, up to the visual cortex. Rather, the retinal image is broken down into significant elements, such as contrast, colour, and movement, while less useful information such as ambient light level, is discounted. The significant elements are transmitted to the visual cortex at two different speeds, the faster providing an outline of the scene while the slower fills in the details. From these elements, and the accumulated experience of sight, a model of the outside world is constructed.

3.3 CONTINUOUS ADJUSTMENTS OF THE VISUAL SYSTEM

In everyday use, the visual system has to be capable of examining objects in different locations, at different distances over a wide range of lighting conditions. These variable requirements indicate the need for the visual system to continually adjust its aim, focus, and sensitivity.

3.3.1 EYE MOVEMENTS

The emphasis given to the fovea by the visual system makes it essential that the visual world is explored by moving the eye so that different points of interest fall on the fovea. There are several different types of eye movement. Searching the visual field is done by a series of fixations and saccades. Fixations are attempts to hold an object of interest steady on the fovea. Movement between fixations is made by saccades. Saccades are very fast, velocities ranging up to 1000 degrees/second depending upon the distance moved. Saccadic eye movements have a latency of about 200 ms, which limits how frequently the line of sight can be moved to about five movements per second. Visual functions are substantially limited during saccadic movements. Figure 3.6 shows the pattern of fixations made while driving on an unlit road at night using headlamps alone.

About the only situation in which saccades rarely occur is in smooth pursuit eye movements. Such movements are relatively slow, up to 40 degrees/second, and are used to keep the image of a smoothly moving object, such as a passing vehicle, on the fovea. The advantage of such movements is that, if successful, the target is always on the fovea, which means that more detailed information about it can be extracted. The smooth pursuit system cannot follow either smoothly moving targets at high velocities or slow but erratically moving targets.

These eye movements all occur in a single eye, but movements in the two eyes are not independent. Rather, they are coordinated so that the lines of sight of the two

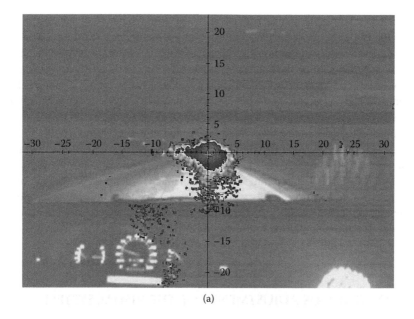

(a)

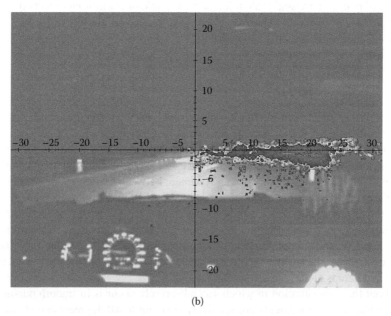

(b)

FIGURE 3.6   The pattern of fixations made while driving along an unlit road at night using headlamps (a) along a straight road and (b) around a curve. The pattern is defined into two groups based on the number of fixation points measured at each pixel. The pixel with the maximum number of fixations is given the value of 100 percent and is marked with a cross. All the other pixels are then allocated to one of two groups, those with 50 percent or more of the maximum number of fixations (dark grey) and those with 1 to 50 percent of the maximum (light grey) (from Stahl 2004).

eyes are both pointed at the same target at the same time. If the lines of sight of the two eyes are not aimed at the same target at the same time, the target may be seen as double. Movements of the two eyes that keep the primary lines of sight converged on a target as it changes in distance, e.g., from the view through the windscreen to the vehicle's instrument panel, are called vergence movements. These movements are very slow, up to 10 degrees/second, but can occur as a jump movement or can smoothly follow a target moving in a fore-and-aft direction. Both types of movement involve a change in the angle between the two eyes.

3.3.2 ACCOMMODATION

The eye has a fixed image distance, so to bring the images of objects at different distances to focus on the retina it is necessary to vary the optical power of the eye. This process is called accommodation. There are three optical components involved in accommodation. The first is the thin film of tears on the cornea. This film cleans the surface of the eye, starts the optical refraction process necessary for focusing objects, and smoothes out small imperfections in the surface of the second optical component, the cornea. The cornea covers the transparent anterior one-fifth of the eyeball (Figure 3.1). With the tear layer, it forms the major refracting component of the eye and gives the eye about 70 percent of its optical power. The crystalline lens provides the remaining 30 percent. The ciliary muscles have the ability to change the curvature of the lens and thereby adjust the power of the eyes' optical system in response to changing target distances.

Accommodation is a continuous process and is always a response to an image of the target located on or near the fovea rather than in the periphery of the retina. Any condition, either physical or physiological, that handicaps the fovea, such as a low light level, will adversely affect the ability to accommodate. As adaptation luminance decreases below 0.03 cd/m^2, the range of accommodation narrows so that it becomes increasing difficult to focus objects near and far from the observer (Leibowitz and Owens 1975). Reduced contrast sensitivity and poor visual acuity are consequences of inappropriate accommodation. When there is no stimulus for accommodation, as in complete darkness or in a uniform luminance visual field such as occurs in a dense fog or a whiteout snowstorm, the visual system accommodates to focus at approximately 70 cm.

3.3.3 ADAPTATION

The range of luminances to which the human visual system can be exposed is very large, from 0.000001 cd/m^2 on a very dark night to 100,000 cd/m^2 on a sunlit beach. To cope with this range, the visual system continuously and unconsciously changes its sensitivity through a process called adaptation. Adaptation involves three different processes:

Change in Pupil Size — The iris constricts and dilates in response to increased and decreased levels of retinal illumination. For young people, the diameter of the pupil can range from about 2 to 8 mm. The amount of light transmitted through the pupil is proportional to its area, so this range of diameters implies a maximum effect of pupil changes on retinal irradiance of 16 to 1. As the visual system can operate over a range of about 1,000,000,000,000 to 1, this indicates that the pupil plays only a minor role in the adaptation of the visual system. Iris constriction is faster (about 0.3 s) than dilation (about 1.5 s).

Neural Adaptation — This is a fast (less than 200 ms) change in sensitivity produced by synaptic interactions in the retina. It is effective over a luminance range of two to three log units.

Photochemical Adaptation — The four types of retinal photoreceptors used by the visual system contain four different photopigments. When light is absorbed by the photoreceptor, the pigment is bleached. In the dark, the pigment is regenerated and is again available to absorb light. The sensitivity of the eye to light is largely a function of the percentage of unbleached pigment. Under conditions of steady retinal illumination, the concentration of photopigment produced by the competing processes of bleaching and regeneration is in equilibrium. When the retinal illumination is changed, as happens when a vehicle with headlamps on passes, pigment is bleached and then regenerated so as to re-establish equilibrium. Exactly how long it takes to re-establish equilibrium after a change in retinal illumination depends on the magnitude of the change, the extent to which it involves different photoreceptors, and the direction of the change. For changes in retinal illumination of about two to three log units, neural adaptation is sufficient so equilibrium should be reached in less than a second. For larger changes, photochemical adaptation is necessary. If the change in retinal illumination is large but still lies completely within the range of operation of the cone photoreceptors, a few minutes will be sufficient for reach equilibrium. If the change in retinal illumination ranges from cone photoreceptor operation to rod photoreceptor operation, tens of minutes may be necessary for adaptation to be completed. As for the direction of change, once the photochemical processes are involved, changes to a higher retinal illumination can be achieved much more rapidly than changes to a lower retinal illumination. It is important to note that these changes in adaptation are not linear with time. The times given above are for equilibrium to be reached, but this is approached asymptotically. The vast majority of the adaptation takes place in much shorter times.

When the visual system is not completely adapted to the prevailing retinal illumination, its capabilities are diminished. This state of changing adaptation is called transient adaptation. Transient adaptation can be significant where sudden changes from high to low retinal illumination occur, such as on entering a long road tunnel on a sunny day.

3.3.4 PHOTOPIC, SCOTOPIC, AND MESOPIC VISION

The process of adaptation can change the spectral sensitivity of the visual system because at different retinal illuminances, different combinations of retinal photoreceptors are operating. The three states of sensitivity are conventionally identified as:

Photopic vision — This state of the visual system occurs at luminances higher than approximately 3 cd/m^2. For these luminances, the retinal response is dominated by the cone photoreceptors. This means that both colour vision and fine resolution of detail are available.

Scotopic vision — This operating state of the visual system occurs at luminances less than approximately 0.001 cd/m^2. For these luminances only the rod photoreceptors respond to stimulation, the cone photoreceptors being insufficiently sensitive to respond to the low level of retinal irradiance. This means that colour is not perceived, only shades of grey, and the fovea of the retina is blind.

Mesopic vision — This operating state of the visual system is intermediate between the photopic and scotopic states, i.e., between about 0.001 cd/m^2 and 3 cd/m^2. In the mesopic state both cones and rod photoreceptors are active. As luminance declines through the mesopic region, the fovea, which contains only cone photoreceptors, slowly declines in absolute sensitivity, without significant change in spectral sensitivity, until vision fails altogether as the scotopic state is reached. In the periphery, the rod photoreceptors gradually come to dominate the cone photoreceptors, resulting in a slow deterioration in colour vision and resolution and a shift in spectral sensitivity to shorter wavelengths.

The relevance of the different operating states for road and vehicle lighting varies. Scotopic vision is largely irrelevant. Any vehicle lighting worthy of the name provides enough light to at least move the visual system into the mesopic state when the driver is looking in the lit area, although parts of the visual scene outside the lit area may be at scotopic luminances. When there is road lighting present as well, there is usually enough to ensure that the visual system operates near the boundary of the photopic and mesopic states (Plainis et al. 2005).

3.4 CAPABILITIES OF THE VISUAL SYSTEM

The human visual system, like every other physiological system, has a limited range of capabilities. These limits are called the thresholds of vision. By convention, a threshold is measured as the value of the stimulus that is detected on 50 percent of the occasions it is presented. Threshold measurements come in many different forms and depend on many different variables, most of which interact (Boff and Lincoln 1988). Threshold measurements are mainly of interest for determining what will not be seen rather than how well something will be seen, but for driving at night, knowing what will not be seen is useful. The intention here is to summarize the thresholds of relevance to driving at night and how they are affected by the characteristics of the human visual system. For these threshold measurements it can be assumed that the observers were fully adapted, that the target was presented on a field of uniform luminance, and that the observers' accommodation was correct.

3.4.1 THRESHOLD MEASURES

The threshold capabilities of the human visual system can conveniently be divided into spatial, temporal, and colour classes.

Spatial threshold measures — These measures relate to the ability to extract a target from its background or to resolve detail within a target. For spatial threshold measures, it is usually assumed that the target does not vary with time. Common spatial threshold measures are threshold luminance contrast and visual acuity. The luminance contrast of a target quantifies its visibility relative to its immediate background. The higher is the luminance contrast, the easier it is to detect the target. There are three different forms of luminance contrast commonly used for uniform luminance targets seen against a uniform luminance background (see Section 1.3.2). Visual acuity is a measure of the ability to resolve detail for a target with a fixed luminance contrast. Visual acuity is most meaningfully quantified as the angle subtended at the eye by the detail that can be resolved on 50 percent of the occasions the target is presented, expressed in minutes of arc. An alternative is decimal acuity, which is the reciprocal of the angle subtended measured in minutes of arc. Using the former measure, the visual acuity corresponding to "normal" vision is taken to be 1 min arc. Unfortunately, for simplicity, a relative measure of visual acuity is used by the medical profession. This is the distance at which a patient can read a given size of letter or symbol relative to the distance an average member of the population with normal vision could read the same letter or symbol. For example, if the person is said to have 6/10 vision it means that the person can only read a given letter at 6 m that an average member of the population with normal vision can read from 10 m.

Temporal threshold measures — These measures relate to the speed of the response of the human visual system and its ability to detect fluctuations in luminance. For temporal threshold measures it is usually assumed that the target is fixed in position and size.

The ability of the human visual system to detect fluctuations in luminance can be measured as the frequency of the fluctuation in Hertz (cycles/second) and the modulation of the fluctuation for the stimulus that can be detected on 50 percent of the occasions it is presented. The modulation is expressed as

$$M = (L_{max} - L_{min}) / (L_{max} + L_{min})$$

where M is the modulation, L_{max} is the maximum luminance (cd/m^2), and L_{min} is the minimum luminance (cd/m^2).

This formula gives modulations that range from 0 to 1. Sometimes, modulation is expressed as a percentage modulation, calculated by multiplying the modulation by 100.

Colour threshold measures — Colour threshold measures are based on the separation in colour space of two colours that can just be discriminated. In principle, the separation can be measured in many different ways but by far the most widely used has been separation on the CIE 1931 chromaticity diagram or the related uniform chromaticity scale diagram (see Section 2.4.1).

3.4.2 Factors Determining Visual Threshold

There are three distinct groups of factors that influence the thresholds of vision: visual system state, target characteristics, and the background against which the target appears.

Important visual system factors are the luminance to which the visual system is adapted, the position in the visual field where the target appears, and the extent to which the eye is correctly accommodated. The luminance to which the visual system is adapted determines which photoreceptor types are operating. The position in the visual field where the target appears determines the size of the receptive field available to the visual system and the type of photoreceptors available. The state of accommodation determines the retinal image quality. As a general rule, the lower the luminance to which the visual system is adapted, the further the target is from the fovea, and the more mismatched the accommodation of the eye is to the viewing distance, the larger will be the threshold values.

Important target characteristics are the visual size and luminance contrast of the target and any colour difference between the target and the immediate background. The effect of these variables on thresholds will be modified by any movement of the target or the observer. Any one of these task characteristics can be the threshold measure of interest but the others will interact with it. For example, the visual acuity of a target will be different for targets of different luminance contrast and colour difference. As a general rule, the closer the other target characteristics are to their own threshold, the greater will be the threshold of the measured variable. When the target is moving, the faster and the more erratic is its movement, the greater will be the increase in the measured threshold because of the difficulty of keeping the retinal image of the target on the fovea.

The important factors for the effect of the background against which the target appears are the area, luminance, and colour of the background. These factors are important because they determine the state of adaptation of the visual system and the potential for interacting with the image processing of the target. As a general rule, the larger the area around the target that is of a similar luminance to the target and neutral in colour, the smaller will be the threshold measure.

3.4.3 SPATIAL THRESHOLDS

About the simplest possible visual task is the detection of a spot of light presented continuously against a uniform luminance background, e.g., the headlamp of a distant bicycle on an unlit rural road. For such a target the visual system demonstrates spatial summation, i.e., the product of target luminance and target area is a constant. This implies that the total amount of energy required to stimulate the visual system so that the target can be detected is the same regardless of whether it is concentrated in a small spot or distributed over a larger area. Spatial summation breaks down when the target is above a given size, called the critical size. The critical size varies with the angular deviation from the fovea. The critical size is about 0.5 degree at 5 degrees from the fovea, and about 2 degrees at 35 degrees from the fovea (Hallett 1963). There is very little spatial summation in the fovea, the critical size being about 6 min arc.

Given that the size of the target is above the critical size, the detection of the presence of a spot of light is determined simply by the luminance contrast. When the luminance of the surround is in the photopic range, Weber's law states that the luminance difference needed for a target to be detected is a constant proportion of the background luminance. Of course, this is for a target of fixed size. A more general picture of the effect of adaptation luminance on threshold contrast for targets of different size is shown in Figure 3.7. The increase in threshold contrast as adaptation luminance decreases is obvious, as is the increase in threshold contrast with

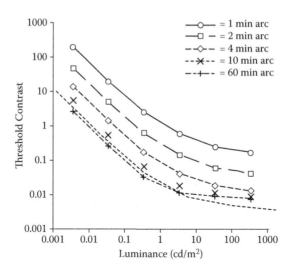

FIGURE 3.7   Threshold contrast plotted against background luminance (cd/m²) for disc targets of various diameters (min arc), viewed foveally. The discs were presented for 1 second and were brighter than the background (after Blackwell 1959).

decreasing target size (Blackwell 1959). These data were obtained using a disc of different sizes presented, foveally, for 1 second. Decreasing the presentation time below 1 second increases the threshold contrast, for all sizes, particularly at lower adaptation luminances (Graham and Kemp 1938).

Figure 3.8 shows the threshold contrast necessary to detect a bar pattern subtending 5 degrees, stationary or moving at 24 degrees/second, against a background luminance of 9.8 cd/m², for different eccentricities (Rogers 1972). It is evident from Figure 3.8 that less threshold contrast is needed to detect a moving target than a stationary target in the fovea and in the peripheral visual field. Interestingly, the threshold contrast for the stationary target increases dramatically with increasing eccentricity but the same is not true for the moving target. For the moving target, threshold contrast changes little with increasing eccentricity.

Turning now to visual acuity, Figure 3.9 shows the variation in visual acuity with adaptation luminance for foveal viewing of the target. As adaptation luminance decreases, visual acuity, measured as the reciprocal of the minimum gap size, decreases (Shlaer 1937).

Figure 3.10 shows visual acuity, measured as the reciprocal of the minimum gap size measured in minutes of arc, plotted against eccentricity, for a range of adaptation luminances. For the adaptation luminance of 3.2 cd/m², i.e., for the conventional photopic/mesopic border, acuity is at about 1 min arc in the fovea and declines rapidly to about 10 min arc as eccentricity increases. For adaptation luminances below 0.006 cd/m², i.e., approaching the scotopic state where the fovea is blind and only the rod photoreceptors are active, visual acuity is best at about 10 min arc, 4 to 8 degrees off axis (Mandlebaum and Sloan 1947). For the mesopic state, the visual acuity is in transition between these two extremes.

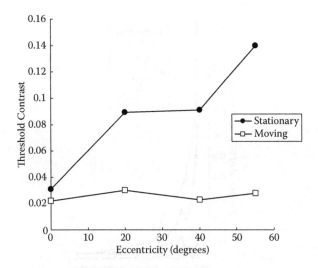

FIGURE 3.8 Threshold contrast plotted against eccentricity (degrees) for stationary and moving (24 degrees/second) bar pattern targets (after Rogers 1972).

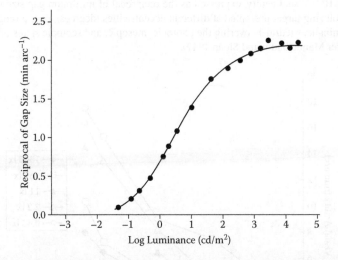

FIGURE 3.9 Visual acuity, expressed as the reciprocal of the minimum gap size (min arc $^{-1}$) for a Landolt ring, plotted against log background luminance (cd/m^2) (after Shlaer 1937).

Figure 3.11 shows the visual acuity for smooth relative movement of target and observer at different velocities under different illuminances (Miller 1958). As would be expected, visual acuity worsens as illuminance falls and angular velocity increases.

In the above discussion, threshold contrast and visual acuity have been considered separately because threshold contrast is usually measured with large targets, without detail, and visual acuity is measured with high-luminance contrast targets, with

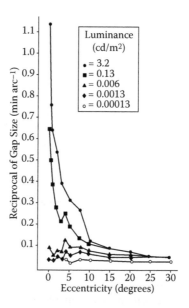

FIGURE 3.10 Visual acuity, expressed as the reciprocal of minimum gap size (min arc^{-1}) for a Landolt ring target presented at different eccentricities (degrees), over a range of background luminances (cd/m^2) covering the photopic, mesopic, and scotopic states of the visual system (after Mandelbaum and Sloan 1947).

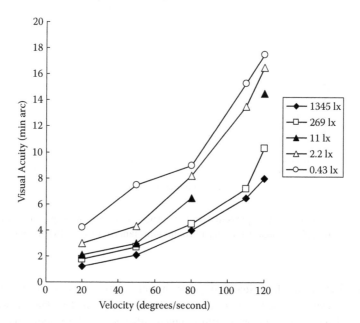

FIGURE 3.11 Visual acuity, expressed as the minimum gap size (min arc), for a Landolt ring target, illuminated to different illuminances (lx), with the observer moving at different angular velocities (degrees/second). Viewing is monocular and for a constant presentation time of 0.5 seconds (after Miller 1958).

detail. But many things of practical interest vary in both luminance contrast and size of detail and these two target characteristics can be expected to interact. The threshold capabilities of the visual system to such targets can be expressed as the contrast sensitivity function. This is a rather grand name for what is essentially a simple piece of information, the frequency response of the visual system to spatial variations in luminance. The contrast sensitivity function of the visual system is measured using sine-wave grating targets of different spatial frequencies and adjustable luminance modulation (see Sections 1.3.1 and 1.3.2). The spatial frequency of the grating consists of the number of cycles of the grating that lie within a 1-degree field of view for the observer, and hence is expressed in cycles per degree. The threshold luminance contrast condition is usually measured as luminance modulation, but it is often displayed as contrast sensitivity, which is the reciprocal of luminance modulation. Figure 3.12 shows contrast sensitivity functions for different adaptation luminances (Van Nes and Bouman 1967).

The value of this apparently esoteric piece of information is that any variation in luminance across a surface can be represented as a waveform, and any waveform can be represented as a series of sine waves of different modulation and frequency. The response of the visual system to sine waves of different modulation and frequency is given by the contrast sensitivity function. Thus, the contrast sensitivity function can be used to determine if and how a complex variation in luminance will be seen. If the luminance pattern has contrast sensitivities at all spatial frequencies that are greater than the threshold contrast sensitivities, i.e., all are above the contrast sensitivity function, the luminance pattern will be invisible. It is only when at least one spatial frequency has a contrast sensitivity below the contrast sensitivity function that the

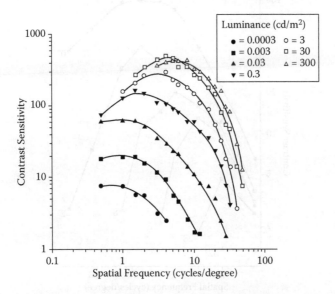

FIGURE 3.12 Contrast sensitivity functions for sine-wave gratings at different levels of background luminance (cd/m^2), covering the photopic, mesopic, and scotopic states of the visual system (after Van Nes and Bouman 1967).

luminance pattern will be visible in some form. The extent to which the luminance pattern will be seen in its entirety depends on the number of spatial frequencies for which the contrast sensitivity lies below the threshold contrast sensitivity; the more spatial frequencies for which this occurs, the more complete is the perception of the luminance pattern. Contrast sensitivity functions can be used for many practical purposes. For example they can be used to determine what size and contrast a road sign needs to be read from a given distance. The distance from which the observer views the luminance pattern is important because changing the viewing distance changes the spatial frequency of the pattern. As viewing distance increases, spatial frequency increases.

Returning now to Figure 3.12, it is apparent that decreasing adaptation luminance decreases both the contrast sensitivity and the maximum spatial frequency detectable, i.e., it produces a higher threshold contrast and a worse, i.e., larger, visual acuity. Also clear is the fact that the contrast sensitivity function changes rapidly below an adaptation luminance of about 30 cd/m^2. The deterioration takes the form of reduced contrast sensitivities at all spatial frequencies and a shift in the spatial frequency at which maximum contrast sensitivity occurs to a lower value.

The effect of eccentricity on the contrast sensitivity function is shown in Figure 3.13. As might be expected from the increase in receptive field size with increasing eccentricity, the contrast sensitivity function shows a dramatic reduction in the highest spatial frequency visible as deviation from the fovea increases, as well as a reduction in peak contrast sensitivity. What this means is that it is not possible

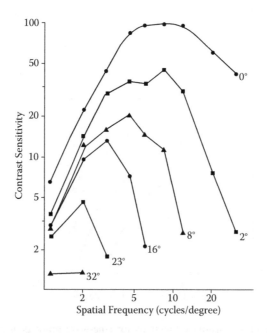

FIGURE 3.13 Contrast sensitivity functions for a 2.5-degree stimulus presented at different eccentricities (degrees) (after Hilz and Cavonius 1974).

to see fine detail more than a few degrees away from the fovea, but large objects are visible far into the periphery.

3.4.4 TEMPORAL THRESHOLDS

The simplest possible form of a temporal visual task is the detection of a spot of light briefly presented against a uniform luminance background. For such a target the visual system demonstrates temporal summation, i.e., the product of target luminance and the duration of the flash is a constant. This implies that the total amount of energy required to stimulate the visual system so that the target can be detected is the same, regardless of the time for which the target is presented. Temporal summation breaks down above a fixed duration, called the critical duration. The critical duration varies with adaptation luminance, ranging from 0.1 s for scotopic luminances to 0.03 s for photopic luminances. For presentation times longer than the critical duration, presentation time has no effect, the ability to detect the flash being determined by the difference in luminance between the flash and the background.

While the ability to detect a single flash is of interest for signaling purposes, an aspect of temporal thresholds of wider relevance to driving is the ability to detect repetitive light fluctuations, such as occur when a turn lamp is activated. Figure 3.14 shows the temporal equivalent of the contrast sensitivity function: the modulation sensitivity function, i.e., the reciprocal of the modulation threshold plotted against the frequency of the oscillation, for different adaptation luminances. These data were collected from a 68-degree-diameter field, uniformly illuminated, the flicker waveform being sinusoidal. It is clear that decreasing the mean luminance decreases

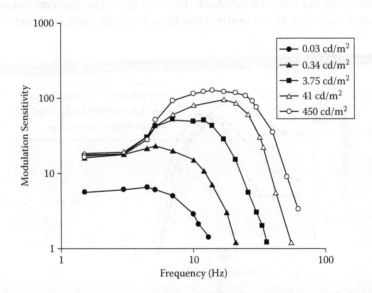

FIGURE 3.14 Modulation sensitivity functions for a large visual field at different average luminances (cd/m²). Modulation sensitivity is the reciprocal of modulation threshold (after Kelly 1961).

the sensitivity to modulation and shifts the frequency for peak sensitivity from about 20 Hz to 8 Hz. The other important point is that apart from the lowest average luminance (0.03 cd/m²), the results for all the other average luminances come to a common curve at low frequencies, but have different curves at high frequencies. This means that at low frequencies the ability to detect flicker is stable at a fixed modulation over a wide range of average luminances. Signaling for drivers usually involves low frequencies.

Figure 3.14 can be used to determine if a light fluctuation will be visible for a large area fluctuation. For a sine wave oscillation, if the modulation at the given frequency is above the curve for the appropriate average luminance, the flicker will not be visible. If it is below the curve, it will be visible. But most vehicle signals and traffic signals are much smaller than the 68-degree target used to obtain the data shown in Figure 3.14 and have to be seen by day and by night. Figure 3.15 shows the modulation sensitivity function for 68-degree and 4-degree targets, both with an average luminance of 100 cd/m², the 68-degree target having an edge blurred to dark over another 18 degrees, while the 4-degree target has a sharp edge to a dark background (Kelly 1959). Also shown is the modulation sensitivity function for a 2-degree target seen against a 60-degree background with the same average luminance (de Lange 1958). What is evident from Figure 3.15 is that peak sensitivity to fluctuations is in the range 8 to 20 Hz for the small targets with both bright and dark backgrounds but for low frequencies, background luminance makes a large difference to sensitivity.

3.4.5 COLOUR THRESHOLDS

Both the spatial and temporal thresholds discussed above have been measured using achromatic targets lit by nominally white light. In the photopic state and to some

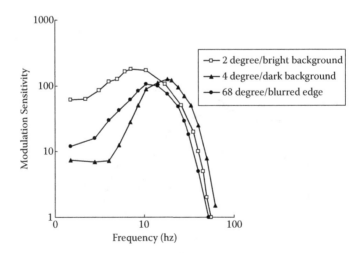

FIGURE 3.15 Modulation sensitivity functions for targets of different visual sizes, with different background brightnesses. Modulation sensitivity is the reciprocal of threshold modulation (after Boff and Lincoln 1988).

extent in the mesopic state, the human visual system has a well-developed ability to discriminate colours. The thresholds for discriminating different colours are given by the MacAdam ellipses plotted in the CIE 1931 chromaticity diagram (MacAdam 1942). Each ellipse represents the standard deviation in the chromaticity coordinates for colour matches made between two small visual fields with the reference field having the chromaticity of the centre point of the ellipse. The MacAdam ellipses were obtained in conditions ideal for comparison (photopic conditions with simultaneous viewing of adjacent small fields by a highly practiced observer) and so represent what is likely to be the finest colour discrimination possible. Brown (1951) showed that the effect of reducing luminance to low photopic levels was to increase the size of the ellipses. Colour discrimination between targets presented successively or between targets in which there is a wide range of colours and patterns present is similarly degraded (Narendran et al. 2000).

3.5 PERCEPTION OF A SCENE

Thresholds define the limits of the capabilities of the human visual system, but how does the visual system function when presented with targets above threshold? Such stimuli are the material used in the study of perception (Purves and Beau Lotto 2003). When considering how we perceive the world when the objects it contains are all above threshold, the overwhelming impression is one of stability in the face of continuous variation. As the eyes move in the head and the head itself moves about, the images of objects move across the retina and change their shape and size according to the laws of physical optics. Despite these variations our perception of objects rarely changes. This invariance of perception is called perceptual constancy. There are four fundamental attributes of an object that are maintained constant over a wide range of lighting conditions. They are:

Lightness — Lightness is the perceptual attribute related to the physical quantity, reflectance. In most lighting situations, it is possible to distinguish between the illuminance on a surface and its reflectance, i.e., to perceive the difference between a low-reflectance surface receiving a high illuminance and a high-reflectance surface receiving a low illuminance, even when both surfaces have the same luminance. It is this ability to perceptually separate the luminance of the retinal image into its components of illuminance and reflectance that ensures that a black cat placed in the brightest part of a headlamp beam is seen as black while a piece of white cardboard outside the beam is seen as white, even when the luminance of the cat is higher than the luminance of the cardboard.

Colour — Physically, the stimulus a surface presents to the visual system depends on the spectral content of the light illuminating the surface and the spectral reflectance of the surface. However, quite large changes in the spectral content of the illuminant can be made without causing any changes in the perceived colour of the surface, i.e., colour constancy occurs. Colour constancy is similar in many ways to lightness constancy. There are two factors that need to be separated, the spectral distribution of the incident light and the spectral reflectance of the surface. As long as the spectral content of the incident light can be identified, the spectral reflectance of the surface, and hence its colour, will be stable.

Size — As an object gets further away, the size of its retinal image gets smaller but the object itself is not seen as getting smaller. This is because by using clues such as texture, motion parallax, and masking, it is possible to estimate the distance to the object and then to compensate, unconsciously, for the decrease in retinal image size.

Shape — as an object changes its orientation in space, its retinal image changes. Nonetheless, in most lighting conditions the distribution of light and shade across the object makes it possible to determine its orientation in space. This means that in most lighting conditions a circular wheel that is tilted will continue to be seen as a tilted circular wheel even though its retinal image is an ellipse.

These constancies represent the application of everyday experience and the integration of all the information about the lighting available in the whole retinal image to the interpretation of a part of the retinal image that is inherently ambiguous. Given this process it should not be too surprising that the constancies can be broken by restricting the information available coincident with the object being viewed. For example, viewing a uniform luminance surface through an aperture that restricts the view to a limited part of the surface will often eliminate lightness constancy, i.e., make it impossible to accurately judge the reflectance of the surface. Likewise, eliminating cues to distance, such as gradients in texture, motion parallax, and overlapping of objects will destroy size constancy; changing cues to the plane in which an object is lying will reduce shape constancy and eliminating information on the spectral content of the illuminant will reduce colour constancy.

The constancies are most likely to be maintained when there is enough light for the observer to see the object and the surfaces around it clearly, the light being provided by an obvious light source that has a spectral power distribution covering all the visible spectrum. In addition, the constancies are most likely to be maintained when there are a variety of surface colours in the scene, including some small white surfaces, and there are no large glossy areas, both factors that help with the identification of the spectral content of the light source (Lynes 1971). Conversely, constancy is likely to break down whenever there is insufficient or misleading information available from the surrounding parts of the visual field. A breakdown of constancy is unlikely during the day but the limited coverage of headlamps and the limited spectral power distribution of some of the light sources commonly used for road lighting make a breakdown more likely when driving at night.

As examples of the problems such breakdowns can cause, consider the driver's perception of absolute and relative speed. When driving along a road, the retina receives a moving pattern called the optic flow in which different parts of the visual field appear to flow around the observer at different speeds in varying directions (Figure 3.16). Analysis of optic flow exposes both the structure of the world around you and your movement through that world (Gibson 1950). But how do you know that you are moving through the world rather than the world moving past you? The answer is that you do not, at least not from optic flow alone. Either other information is necessary to conclude that you are moving through the world or you have to make an assumption. Among the other sources of information are changes in your vestibular system in your inner ear when you are accelerating or braking and vibration felt through the vehicle's contact with the road. If such information is not enough, then the usual assumption is that the larger surrounding object (the world) is stationary and

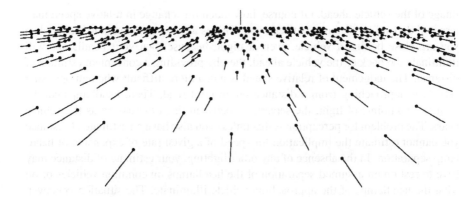

FIGURE 3.16 An optic flow pattern for an observer moving straight across and parallel to a plane in a direction given by the short line above the horizon. Each black dot represents an element on the plane and the line attached to each dot shows the associated velocity vector. The length of the line represents the instantaneous velocity and the direction of the line represents the direction of movement of the element (from Sekular and Blake 1994).

the smaller enclosed object (you) is moving. This is the explanation for the common movement illusion experienced when seated on a train and the adjacent train filling the window starts to move. The initial perception is that the train you are on is moving, until the absence of vestibular and vibration information tells you otherwise.

Given that you have determined that you are moving through the world, two other pieces of information are useful for the driver, his direction of movement and his speed. Gibson (1950) suggested that the direction of movement could be obtained by changes in the pattern of optic flow. When you are fixating the point that you are moving towards, the pattern of optic flow will be seen to expand if you are moving forward and to contract if you are moving back. Things get more complicated if you are not looking at the point towards which you are attempting to move because then the pattern of optical flow changes every time you make an eye movement (Regan and Beverley 1982). There is no doubt that the visual system has a mechanism for extracting the effect of eye movements but exactly what that is is uncertain.

As for speed, absolute speed cannot be obtained from optic flow alone because to judge speed you need to be able to estimate distance and distance cannot be obtained from optic flow. There are many other visual cues to distance in the retinal image such as perspective, relative sizes of familiar objects, texture gradients, shading, and masking of one object by another. Of course, much of this information is only available when the world through which you are moving is illuminated. Driving at night on headlamps alone limits the amount of distance information available and hence makes it difficult to judge absolute speed visually.

While absolute speed is of interest when driving on an empty road, when there are other vehicles on the road relative speed is of interest. For example, if you are following another vehicle, then as long as the retinal image size is constant, you are both travelling at the same speed and you are neither increasing nor decreasing your separation. If the retinal image starts to expand you are closing on the vehicle ahead and the rate at which you are closing is related to the rate of expansion of the retinal image of the vehicle ahead. Of course, how much this change in relative speed matters will depend on distance between you. This is an easy judgement to make when you are close behind a vehicle, even at night, because then your headlamps will illuminate the back of the vehicle ahead, thereby providing a convenient estimate of distance. The judgement of relative speed is much more difficult when an opposing vehicle is approaching from a distance on an unlit road. Then, headlamps can be seen as two points of light, the separation between them increasing as the vehicle nears. The problem for perception is that unless you also have an estimate of distance you cannot estimate the implication for speed of a given rate of expansion of headlamp separation. In the absence of any other lighting, your estimate of distance may have to rest on an assumed separation of the headlamps on common vehicles or on what the headlamps of the approaching vehicle illuminate. The situation gets even more difficult for a motorcycle when there is only one headlamp. Then, if you want to estimate the approach speed you have to detect the increase in size of the single headlamp as well as judge the distance.

Clearly, lighting has a role to play in stabilizing perception when driving at night, particularly road lighting. Road lighting provides a much more extensive view of the scene than headlamps alone, but its value in stabilizing perception is rarely considered. Even when road lighting is provided, enthusiasts for energy savings tend to want the light distribution restricted to the road surface. This discussion suggests that such a restriction would be unwise as providing some light on the surroundings of the road will help to set the road in context and provide a richer pattern of optic flow for the visual system to work on. Land and Horwood (1995) have shown that the edges of the road or lane markings detected in peripheral vision provide guidance for keeping the vehicle in the lane. When headlamps alone are in use, particularly low-beam headlamps, the amount of information available from the visual scene is limited, as is the pattern of optic flow. This suggests that a low level of light spread over a larger area with the aim of enlarging the optic flow pattern should be one objective for the designers of headlamp systems.

3.6 VISIBILITY AND ACCIDENTS

While there is little doubt that good visibility is a necessary condition for safe driving, alone it is not sufficient to guarantee safety. This is evident from studies of the factors contributing to accidents. Figure 3.17 shows the percentage of accidents in which, in the opinion of the investigating team, each of the listed factors was involved (Sabey and Staughton 1975; Hills 1980). These percentages are based on on-the-spot investigations of 2,036 road accidents in a rural area of the UK. The percentages do not sum to 100 percent because road accidents rarely have a single cause, usually a combination of causes is involved. Even so, Figure 3.17 declares that in 95 percent of the accidents studied, the road user contributed to the accident. Further, 44 percent of the accidents involved what is termed perceptual error. Included under perceptual error are distraction and lack of attention, incorrect interpretation, misjudgement of speed and distance, and a category intriguingly called "looked but failed to see."

A more detailed examination of the class of perceptual error can be used to explore the significance of visibility for driving. The "looked but failed to see" category

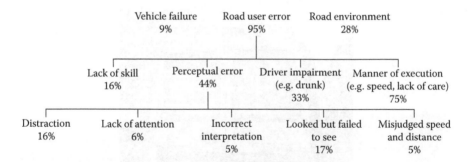

FIGURE 3.17 Apparent contributory factors in road accidents (after Hills 1980).

makes the largest contribution, a status supported by Cairney and Catchpole (1996) who found it to be frequently involved in accidents in urban areas. A classic example of looking but failing to see is a driver who pulls out from a side road straight into the path of a motorcyclist on the main road (Wulf et al. 1989). The driver looked along the main road but failed to see the motorcyclist although measurements of luminance contrast would indicate that the motorcyclist was clearly visible. How can this be? How can a driver look in the direction of a clearly visible vehicle and not see it? The answer lies in the fact that the total amount of information available through the visual system and through the other senses at any instant is too much for the brain to deal with simultaneously (Wolfe et al. 2006). To overcome this problem, the brain uses a number of mechanisms to select which stimuli to examine in detail and which to ignore. This process of selection is unconscious and is called attention. It represents an allocation of neural resources to a specific activity at a specific time and a consequent neglect of other information arriving at that time. Thus, if the driver is paying attention to another part of the visual field or input from another sense organ, or is attending to some internal brain activity, it is only too likely that a highly visible object will not be seen, even when viewed directly. This explains two of the other perceptual errors shown in Figure 3.17. Distraction occurs when attention is directed towards some other external but irrelevant source of information. Lack of attention occurs when attention is directed internally.

Given that some selection amongst the incoming information is necessary, how does a driver allocate neural resources? Four factors have been found to influence the focus of attention: conspicuity, expectation, mental workload, and capacity.

Conspicuity refers to the extent to which an object stands out from whatever is around it. Visibility is a necessary condition for detection but it is not a sufficient condition to make an object conspicuous. To make an object conspicuous it has to have some feature that easily differentiates it from other objects around it. What that feature may be can vary widely, depending on the background against which the object is usually seen. Vehicles that are likely to be stopped by the roadside in unusual places are often painted in dramatic patterns that vary in colour and reflectance to increase their conspicuity by day and equipped with flashing lights to enhance their conspicuity at night (Figure 3.18). The choice of the feature or features to use to enhance conspicuity should be made on the basis of the visibility of the

FIGURE 3.18 An Environment Agency vehicle painted and lit to be conspicuous by day and night. The painted chevrons are alternately red and white.

feature and its rarity. To be conspicuous, it is necessary to be both visible and different. The higher the level of conspicuity, the less likely it is that a driver will look but fail to see.

Expectation is the extent to which the presence of the object can be predicted. A clearly marked pedestrian crossing sets up an expectation that there may be pedestrians on the road and encourages the driver to look directly at it. The greater is the expectation that a specific object will be present, the less likely it is that the object will be looked at but not seen. Experience is the basis of expectation, both in general and in localities. Experienced drivers are less likely than novices to look but not see. Drivers who are familiar with a locality know what to expect and hence what sort of hazards are most likely to be occur.

Mental workload refers to the amount of information that has to be handled at any moment in time. It is important to appreciate that mental workload extends beyond the information arriving at the senses to the handling of past information using memory and decision-making capabilities. Given that attention is a process for allocating neural resources and that, for an individual, such resources are fixed, it is obvious that the greater the mental workload, the more likely it is that something of significance will be ignored, i.e., looking but failing to see is more likely as mental workload increases. Increased mental workload has been shown to be related to worse driving performance (Chaparro et al. 2005).

Capacity refers to the ability to handle information flow. Different people have different capacities. Further, an individual's capacity to handle information flow can be decreased by age, fatigue, alcohol, and drugs. Again, the more restricted the individual's capacity, the more likely it is that the individual will look but fail to see.

To summarize, unexpected, inconspicuous objects that occur in front of drivers with limited capacity at times of high mental load are most likely to be looked at but not seen, even when they have a high visibility.

The remaining two categories of accident causes classified as perceptual error are incorrect interpretation and misjudgement of speed and distance. Incorrect interpretation is probably related to the limited time drivers have available to gather information. Frequently, drivers have only one or two seconds in which to decide whether to make a particular maneuvre. In the few seconds available the driver will only be able to fixate and accommodate on a few places. Limited information will sometimes lead to misinterpretation of the situation facing the driver.

Judgements of speed and distance are both involved in two maneuvres that produce many serious accidents, turning across traffic and overtaking, these maneuvres requiring the driver to move from the correct side of the road and go across or into the opposing traffic (Caird and Hancock 2002). To do this safely it is necessary to make accurate judgements of how far off an approaching vehicle is and at what speed it is approaching. Experimental evidence (Jones and Heimstra 1964) suggests that people are not very good at making either of these judgements, particularly when the judgement has to be made at a long distance and the approaching vehicle is moving directly towards the driver, so the change in size and the perceived relative motion are small.

By now it will be appreciated that the driver's task is a complex one, involving both visual and cognitive factors (Wierda 1996). Within a very limited time the driver has to interpret what is likely to happen on the road ahead. To do this the driver has developed a series of expectations of other drivers' behaviour and of what are the appropriate locations to examine. The driver will be faced with objects of different degrees of visibility and conspicuity and will have to make judgements for which the visual system is not well suited. It is this context that vehicle and roadway lighting has to operate.

3.7 SIGHT AND DRIVING

Because sight is considered an essential sense for driving, most countries require some sort of vision test before a licence to drive is issued. These vary from country to country. For most drivers these tests involve little more than a measurement of visual acuity made under photopic conditions with unlimited time, although more stringent tests involving field of view and colour vision may be required for drivers of commercial and public service vehicles (see Section 12.5.1). Despite these requirements, attempts to find a link between simple visual functions and the accident record of drivers have proved largely fruitless. Burg (1967) measured the static and dynamic visual acuity, the field of vision, the extent of misalignment of the two eyes, and the rate of recovery from glare for 17,500 drivers who between them had had over 5200 accidents in the previous three years. He was able to find only very weak correlations between the measured visual capabilities and the drivers' accident records. Charman (1997) has confirmed that simple visual capabilities are at best weakly correlated with accident occurrence. This failure can be explained in two ways. First, it is possible that drivers with worse visual capabilities are aware of their limitations and drive within them (see Section 12.4.5). Second, there is the point that the visual capabilities measured were very simple and had no time limit. This is a problem because the driver often has to detect the presence and/or movements of a number of different obstacles, recognize them and their characteristic patterns of movement, and then relate them to each other, all the while maintaining an understanding of the more slowly changing features of the road, and all in a short time before making a maneuvre.

Somewhat more successful have been attempts to relate limited visual function to driving performance using younger drivers with their visual capabilities artificially constricted. Troutbeck and Wood (1994) restricted the driver's field of view to 40 degrees. As a result, the drivers drove more slowly and had difficulty in detecting and identifying road signs, avoiding obstacles, and driving through narrow gaps. Brooks et al. (2005) used a driving simulator on which young drivers drove around a winding course at 55 mph and tried to detect up to six pedestrians standing at the edge of the road. While doing this, the driver's retinal image quality was either reduced by defocusing lenses, or their visual field was restricted, or the luminances of the central road markings were varied over a wide range. Figure 3.19 shows how the drivers' visual acuity, the percentage of time the whole vehicle was within the correct lane, and the percentage of pedestrians detected varied with blur, field size, and road marking luminance.

It is clear from Figure 3.19 that while visual acuity is very sensitive to a blurred retinal image, the ability to steer a car is not. What is also somewhat sensitive to blur is the ability to detect pedestrians. The same is true for reducing the luminance of road markings but not for reducing the visual field size. Both steering accuracy and the ability to detect pedestrians are sensitive to reduced visual field size. These results are consistent with what is known as the selective degradation hypothesis (Leibowitz and Owens 1977). This hypothesis maintains that the reason for the increased probability of accidents at night is that all visual capabilities do not decline equally as luminance is reduced, but that drivers are not aware of this. Specifically, while the ability to see detail in the fovea deteriorates with reduced luminance, drivers are not aware how great the deterioration is because they experience little difficulty in steering using peripheral information and because many objects that need to be seen, such as retroreflective road signs, are designed to be highly visible. It is only when the driver comes across an unexpected and inconspicuous object, such as a deer, that the deterioration in foveal capabilities becomes apparent.

It should now be apparent why it has been difficult to establish much of a correlation between simple visual functions measured under advantageous conditions and road accidents. The basic problem is that driving involves everything from simple detection and resolution of detail to high-level perception using different parts of the visual field, all interacting with cognitive abilities and performed under time pressure. The failure to relate simple visual functions to accidents does not mean that visual capabilities and the lighting conditions that support them are unimportant. Rather, the evidence of the effect of visual capabilities on driving performance means that visual capabilities and the lighting conditions that support them are part of the problem and part of the solution but not the whole problem, nor the whole solution.

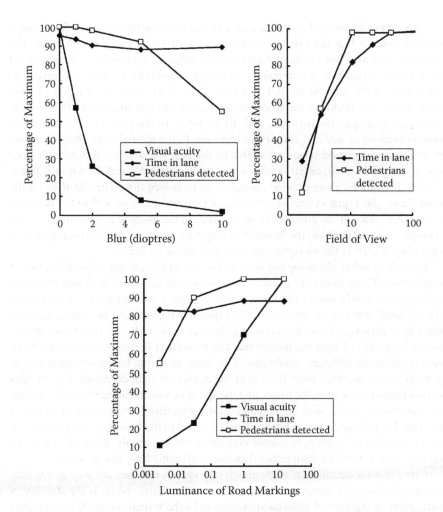

FIGURE 3.19 The percentage of the maximum possible value for visual acuity, time wholly in lane, and pedestrians detected with different degrees of blur (dioptres), different fields of view (degrees), and different road marking luminances (cd/m²). Visual acuity measurements were not made for the restricted visual field variable (after Brooks et al. 2005).

3.8 SUMMARY

The visual system consists of two parts, an optical system that produces an image on the retina of the eye and an image processing system that extracts different aspects of that image at various stages of its progress up the optic nerve to the visual cortex, while preserving the location where the information came from. The visual system devotes most of its resources to analyzing the central area of the retina, particularly the fovea. This implies that peripheral vision is mainly devoted to identifying something that should be examined in detail by turning the head and eyes so the image of whatever it is falls on the fovea.

This emphasis on the fovea, together with the fact that the eye has a fixed image distance, means that the eye is continuously adjusting its fixation and focus. The visual system also has to operate over a wide range of luminances, from sunlight to starlight. To do this it continually adjusts its sensitivity to light, increasing its sensitivity as the amount of light available falls. Decreasing the amount of light from daylight to darkness takes the visual system through three distinct operating states, the photopic, the mesopic, and the scotopic. In the photopic condition, fine discriminations of size and colour can be made. In the mesopic, the ability to make these discriminations deteriorates so that by the time the scotopic is reached, colour can no longer be seen, detail is impossible to discriminate, and the fovea is blind. Vehicle headlamps alone provide enough light to ensure that, when fixating on the road ahead, the visual system is operating in the photopic state within the beam, in the mesopic state at the edge of the beam, and, in the absence of moonlight, in the scotopic state well outside the beam. Road lighting ensures that the visual system is operating at least in the mesopic state over a much larger area.

Like every other physiological system, the visual system has a limited range of capabilities. These limits are expressed by the thresholds of vision. There are many different thresholds, one of the most common being visual acuity, i.e., the smallest size of detail that can be resolved. Others quantify the smallest luminance contrast that can be detected, the smallest colour difference than can be discriminated, and the lowest frequency of light fluctuation that can be seen as flickering. Different thresholds occur under different conditions of lighting and stimulus presentation but, in general, vision becomes more limited as the amount of light decreases, the stimulus occurs further away from the fovea, and the speed of movement increases. Threshold measurements provide well-defined and sensitive metrics to explore the operation of the visual system and so have been extensively used in the field of vision science, but for the practice of lighting, threshold measurements are mainly of interest for determining what will not be seen rather than how well something will be seen.

Given that the details of a scene are clearly visible, i.e., they are well above their threshold values, the dominant characteristic of the visual system is the stability of perception in the face of continuous variation in the retinal image. Given lighting conditions that provide enough light with a wide spectral distribution in such a way that how the space is lit can be easily understood, the lightness, colour, size, and shape of objects in the space remain constant no matter how they are viewed. It is only when the information about the space and the way it is lit is restricted or misleading that these perceptual constancies will break down.

While there is little doubt that good visibility is a necessary condition for safe driving, alone it is not sufficient to guarantee safety. This is evident from studies of the factors contributing to accidents. Almost half of accidents involve what is termed perceptual error, a category that includes distraction, lack of attention, incorrect interpretation, misjudgement of speed and distance, and something called "looked but failed to see." In accidents in which looked but failed to see is a factor, the driver reportedly looked at something that measurements define as being visible but failed to see it. The reason for this unexpected behaviour is that the total amount of information available through the visual system and through the other senses at any instant is too much for the brain to deal with simultaneously. To overcome this problem, the brain has to select which stimuli to examine in detail and which to ignore. This process of selection is unconscious and is called attention. Attention is influenced by conspicuity, expectation, mental load, and capacity. Unexpected, inconspicuous objects that occur in front of drivers with limited capacity at times of high mental load are most likely to be looked at but not seen, even when they have a high visibility.

The belief that vision is important to driving safely is the reason why measurements of visual capability are an integral part of the test for the issuing of a driving licence in most countries. Despite these requirements, attempts to find a link between simple visual functions such as acuity and the accident record of drivers have proved largely fruitless. This may be because drivers with worse visual capabilities are aware of their abilities and drive within them. Alternatively, it may be that the visual capabilities measured are too simple. The fact is, the driver's task is a complex one, involving both visual and cognitive factors. Within a very limited time the driver has to interpret what is likely to happen on the road ahead. To do this the driver has developed a series of expectations of other drivers' behaviour and of what are the appropriate locations to examine. The driver will be faced with objects of different degrees of visibility and conspicuity and will have to make judgements for which the visual system is not always well suited. It is in this context that vehicle and roadway lighting has to operate.

4 Road Lighting

4.1 SOME HISTORY

Paved roads have existed in Europe since Roman times and in South America since the time of the Inca empire, but road lighting is much more recent. Lighting designed specifically for enhancing the safety of the driver began to appear in the 1930s. Three factors converged to make road lighting possible and desirable at this time. The first was the availability of the necessary technologies in the form of an extensive electricity distribution network together with suitable lamps and luminaires. The second was the establishment of official systems for regulating the design and use of vehicles and the control of traffic. The third was the growth in the number of vehicles on the roads and the speeds those vehicles could sustain. Despite this convergence, the growth in road lighting was slow. Indeed, the autobahn system in Germany, introduced in the 1930s, and the motorway systems in Belgium and Britain, introduced in the 1950s, all designed for high-speed traffic, were opened without lighting. However, as traffic densities and traffic speeds have continued to increase, road lighting has become more extensive until it is now routinely considered as an important component of any road scheme.

The road lighting discussed above was for roads outside towns and cities, i.e., in locations where the main concern was for the safety of the driver. The lighting of roads in urban areas began much earlier for reasons of public safety (Painter 1999, 2000). In Paris, in the 15th century, it was decreed "during the months of November, December and January, a lantern is to be hung out under the level of the first floor window sills before 6 o'clock every night. It is to be placed in such a prominent position that the street receives sufficient light" (Schivelbusch 1988). As cities grew, there was an increased demand for some form of public lighting at night. This demand was first widely met by the introduction of gas lighting. In London, by 1823, the gas lighting system had grown to such an extent that 39,000 gas lamps provided lighting for 215 miles of urban road (Chandler 1949). Gas was the major source for urban exterior lighting at night for about 100 years, although the first exterior electric lighting installations, using arc lamps, were installed in the 1850s. It was not until the early 20th century that gas began to give way to electricity as the primary means of providing light at night. Today, urban road lighting is designed to meet a number of objectives such as reducing the fear of crime, improving the safety of drivers, cyclists and pedestrians, and enhancing the attractiveness of the environment. Of these, making life safer for drivers, cyclists, and pedestrians is paramount because it is always relevant while the others may or may not be, depending on the site.

4.2 THE TECHNOLOGY OF ROAD LIGHTING

The technology of road lighting consists of light sources and luminaires arranged in different ways and controlled by different means.

Light sources: The original electric light source used for road lighting was the incandescent lamp but over the decades the main light sources used have ranged through mercury vapour, fluorescent, low-pressure sodium, high-pressure sodium, and metal halide discharge. The characteristics of these light sources are summarized in Section 2.5. In the UK, the light source most widely used for road lighting is low-pressure sodium with high-pressure sodium being dominant in urban areas. In the US, high-pressure sodium is the light source most widely used for road lighting, with metal halide becoming more common in urban areas. The characteristics that are deemed most important for road lighting are first cost, for obvious reasons; luminous efficacy, because that affects energy costs; lamp life, because that affects maintenance costs; and colour rendering, but only in urban areas where the appearance of people and buildings is given greater weight. In the future, it is likely that greater interest will be shown in compact fluorescent and LED light sources.

Luminaires: The luminaires used for road lighting can be classified in several different ways. The most complete is the full luminous intensity distribution. There are a large number of luminous intensity distributions available, some symmetrical and some asymmetrical. Further, many luminaires allow for on-site adjustments in light source position within the luminaire to modify the luminous intensity distribution. Different light distributions are necessary because roads of different widths and layouts require different light distributions if the light is to be directed onto the road surface and not wasted. The exception to this concern takes the form of high mast lighting where the luminaires are mounted thirty metres or more above the ground. High mast lighting is used for lighting complex road junctions where the waste inherent in illuminating large areas of grass between roads is more than offset by the cost savings produced by minimizing the number of columns and simplifying the electricity distribution network.

A simplified system of luminaire classification based on the luminous intensity distribution just above and below the horizontal plane through the luminaire is also used. Limiting the amount of light emitted just above the horizontal plane through the luminaire is intended to reduce light pollution (see Section 13.7), while limiting the amount of light emitted just below the horizontal is done to control glare to the driver. The Illuminating Engineering Society of North America (IESNA) uses such a system, the system having four classes: full cutoff, cutoff, semi-cutoff, and non-cutoff (IESNA 2005a).

A full cutoff luminous intensity distribution is defined as having zero luminous intensity at or above 90 degrees from the downward vertical and no luminous intensity in the range 80–90 degrees from the downward vertical greater than 10 percent of the light source luminous flux.

A cutoff luminous intensity distribution is defined as having no luminous intensity at or above 90 degrees from the downward vertical greater than 2.5 percent of the light source luminous flux and no luminous intensity in the range 80–90 degrees from the downward vertical greater than 10 percent of the light source luminous flux.

A semi-cutoff luminous intensity distribution is defined as having no luminous intensity at or above 90 degrees from the downward vertical greater than 5 percent of the light source luminous flux and no luminous intensity in the range 80–90 degrees from the downward vertical greater than 20 percent of the light source luminous flux.

A noncutoff luminous intensity distribution is defined as a luminous intensity distribution where there is no limitation of luminous intensity above the angle having the maximum luminous intensity.

The luminaire classification system used in the UK (BSI 2003a) has six classes defined by the luminous intensity of the luminaire, in candelas/1000 lumens of light source luminous flux, at 70, 80, and 90 degrees from the downward vertical, in any direction, and the luminous intensity above 95 degrees, in any direction (see Section 4.4.2). This system is aimed primarily at controlling disability glare to the driver.

To be effective, both classifications assume that the luminaire is mounted horizontally but this is not necessarily the case. Road lighting luminaires may be tilted up as much as 15 degrees from the horizontal. The IESNA system has also been criticized because of the anomalies that arise from specifying a limit on luminaire luminous intensity in terms of light source luminous flux (Bullough 2002a). One example of an anomaly relevant to light pollution is the fact that the luminous flux emitted above a horizontal plane through a cutoff luminaire, as defined above, can, in principle, vary from 0 to 16 percent of the light source luminous flux, while the corresponding figures for a semi-cutoff luminaire are 0 to 31 percent. As a result of such anomalies, people concerned about light pollution tend to use an alternative term, fully shielded, to describe luminaires that emit no light directly above the horizontal plane through the luminaire. Fully shielded luminaires typically have the aperture through which light is emitted sealed by a transparent, flat lens. Luminaires where the lens drops below the plane of the aperture are not fully shielded. Broadly, such drop luminaires can be divided into shallow bowl and deep bowl types depending on the distance the lens extends below the aperture through which light escapes from the luminaire.

The IESNA is in the process of developing a new outdoor luminaire classification system based on the percentage of light source luminous flux emitted in a number of zones about the luminaire (IESNA 2007). Figure 4.1 shows a sphere centred on the luminaire, divided into three zones. The luminous flux emitted into the hemisphere above the luminaire is the uplight. The luminous flux emitted into the quarter of the sphere in front of the luminaire and below the horizontal plane is the forward light, and that emitted into the quarter of the sphere behind the luminaire and below the horizontal plane is the backlight. The remaining zone not evident in Figure 4.1 is the trapped light, which is the luminous flux emitted by the light source that does not get out of the luminaire. Uplight, forward light, backlight, and trapped light are all expressed as percentages of the light source luminous flux. The uplight, forward light, and backlight solid angles are all subdivided into different zones according to the angle from the downward vertical from the luminaire, uplight into two classes and forward light and backlight each into four classes. These subdivisions are introduced so as to identify the percentage of luminous flux emitted by the luminaire relevant to different effects such as glare, light trespass, and sky glow (see Section 13.7).

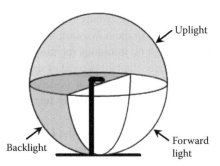

FIGURE 4.1 Three zones around a luminaire according to the IESNA outdoor luminaire classification system being developed (from IESNA 2007).

The new IESNA proposal is not yet a full classification system in that no limiting values are given for assigning a specific luminaire to a class. Rather, it is a way of systematically dividing up the luminous flux emitted by a light source from which different luminaire classes might be developed. Nonetheless, it is useful in exposing the typical distributions for common types of road lighting luminaires. Table 4.1 gives the percentage light source luminous flux occurring in each of the subdivisions of the zones for four luminaires, two luminaires used for lighting traffic routes, and two used for lighting urban areas that include roads. It is clear from Table 4.1 that the luminaires used to light traffic routes, i.e., road lighting luminaires, are very similar in their distribution of light source luminous flux although the drop lens luminaire has less trapped light and slightly more uplight than the flat lens luminaire. Where large differences occur is in the urban area lighting luminaires particularly in the amount of upward luminous flux. It is also apparent that the road lighting luminaires provide more forward light than the urban area lighting luminaires, these latter having approximately equal forward light and backlight.

The luminous flux distribution is not the only important characteristic of road lighting luminaires. Others are the light output ratio, the protection provided against the ingress of dirt and moisture, and the physical size. The luminaire light output ratio, or the sum of uplight, forward light, and backlight in the IESNA outdoor luminaire classification system, quantifies the proportion of light emitted by the light source that gets out of the luminaire. Light output ratios for road lighting luminaires tend to be about 0.8. The level of ingress protection is given by the IP number (see Table 2.5). Typically, road lighting luminaires are IP65, meaning they are strongly protected against the ingress of dust and driving rain. This implies that the interior of the luminaire should remain clean although the outside will still require regular cleaning if a marked deterioration in light output is to be avoided. As for physical size, this is relevant to the amount of leverage applied to a lighting column in high winds. The smaller the physical size of the luminaire and the more aerodynamic its shape, the less will be the leverage.

Columns: Most road lighting installations use columns to carry the luminaire or luminaires, although some installations use wire suspensions between buildings to light urban streets or catenary suspensions along traffic routes. Columns vary in

TABLE 4.1

Percentage of Light Source Luminous Flux in the Different Zones of the Developing IESNA Outdoor Luminaire Classification System for Two Road Lighting Luminaires and Two Urban Area Lighting Luminaires

Zone	Zone Subdivision (Degrees from Downward Vertical)	150 W HPS Drop Lens Road Lighting Luminaire (%)	150 W HPS Flat Lens Road Lighting Luminaire (%)	175 W MH Glass, Tear-Drop Urban Area Lighting Luminaire (%)	175 W MH Acrylic, Acorn, Urban Area Lighting Luminaire (%)
Forward Light	0–30	11.5	5.0	3.8	0.5
	30–60	27.6	27.3	14.9	4.0
	60–80	17.1	20.6	12.9	13.2
	80–90	1.3	0.8	0.7	8.4
	Total	57.4	53.7	32.3	26.1
Backlight	0–30	5.2	4.1	3.8	0.5
	30–60	11.8	12.3	14.9	4.0
	60–80	6.0	4.4	12.9	13.2
	80–90	0.9	0.8	0.7	8.4
	Total	24.0	21.5	32.3	26.1
Uplight	90–100	1.1	0.0	0.1	8.9
	100–180	1.3	0.0	0.1	23.0
	Total	2.4	0.0	0.2	31.8
Trapped Light	Total	16.2	24.8	35.2	16.0

HPS = high pressure sodium, MH = metal halide.

From IESNA (2007).

height and materials used. Column heights above ground level typically range from 3.5 m to 14 m, although high mast installations use columns of 30 m and taller. Materials are usually aluminium or steel, although concrete was used in the past and plastic composites are of current interest. In rural areas where the electricity network uses overhead distribution, the poles carrying the electricity lines are often used for mounting the luminaires, regardless of the distortions this imposes on the lighting conditions achieved.

Lighting columns themselves are a hazard for the driver, which is one reason why concrete columns have fallen into disuse. Measurements in The Netherlands suggest that 20 percent of accidents causing death or injury involve single vehicle accidents where the vehicle leaves the road and collides with objects on or near the road. In 11 percent of such accidents the object collided with is a lighting column (Schreuder 1998). There are three approaches to reducing the hazard of the lighting column; to set the column back from the road edge so as to provide an obstacle-free zone alongside the road, to shield the column with a crash barrier of some sort, or to design the column to either bend or break away on impact. Each of these approaches has limitations. The set-back required on high-speed roads is large and may not always be available. Crash barriers are expensive although they may be necessary for other reasons. Columns that bend are restricted to shorter columns, usually less than 7 m, that are made of aluminium or plastic composite. Columns that break away on impact may cause more accidents because it is not possible to predict where they will fall.

The choice of light source, luminaire, and column will be guided by many factors, among them the desired layout of luminaires. The layout of the luminaires for single carriageway roads is usually single-sided, staggered, or opposite. In a single-sided installation all the luminaires are located on one side of the carriageway. The single-sided layout is used when the width of the carriageway is equal to or less than the mounting height of the luminaires. The luminance of the lane on the far side of the carriageway is usually less than that on the near side. In a staggered layout, luminaires are arranged alternately on opposite sides of the carriageway. Staggered layouts are typically used where the width of the carriageway is between 1 to 1.5 times the mounting height of the luminaires. With this layout, care is necessary to ensure that the luminance uniformity criteria are met. In the opposite layout, pairs of luminaires are located symmetrically opposite each other on the sides of the carriageway. This layout is typically used when the width of the carriageway is more then 1.5 times the mounting height of the luminaires.

The layout of luminaires for dual carriageways and motorways is usually central twin, central twin and opposite, or catenary. In a central twin layout, pairs of luminaires are located on a single column in the central reservation. This layout can be considered as a single-sided layout for the two carriageways. Where the overall width of the road is wider, either because the central reservation is wide or there are several lanes, the central twin and opposite layout can be used. In this, the central twin luminaires alternate with the opposite luminaires to form a staggered layout. In the catenary layout, luminaires are suspended from a catenary cable along the central reservation. The catenary layout offers good luminance uniformity, less glare, because the luminaires are viewed axially, and excellent visual guidance.

The above layouts apply to continuous linear stretches of roads not to sharp bends or conflict areas such as intersections and roundabouts. Layouts for these areas are bespoke, although some typical layouts are given in road lighting recommendations (BSI 2003b; IESNA 2005a).

Controls: The control of road lighting is usually based either on time or on the amount of daylight available and is applied to individual luminaires or to groups of luminaires. In the past, most road lighting was controlled by time switches to be on from half an hour after sunset to half an hour before sunrise, although enthusiasm for energy savings meant that some installations were partially or totally switched off at midnight. However, the use of time switches makes it difficult to deal with unexpected meteorological conditions. Today, the most common control system is based on a photoelectric cell, this being used to detect the amount of daylight available and thus to ensure that the road lighting is only used when necessary. In the future, this relatively simple control system is likely to become much more sophisticated. Developments in light source technology and electronic control gear are making dimming of the discharge light sources commonly used for road lighting feasible over a useful range. Further, developments in computer networking using mains signaling and wireless communication are making it possible to control many individual luminaires from a remote site and hence to manage their operation (see Section 14.4.2). These technical advances are behind the current interest in dimming road lighting according to traffic flow and weather conditions (Guo et al. 2007). In addition, the ability to dim road lighting can be used to compensate for the decline in light output that occurs over light source life. At present, designers tend to anticipate this decline so that, when new, most road lighting installations produce more than the required road surface luminance. This excess can be eliminated by dimming; the level of dimming being reduced as the light output of the light source decreases. Depending on the light source, such compensating dimming can lead to energy savings of about 10 percent.

4.3 METRICS OF ROAD LIGHTING

Given the need to make a choice between a wide variety of light sources, luminaire types, column heights, road layouts, and control systems it is necessary for the designer of road lighting to have a target to aim at. What will be discussed here are the attempts that have been made to provide single number metrics suitable for quantifying the effectiveness of the choices made.

The primary purpose of road lighting is to make people, vehicles, and objects on the road visible without causing discomfort to the driver. Road lighting makes people, vehicles, and objects on the road visible by producing a difference between the luminance of the person, vehicle, or object and the luminance of its immediate background, which is usually the road surface or its edges (BSI 2003b). In principle, this may present a problem because, depending on the luminance of the people, vehicle, or object and its immediate background, the luminance difference can be positive or negative, above or below threshold. Both positive and negative luminance differences that are below threshold are unlikely to be seen. In practice, this is seldom a problem because people, vehicles, and objects rarely have a single, uniform luminance and neither do their backgrounds. Figure 4.2 shows a view of a vehicle on a road lit by road lighting. The vehicle varies in luminance as does the background against which it is seen, which means that at any instant the vehicle presents a number of luminance differences to the observer. This is a good thing because it means that it is unlikely that the whole vehicle will disappear.

While the existence of multiple luminance differences is useful for detection, they may make identification difficult, particularly when the pattern of luminance differences changes dramatically as the vehicle moves. A closer examination of Figure 4.2 shows that road lighting and vehicle lighting provide some stability to the pattern of luminance differences. Specifically, the road lighting produces a pattern of shadow on the road around the vehicle while the headlamps of the vehicle increase the luminance of the road immediately in front of the vehicle. The shadow pattern around the vehicle fluctuates as the vehicle moves beneath successive road lighting luminaires, but is always present. Similarly, the luminance differences between the vehicle pro-

FIGURE 4.2   A vehicle on a lit road with its headlamps on low beam. Note the variations in luminance within the vehicle and between the vehicle and the immediate background, especially the road surface.

file and the road immediately ahead will vary as the vehicle moves under the road lighting but are always present.

Despite this variability in luminance differences for real objects, the metrics that have been developed to quantify the ability of road lighting to make objects visible have been based on the luminance of planar surfaces. The first such metric was called revealing power (Waldram 1938). To calculate the revealing power of a road lighting installation at a specific point requires four pieces of information: the road surface luminance, the vertical illuminance on the critical object, the luminance difference threshold of the critical object, and the cumulative frequency distribution of the reflectances associated with critical objects. The last two requirements depend on the definition of the critical object. The critical object is usually taken to be a diffusely reflecting square plate of 20 cm side, placed vertically on the road surface (Narisada 1995). The justification for the choice of an object of this size is that an object 20 cm high will just pass beneath most vehicles without hitting the underside. While the physical dimensions of the critical object are 20 cm, for determining the luminance difference threshold it is necessary to define the visual size of the critical object, i.e., the angle the critical object subtends at the eye. This, in turn, requires the specification of a distance at which the critical object is to be seen. The distance usually specified is 100 m on the basis that if the critical object can be seen at 100 m, it can be avoided (de Boer et al. 1959). Given the visual size of the critical object and the luminance of the road surface forming the background to the object, the luminance difference threshold can be determined from the psychophysical literature (Boff and Lincoln 1988). Given the luminance difference threshold and the road surface luminance, it is possible to identify the upper and lower luminances of the critical object necessary for it to be above threshold and hence visible. Once these luminances have been calculated, then with knowledge of the vertical illuminance on the critical object, the upper and lower reflectances that are necessary for the critical object to be visible can be calculated. These are known as the critical reflectances. The final step is to apply these critical reflectances to the cumulative frequency distribution of reflectances that are likely to seen on the road (Smith, 1938; Hansen and Larsen, 1979). Figure 4.3 shows such a cumulative frequency curve for pedestrian clothing reported by Smith (1938). As an example, assume that the upper and lower critical reflectances are 0.15 and 0.20 respectively. Critical objects with reflectances of 0.15 and below will be seen in negative contrast against the road while critical objects with reflectances of 0.20 and above will be seen in positive contrast. Examination of Figure 4.3 reveals that 80 percent of pedestrian clothing will fall below the lower critical reflectance and 10 percent will be above the upper critical reflectance, making the total revealing power at the point considered to be 90 percent. Of course, to cover the whole road it is necessary to make these calculations for a regular array of points. This will give a distribution of revealing power values. The general level can then be expressed as the area ratio, this being the percentage of the road area where the revealing power is above a fixed level, such as 70 percent (Narisada et al. 2003).

This somewhat convoluted procedure has the advantage of quantifying the effectiveness of road lighting in terms of visibility and emphasizing the importance of vertical illuminance. It has been used to generate equi-revealing power contours for different road lighting luminaire layouts (Narisada and Karasawa 2001) and for identifying significant design parameters (Narisada et al. 2003). However, it does suffer from some limitations. The most severe is that the data on the cumulative frequency distribution of reflectances are confined to pedestrian clothing. Although

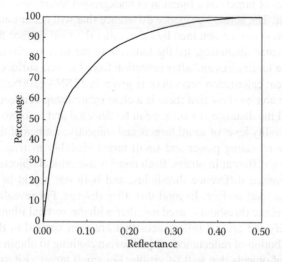

FIGURE 4.3 The cumulative frequency of the reflectances of clothing worn by pedestrians (after Smith 1938).

it is certainly true that pedestrians are still highly exposed to the risk of injury or death when moving in or close to traffic (see Section 1.5), they are not the only objects that need to be seen on the road. More up-to-date data on the reflectances of all objects likely to be found on roads are needed if revealing power calculations are to be meaningful today. Another limitation is the fact that the luminance increment threshold data in the literature has usually been obtained using uniform luminance fields to which the observers were fully adapted. The luminance distribution in front of the driver is rarely uniform and may even be diverse enough for the driver not to be fully adapted following some changes on fixation, although Narisada (1995) claims to have found a way around this problem. These limitations, together with the assumptions made about the critical object, have been enough to ensure that revealing power has been a subject of academic interest rather than a topic of practical relevance.

Revealing power has not been the only attempt to quantify the effectiveness of road lighting using visibility. In the US, an approach called small target visibility has been developed. Small target visibility is defined as the weighted average of the visibility levels of a number of flat, 18 cm side, square targets, all with a diffuse reflectance of 0.5, mounted vertically on the road surface (IESNA 2005a). The targets are arranged in a regular matrix of points across the road surface. The target is always 83 m ahead of the observer, oriented normal to the observer-to-target sight line, which is parallel to the centreline of the road. This distance and viewing direction convention maintains constant both target size and target/observer geometry. The observer is assumed to be 60 years of age, with normal vision, and using a fixation time of 200 ms. The visibility level of the target is defined as the ratio of the actual luminance difference of the target to the luminance difference threshold of the target; the higher the ratio, the more visible the target. Adrian (1989) has provided a quantitative model that allows the calculation of luminance difference thresholds for different sizes of targets as a function of background luminance, for positive and negative contrast. The actual luminance difference that will occur under a proposed lighting installation can be obtained by calculating the road surface luminance and the target luminance, including, for the latter, both the light received on the target directly from the luminaires and after reflection from the road surface (Adrian et al. 1993). A complete calculation procedure is given in IESNA (2005a). Janoff (1990, 1992a) has been able to show that there is a clear relationship between the visibility of the target and the distance at which it can be detected and recognized, and a link between the visibility level of small targets and subjective ratings of their visibility.

It is clear the revealing power and small target visibility metrics are similar in some respects but different in others. Both metrics use critical objects, both involve the use of luminance difference thresholds, and both use a grid of measurement points across the road surface. Beyond this, they diverge. For revealing power, the luminance difference threshold is used together with the vertical illuminance on the object, to identify the critical reflectances that are then applied to the cumulative frequency distribution of reflectances of pedestrian clothing to obtain an estimate of the percentage of objects that will be visible. For small target visibility, the critical object has a fixed reflectance and the merit of the road lighting installation is judged simply on the weighted average of the calculated visibility levels. This eliminates the concerns expressed above about the relevance of the distribution of reflectances but leaves unanswered concerns about basing the assessment of road lighting on a single reflectance at the extreme of the cumulative frequency of pedestrian clothing (Figure 4.3). Small target visibility has been adopted as a metric for the quality of road lighting in the US (IESNA 2005a), but only half-heartedly in the sense that it is one of three alternative sets of recommendations, the other two being based on conventional photometric measures (see Section 4.4.1).

Both revealing power and small target visibility are attempts to develop metrics for road lighting that are closer to its ultimate purpose than conventional photometric measures, namely to make objects visible. Unfortunately, both are probably doomed to failure because the visual task of driving is multifaceted and both revealing power and small target visibility address only one facet. The driver certainly has to be able to detect small targets on-axis but also needs to be able to detect objects on the road but off-axis, to detect movement at the edges of the roadway, to judge the relative speeds and direction of movement of other vehicles, and to identify many different signs and signals. Further, the revealing power and small target visibility metrics are only really applicable in low traffic densities where the driver can see a considerable distance down the road. In heavy traffic, detection of small obstacles at a distance is not an issue, much more importance being attached to the relative movements of nearby vehicles. Even in the absence of other traffic where the driver can see far down the road, there can be no guarantee that all objects will be small and when they are, they are of marginal significance for accidents. Data from The Netherlands suggest that only 0.6 percent of accidents causing death or injury involve collision with small, loose objects on or near the road (Schreuder 1998).

This is a practical objection to the use of revealing power or small target visibility as metrics for road lighting effectiveness. A more fundamental objection arises from the fact that most objects on and near roads that have to be seen have multiple luminance differences, both within the object and between the object and the background (Figure 4.2). This poses a problem for the use of a single visibility level for an object, because when it has multiple luminance differences, the object has multiple visibility levels, a difficulty experienced by Blackwell et al. (1964) in their early work on visibility measurement. Attempts have been made to overcome this problem by using multifaceted hemispherical objects rather than flat plates (Lecocq 1993). This approach has been shown to produce good predictions of perceived visibility on the road but so has the STV approach (Bacelar et al. 1999). As a result, the use of hemispherical targets for visibility calculations has not been pursued. This is a pity because they are more representative of the complexity of real targets. It is not until this complexity is recognized and the very diverse nature of the driver's tasks is accepted that the visibility approach to quantifying the quality of road lighting will be appreciated.

So far, this discussion of the metrics of road lighting has been focussed on the effects on visibility of the light that illuminates people, vehicles, and objects on the road. However, light emitted by road lighting luminaires can have other effects on visibility, as well as visual discomfort, when they deliver significant amounts of light directly to the eyes of the driver. When these effects are noticeable they are called glare. There are two forms of glare that are relevant to road lighting, disability glare, and discomfort glare. Disability glare, as its name implies, produces a measurable change in visibility. This change is caused by light scattered in the eye, the scattered light forming a luminous veil over the retinal image of the road (Vos 1984). The amount of disability glare produced in a given situation can be measured by comparing the visibility of an object seen in the presence of the glare source with the visibility of the same object seen through a uniform luminous veil. When the visibilities are the same, the luminance of the veil is a measure of the amount of disability glare produced by the glare source, and is called the equivalent veiling luminance. Numerous studies have lead to several different empirical methods for predicting the equivalent veiling luminance (Holladay 1926; Stiles 1930; Stiles and Crawford 1937). Based on this work an equation has been developed to predict the equivalent veiling luminance from directly measurable variables. It is

$$L_v = 10\Sigma E_n \Theta_n^{-2}$$

where L_v is the equivalent veiling luminance (cd/m^2), E_n is the illuminance at the eye from the nth glare source (lx), and Θ_n is the angle between the line of sight and the nth glare source (degrees).

This formula is adequate for glare sources between about 1 and 30 degrees from the line of sight and for young people. However, a series of modifications of the formula have been suggested to extend the range over which accurate predictions can be made from 0.1 to 100 degrees, for age ranges up to 80 years and for eye iris colour (Vos 1999; CIE 2002).

The CIE (1995b) recommendations for road lighting limit disability glare by setting a maximum percentage increase in the luminance difference threshold occurring under the roadway lighting when the equivalent veiling luminance is allowed for. This criterion is known as the threshold increment. The percentage threshold increment (TI) can be obtained from the formula

$$TI = 65 \, (L_v / L^{0.8})$$

where TI is the threshold increment, L_v is the equivalent veiling luminance (cd/m^2), and L is the average road surface luminance (cd/m^2).

The IESNA (2005a) recommendations for road lighting attempt to control disability glare by setting maximum values for the ratio of the equivalent veiling luminance to the average road surface luminance. Where the small target visibility recommendations are followed, equivalent veiling luminance is included in the estimation of adaptation luminance used to determine the luminance difference threshold. Regardless of whether the percentage threshold increment or the veiling luminance ratio metrics are used, the importance of disability glare diminishes rapidly as the angle between the line of sight and the luminaire increases. Disability glare from road lighting luminaires is a minor matter compared to the disability glare produced by vehicle headlamps.

The other aspect of glare from road lighting luminaires that needs to be considered is discomfort glare. The study of discomfort glare from road lighting has a long and tortuous history, starting with Hopkinson (1940), and continuing until the present day, but without reaching any very definite conclusion. Three alternative metrics have been developed at various times: the luminaire luminance corresponding to the boundary between comfort and discomfort (de Boer et al. 1959), the glare mark (Adrian and Schreuder 1970), and the cumulative brightness evaluation (IESNA 1980). Adrian (1991) examined each of these formulations and showed that despite their apparent differences they actually had a very similar form. Specifically, discomfort glare from road lighting increases with increasing luminaire luminance and the solid angle subtended by the luminaire at the driver's eye, and decreases with increasing road surface luminance and increasing angle between the road lighting luminaire and the line of sight. There are differences in the exponents used in the three metrics but Adrian concluded that they all had the same characteristics and would yield comparable results when used to evaluate different road lighting installations. Despite this conclusion, the latest CIE recommendations for road lighting (CIE 1995b) have dropped the previously used glare mark system and replaced it with a simple assertion that experience suggests that road lighting installations designed to limit threshold increments to less than 10 to 15 percent are generally acceptable as regards discomfort glare. Similarly, the IESNA (2005a) road lighting recommendations have not adopted any specific system for limiting discomfort glare. This lack of enthusiasm for limiting discomfort glare from road lighting suggests that, in practice, such glare is insignificant compared to that experienced when facing oncoming headlamps.

4.4 ROAD LIGHTING STANDARDS

Road lighting standards vary in detail from country to country but they do have some common features. One such feature is the division of the road network into different classes, according to the type of users and the road geometry (Schreuder 1998). Different lighting metrics are used to form standards for the different classes and these metrics are then varied to match the subdivisions of each class. Another common feature is the aspects of road lighting dealt with. Most lighting standards contain metrics for the amount of light on the road, the distribution of light on the road, and the extent of disability glare produced by the road lighting. Two rather different sets of road lighting standards will be examined here, namely, those used in the US and the UK.

4.4.1 ROAD LIGHTING STANDARDS USED IN THE US

Road lighting standards in the US vary from state to state but many states have adopted the IESNA (2005a) recommended practice as a basis for their standards. The road classes used in IESNA (2005a) and their defining characteristics are summarized in Table 4.2.

Another classification used in IESNA (2005a) is the pedestrian conflict area. This has three classes: high, medium, and low. Assignment to a class is based on the number of pedestrians walking on both sides of the road plus those crossing the road anywhere other than at an intersection over a 200 m section of road during the first hour after dark. Specifically, a low-conflict area will have 10 or fewer pedestrians, a

TABLE 4.2

The Road Classification Used in Many Parts of the US

Class	Characteristics
Freeway A	A divided major road with full control of access; a road with high visual complexity and high traffic volumes, typically found in major urban areas and used only by vehicles
Freeway B	A divided major road with full control of access and used only by vehicles
Expressway	A divided major road for through traffic with partial control of access but some intersections, used mainly by vehicles
Major	Roads connecting areas of traffic generation and roads leaving a city; the part of the road network that serves most of the through traffic flow. Average daily traffic volume more than 3,500 vehicles
Collector	Roads servicing traffic between major and local class roads. These roads are used for traffic within residential, commercial and industrial areas, so motor vehicles, bicycles, and pedestrians will all be users. Average daily traffic volume 1,500 to 3,500 vehicles
Local	Roads for access to residential, commercial and industrial property so motor vehicles, bicycles, and pedestrians will all be users. Average daily traffic volume less than 1,500 vehicles

Source: IESNA (2005a).

medium-conflict area will have 11 to 100 pedestrians, and anywhere with more than 100 pedestrians will be a high-conflict area.

There are three different metrics recommended for the design of continuously lit roads and for isolated traffic conflict areas such as intersections on otherwise unlit roads: the illuminance on the road, the luminance of the road as seen by the driver, and the small target visibility for the driver. All are minima that should be maintained over the life of the installation. Table 4.3 gives the recommendations for the different combinations of road and pedestrian conflict classes based on illuminance. As the effect of the illuminance on visibility depends on the reflection characteristics of the road surface, there are different illuminance recommendations depending on the reflection properties of the assumed road surface (see Section 4.5). There are no pedestrian conflict areas for either type of freeway because pedestrians are excluded from such roads. The illuminance recommendations take two forms, both based on the illuminances calculated or measured at a regular array of test points spread over the road. The array should have at least two points per lane across the road and a maximum separation between points along the road of 5 m, for one luminaire cycle, which is the distance between two adjacent luminaires on the same side of the road. The two criteria are the amount of light expressed as the average of the illuminances calculated or measured at the array of test points and the illuminance uniformity expressed as the ratio of the minimum illuminance at any test point to the average illuminance of all the test points.

There are three interesting points about Table 4.3. The first is that the highest illuminances are recommended not for freeways, where traffic speeds are likely to be highest, but for major roads with high pedestrian conflict areas. The second is that the recommended illuminances for any class of road decreases as the pedestrian

TABLE 4.3

Illuminance Criteria for Different Road Classes Used in Parts of the US

Road Class	Pedestrian Conflict Class	Minimum Maintained Average Illuminance for Road Surface R1 (lx)	Minimum Maintained Average Illuminance for Road Surfaces R2 and R3 (lx)	Minimum Maintained Average Illuminance for Road Surface R4 (lx)	Minimum Illuminance Uniformity Ratio (Minimum/ Average)
Freeway A	—	6	9	8	0.33
Freeway B	—	4	6	5	0.33
Expressway	High	10	14	13	0.33
Expressway	Medium	8	12	10	0.33
Expressway	Low	6	9	8	0.33
Major	High	12	17	15	0.33
Major	Medium	9	13	11	0.33
Major	Low	6	9	8	0.33
Collector	High	8	12	10	0.25
Collector	Medium	6	9	8	0.25
Collector	Low	4	6	5	0.25
Local	High	6	9	8	0.17
Local	Medium	5	7	6	0.17
Local	Low	3	4	4	0.17
Isolated traffic conflict area		6	9	8	0.25

Source: IESNA (2005a).

conflict class changes from high to low. The third is that the illuminance uniformity is relaxed as the road becomes more residential.

Table 4.4 gives the minimum maintained luminance recommendations for the same combinations of road and pedestrian conflict classes. There are no divisions of the recommendations according to the road surface reflectance because this is taken into account when calculating the road surface luminance. However, there are two different measures of luminance uniformity, minimum to average luminance ratio and minimum to maximum luminance ratio. These luminances are based on the same regular array of test points described above but for the luminance recommendations the observer is assumed to be positioned 1.45 m above the road surface with a line of sight 1 degree down, resulting in a constant viewing distance for each test point of 83 m. The luminance recommendations have similar features to the illuminance recommendations, namely, the highest maintained average luminance is recommended for major roads with high pedestrian conflict areas, the recommended luminances for any class of road decrease as the pedestrian conflict class changes

TABLE 4.4

Luminance Criteria for Different Road Classes Used in Parts of the US

Road Class	Pedestrian Conflict Class	Minimum Maintained Average Road Surface Luminance (cd/m²)	Minimum Luminance Uniformity Ratio (Minimum/ Average)	Minimum Luminance Uniformity Ratio (Minimum/ Maximum)
Freeway A	—	0.6	0.29	0.17
Freeway B	—	0.4	0.29	0.17
Expressway	High	1.0	0.33	0.20
Expressway	Medium	0.8	0.33	0.20
Expressway	Low	0.6	0.29	0.17
Major	High	1.2	0.33	0.20
Major	Medium	0.9	0.33	0.20
Major	Low	0.6	0.29	0.17
Collector	High	0.8	0.33	0.20
Collector	Medium	0.6	0.29	0.17
Collector	Low	0.4	0.25	0.13
Local	High	0.6	0.17	0.10
Local	Medium	0.5	0.17	0.10
Local	Low	0.3	0.17	0.10
Isolated traffic conflict area		0.6	0.29	0.17

Source: IESNA (2005a).

from high to low, and the luminance uniformity measures are relaxed as the road becomes more residential.

Table 4.5 sets out the small target visibility criteria for different combinations of road class and pedestrian conflict class. Why there is no subdivision for expressways according to the pedestrian conflict class for this metric is not explained. The weighted average visibility levels are based on the same regular array of test points viewed in the same way as described above. The first question that should occur to anyone examining Table 4.5 is why are there luminance criteria as well as small target visibility values? The answer is that the small target visibility assumes an empty road. Where there is opposing traffic, disability glare from the headlamps of approaching vehicles can reduce the visibility of low-contrast objects significantly. To avoid any problem this may cause, minimum values of maintained road surface luminance and luminance uniformity are given, the actual values depending on the separation between opposing traffic streams. The separation between traffic streams is important because the greater the separation, the less the disability glare. The fact that supplementary luminance criteria are necessary when using the small target visibility metric may explain why it has been difficult to get people to use it. The small target visibility (STV) recommendations have similar features to the illuminance and luminance recommendations, namely, the highest

TABLE 4.5

Small Target Visibility Criteria and the Ancillary Luminance Criteria for Different Road Classes Used in Parts of the US

Road Class	Pedestrian Conflict Class	Minimum Maintained Weighted Average Road Surface Visibility Level	Minimum Maintained Average Road Surface Luminance (cd/m²) where the Width of the Median is < 7.3 m	Minimum Maintained Average Road Surface Luminance (cd/m²) where the Width of the Median is > 7.3 m	Minimum Luminance Uniformity Ratio (Minimum/ Maximum)
Freeway A	—	3.2	0.5	0.4	0.17
Freeway B	—	2.6	0.4	0.3	0.17
Expressway	—	3.8	0.5	0.4	0.17
Major	High	4.9	1.0	0.8	0.17
Major	Medium	4.0	0.8	0.7	0.17
Major	Low	3.2	0.6	0.6	0.17
Collector	High	3.8	0.6	0.5	0.17
Collector	Medium	3.2	0.5	0.4	0.17
Collector	Low	2.7	0.4	0.4	0.17
Local	High	2.7	0.5	0.4	0.10
Local	Medium	2.2	0.4	0.3	0.10
Local	Low	1.6	0.3	0.3	0.10
Isolated traffic conflict area		2.6	0.5	0.4	0.17

Source: IESNA (2005a).

maintained STV is recommended for major roads with high pedestrian conflict areas and the recommended STV for any class of road decreases as the pedestrian conflict class changes from high to low. As for the supplementary average maintained luminance criteria, a comparison of Tables 4.4 and 4.5 show that, for the same road/pedestrian conflict class combination, they are the same or less than when luminance criteria are used alone.

Disability glare from the road lighting luminaires is taken into account for all three forms of the recommendations. For the illuminance and luminance recommendations, an explicit metric is used, this being the maximum veiling luminance ratio, which is defined as the ratio of the veiling luminance to the average road surface luminance. For both metrics, for freeways, expressways, and major roads, regardless of the pedestrian conflict class, the maximum veiling luminance ratio is 0.3. For collector and local roads, regardless of the pedestrian conflict class, the maximum veiling luminance ratio is 0.4. For the small target visibility criteria, disability glare from the road lighting luminaires is implicit as it is included in the calculation of visibility level.

TABLE 4.6

Illuminance Recommendations for Intersections of Continuously Lit Roads Used in Parts of the US

Intersection	Minimum Maintained Average Road Surface Illuminance (lx)			Minimum Illuminance Uniformity Ratio (Minimum/ Average)
	High Pedestrian Conflict Area	Medium Pedestrian Conflict Area	Low Pedestrian Conflict Area	
Major/Major	34	26	18	0.33
Major/Collector	29	22	15	0.33
Major/Local	26	20	13	0.33
Collector/Collector	24	18	12	0.25
Collector/Local	21	16	10	0.25
Local/Local	18	14	8	0.17

Source: IESNA (2005a).

An area requiring special attention is where continuously lit major, collector, or local roads intersect. There are no intersections of this type on freeways and where intersections occur on expressways they are likely to be controlled by traffic signals. Table 4.6 sets out the recommended average maintained illuminances for intersections in low, medium, and high pedestrian conflict areas, involving all possible combinations of local, collector, and major roads, together with the minimum illuminance uniformity ratio. The principle behind these recommendations is that the average maintained illuminance should be equal to the sum of the recommended values for the two intersecting roads, assuming a R2 or R3 road surface.

4.4.2 ROAD LIGHTING STANDARDS USED IN THE UK

The road lighting recommendations used in the UK (BSI 2003a,b) identify three distinct situations: traffic routes where vehicles are dominant, conflict areas where streams of vehicles intersect with each other or with pedestrians and cyclists, and residential roads where the lighting is primarily intended for pedestrians and cyclists.

The criteria used to define lighting for traffic routes are the average road surface luminance, the overall luminance uniformity, the longitudinal luminance uniformity, the threshold increment, and the surround ratio. Again, these metrics are based on calculations or measurements at a grid of test points arranged across and along the road. The average road surface luminance is the average of the luminances of the road surface at all the test points. The overall luminance uniformity is the ratio of the lowest luminance at any test point to the average road surface luminance. The longitudinal luminance uniformity is the ratio of the lowest to the highest luminance found at test points on a line along the centre of a single lane. For the whole carriageway, the value taken is the lowest longitudinal luminance uniformity found for any of the lanes of the carriageway.

Traffic routes are divided into different classes. The different classes are based on the type of road, the average daily traffic flow (ADT), the speed limits, the frequency of conflict areas, any parking restrictions, and the presence of pedestrians. Table 4.7 specifies the different classes. Details of the recommended lighting criteria for dry roads of each class are given in Table 4.8. Where roads are likely to be damp or wet for a significant part of the night, a different requirement for overall luminance uniformity can be used. The lighting criteria for this condition are given in Table 4.9. For both dry and wet roads, the minimum surround ratio criterion is only applied where there are no areas with their own criteria adjacent to the carriageway, e.g., a separate cycle path. The surround ratio is the average illuminance just outside the edge of the road to the average illuminance just inside the edge of the road.

Disability glare from road lighting is limited by the use of a maximum threshold increment. Threshold increment is defined in Section 4.3. The values of percentage threshold increment given in Tables 4.8 and 4.9 are for high-luminance light sources such as high-pressure sodium and metal halide. They can be increased by five percentage points when low-luminance light sources such as low-pressure sodium or compact fluorescent are used. An alternative method to limit disability glare is to select a luminaire according to the luminaire classes given in Table 4.10. The different classes are defined by the luminous intensity of the luminaire/1000 lumens of light source luminous flux, at 70, 80, and 90 degrees from the downward vertical, in any direction, and the luminous intensity above 95 degrees, in any direction. Class G3 approximates to a cutoff luminaire in the IESNA (2005a) luminaire classification system. Class G6 approximates to a full-cutoff luminaire.

All traffic routes eventually reach a conflict area. A conflict area is a place where traffic flows merge or cross, e.g., at intersections or roundabouts, or where vehicles, cyclists, and pedestrians are in close proximity, e.g., on a shopping street. Lighting for conflict areas is intended for drivers rather than pedestrians. The criteria used for the lighting of conflict areas are the average road surface illuminance and the overall illuminance uniformity, which is the ratio of the lowest illuminance at any point on the carriageway to the average illuminance of the carriageway. The recommendations for the different lighting classes for conflict areas are given in Table 4.11.

It is obviously important that the lighting of conflict areas should be coordinated with that of the traffic routes leading into them. Table 4.12 indicates the compatible lighting classes for conflict areas on traffic routes. Where two traffic routes lit to different classes lead into a conflict area, the match should be made to the higher traffic route class.

For residential roads, pavements, and cycle paths, an appropriate lighting criterion should be selected from Table 4.13. The photometric measure used for these areas is illuminance as there are likely to be many different directions of view. To ensure adequate illuminance uniformity, the actual maintained average horizontal illuminance should not be more than 1.5 times greater than the minimum maintained average horizontal illuminance.

TABLE 4.7

Lighting Classes for Traffic Routes Used in the UK. ADT is the Average Daily Traffic Flow

Road Name	Detailed Description	ADT	Lighting Class
Motorway	Routes for fast moving, long distance traffic. Fully grade separated and with restrictions on use		
	Main carriageway in complex interchange areas	< 40,000	ME1
		> 40,000	ME1
	Main carriageway with interchanges at < 3 km	< 40,000	ME2
		> 40,000	ME1
	Main carriageways with interchanges > 3 km	< 40,000	ME2
		> 40,000	ME2
	Emergency lanes		ME4a
Strategic route	Routes for fast moving, long distance traffic with little frontage access or pedestrian traffic. Speed limits are usually in excess of 40 mph and there are few junctions. Pedestrian crossings are either segregated or controlled and parked vehicles are usually prohibited		
	Single carriageway	< 15,000	ME3a
		> 15,000	ME2
	Dual carriageway	< 15,000	ME3a
		> 15,000	ME2
Main distributor	Routes between strategic routes and linking urban centres to the strategic network with limited frontage access. In urban areas, speed limits are usually 40 mph or less, parking is restricted at peak times and there are positive measures for pedestrian safety		
	Single carriageway	< 15,000	ME3a
		> 15,000	ME2
	Dual carriageway	< 15,000	ME3a
		> 15,000	ME2
Secondary distributor	Rural areas. These roads link larger villages and heavy goods vehicle generators to the strategic and main distributor network	< 7,000	ME4a
		7,000–15,000	ME3b
		> 15,000	ME3a
	Urban areas. These roads have 30 mph speed limits and very high levels of pedestrian activity with some crossing facilities including zebra crossings. On-street parking is generally unrestricted except for safety reasons	< 7,000	ME3c
		7,000–15,000	ME3b
		> 15,000	ME2

TABLE 4.7 (continued)

Lighting Classes for Traffic Routes Used in the UK. ADT is the Average Daily Traffic Flow

Road Name	Detailed Description	ADT	Lighting Class
Link road	Rural areas. These roads link smaller villages to the distributor network. They are of varying width and not always capable of carrying two-way traffic	Any	ME5
	Urban areas. These roads are residential or industrial interconnecting roads with 30 mph speed limits, random pedestrian movements, and uncontrolled parking	Any	ME4b or S2
		Any with high pedestrian or cyclist traffic	S1

Source: BSI (2003b).

4.4.3 SIMILARITIES AND DIFFERENCES

A comparison of the road lighting standards used in parts of the US and in the UK will reveal that there are some similarities but also a lot of differences. The similarities are most evident in the classifications used for different road types. The US classifications of freeways, expressways, major, collector, and local roads can be roughly matched to the UK classifications of motorway, strategic, main distributor, secondary distributor, and link roads. Another similarity lies in the division of lit areas into traffic routes and conflict areas. Although the US provides recommendations for continuously lighted roads using three different metrics, illuminance, luminance, and small target visibility, and the UK uses only one, luminance, both the US and the UK use illuminance as the metric for lighting conflict areas rather than luminance. The reason for this is that in conflict areas there will be multiple directions of view as drivers, cyclists, and pedestrians attempt to negotiate the intersection or roundabout. A similar argument can be used to explain the use of illuminance as a metric for residential roads. Luminance is only useful as a metric for road lighting where there are a limited number of directions of view. This is the situation for traffic routes where there are few pedestrians and all the drivers are, hopefully, on the road and looking along it.

The differences between the US and UK recommendations become more apparent when the actual values are compared. For traffic routes this is done on the basis of luminance. A comparison of Tables 4.4 and 4.8 shows that the UK recommendations cover a wider range (0.3 to 2.0 cd/m^2) than the US recommendations (0.3 to 1.2 cd/m^2). Interestingly, the highest average road surface luminance in the UK recommendations occurs for motorways while in the US recommendations the maximum occurs for major roads with high pedestrian conflict areas. In fact, the average road

TABLE 4.8

Lighting Recommendations for Dry Traffic Routes in the UK

Lighting Class	Minimum Maintained Average Road Surface Luminance (cd/m²)	Minimum Overall Luminance Uniformity Ratio (Minimum/ Average)	Minimum Longitudinal Luminance Uniformity Ratio for the Carriageway	Maximum Threshold Increment (%)	Minimum Surround Ratio
ME1	2.0	0.40	0.70	10	0.50
ME2	1.5	0.40	0.70	10	0.50
ME3a	1.0	0.40	0.70	15	0.50
ME3b	1.0	0.40	0.60	15	0.50
ME3c	1.0	0.40	0.50	15	0.50
ME4a	0.75	0.40	0.60	15	0.50
ME4b	0.75	0.40	0.50	15	0.50
ME5	0.50	0.35	0.40	15	0.50
ME6	0.30	0.35	0.40	15	None

Source: BSI (2003a).

TABLE 4.9

Lighting Recommendations for Traffic Routes where Roads Are Frequently Wet in the UK

Lighting Class	Minimum Maintained Average Road Surface Luminance when Dry (cd/m²)	Minimum Overall Luminance Uniformity Ratio when Dry	Minimum Overall Luminance Uniformity Ratio when Wet	Minimum Longitudinal Luminance Uniformity for the Carriageway when Dry	Maximum Threshold Increment (%)	Minimum Surround Ratio
MEW1	2.0	0.40	0.15	0.60	10	0.50
MEW2	1.5	0.40	0.15	0.60	10	0.50
MEW3	1.0	0.40	0.15	0.60	15	0.50
MEW4	0.75	0.40	0.15	None	15	0.50
MEW5	0.50	0.35	0.15	None	15	0.50

Source: BSI (2003a).

TABLE 4.10

Luminaire Classes for the Control of Disability Glare Used in the UK

Lighting Class	Maximum Luminous Intensity/1000 Lumens at 70° (cd/1000 lm)	Maximum Luminous Intensity/1000 Lumens at 80° (cd/1000 lm)	Maximum Luminous Intensity/1000 Lumens at 90° (cd/1000 lm)	Luminous Intensity above 95° (cd)
G1	—	200	50	—
G2	—	150	30	—
G3	—	100	20	—
G4	500	100	10	0
G5	350	100	10	0
G6	350	100	0	0

Source: BSI (2003a).

TABLE 4.11

Lighting Recommendations for Conflict Areas in the UK

Lighting Class	Minimum Maintained Average Road Surface Illuminance (lx)	Minimum Overall Illuminance Uniformity Ratio
CE0	50	0.4
CE1	30	0.4
CE2	20	0.4
CE3	15	0.4
CE4	10	0.4
CE5	7.5	0.4

Source: BSI (2003a).

TABLE 4.12

Compatible Lighting Classes for Conflict Areas on Traffic Routes in the UK

Traffic Route Lighting Class	Conflict Area Lighting Class
ME1	CE0
ME2	CE1
ME3	CE2
ME4	CE3
ME5	CE4

Source: BSI (2003b).

TABLE 4.13

Lighting Recommendations for Residential Roads, Pavements, and Cycle Paths in the UK

Lighting Class	Minimum Maintained Average Ground Level Illuminance (lx)	Minimum Maintained Ground Level Illuminance (lx)
S1	15	5
S2	10	3
S3	7.5	1.5
S4	5	1
S5	3	0.6
S6	2	0.6

Source: BSI (2003a).

surface luminance for freeways in the US (0.6 cd/m^2) is less than a third of the average road surface luminance for motorways recommended in the UK (2 cd/m^2). A similar difference occurs in the illuminances recommended for the lighting of conflict areas (c.f. Tables 4.6 and 4.11). The range of illuminances for intersections in the US recommendations is 8–34 lx, but for conflict areas in the UK, which include intersections, the illuminance range is 7.5–50 lx.

Another comparison can be made on the basis of overall luminance uniformity, defined as the minimum to average luminance ratio in both sets of recommendations (c.f. Tables 4.4 and 4.8). The UK recommendations call for more uniform lighting than the US recommendations, the range of overall luminance uniformity for traffic routes being 0.35 to 0.40 for the former and 0.17 to 0.33 for the latter. A similar difference is evident for the lighting of conflict areas in the two sets of recommendations (c.f. Tables 4.6 and 4.11).

Other differences between the UK and US recommendations lie in the various metrics used for what are the same phenomena. There are additional but different luminance uniformity criteria in both sets of recommendations, as well as different approaches to limiting disability glare. There are also differences in the extent to which recommendations for traffic routes, adjacent areas, and conflict areas are coordinated. Of these, the most notable is the use of the surround ratio criterion for adjacent areas in the UK recommendations, a criterion that makes it more likely that drivers will detect people approaching the road (van Bommel and Tekelenburg 1986). In addition, the general US recommendations make no mention of changing values when roads are frequently wet, although individual states may, whereas the UK recommendations specifically consider wet roads. Finally, both UK and US recommendations ignore one application of increasing importance, the somewhat ungainly named city beautification. Essentially, this refers to city and town centres where lighting is part of an economic development plan to enhance the use of the city at night. The role of lighting in city beautification is to make the location an exciting place to be, primarily for pedestrians. Consequently, conventional road lighting is inadequate although some provision for the driver has to be made. Advice on lighting for city beautification can be found in CIBSE (1992), ILE (1995), and IESNA (1999).

In summary, road lighting recommendations in the UK and US are similar in concept but differ in detail. The UK recommendations should lead to brighter, more uniform lighting on roads with faster traffic than the US recommendations. However, the allocation of the highest average road surface luminance to the major roads in high pedestrian conflict areas rather than freeways in the US is consistent with putting most light where it will have the most beneficial effect in reducing fatalities and personal injury accidents (see Section 1.5).

4.5 ROAD LIGHTING IN PRACTICE

The extent to which road lighting standards are achieved in practice depends on the assumptions that have to be made. The lighting criteria requiring the fewest assumptions are those based on illuminance. Given knowledge of the position of a luminaire relative to the road and the luminous intensity distribution of the luminaire, the horizontal illuminance on the road surface can be calculated. Figure 4.4 shows the relevant geometry for an observer looking down the long axis of the road.

The formula for calculating the illuminance at a specific point, P, on the road surface received from a specific luminaire is

$$E = I\,(\varphi, \gamma)\,\cos\gamma^3\,/\,h^2$$

where E_n is the horizontal illuminance at point P from the nth luminaire (lx), $I(\varphi,\gamma)$ is the luminous intensity in the specified direction for the nth luminaire (cd), φ is angle between the transverse axis across the road and the line joining the intersection of the downward vertical through the luminaire with the road and point P, γ is the angle of emission from the downward vertical through the luminaire, and h is the height of the nth luminaire above the road surface (m).

The illuminance contributions at point P from all the surrounding luminaires are summed. Similar calculations are then made for a grid of points arranged across and along the road so that the average road surface illuminance and the illuminance uniformity criteria can be calculated.

Today, these calculations are almost always made using computer software but assumptions are still present. For a start, it is always assumed that the road is level. Then there is the fact that the average road surface illuminance recommended is a maintained value. This means the average road surface illuminance should not fall below the recommended value at any time during the life of the installation. This calls for assumptions about the change in the luminous flux of the light source and the decline in light output of the luminaire over time. Guidance on suitable values for these variables is given by manufacturers and in design guidance (BSI 2003b; IESNA 2005a), but they are still assumptions for any specific site.

Road lighting criteria based on luminance include all these assumptions plus others concerned with the road surface reflection properties. The luminance of any point on a road surface lit by a single luminaire and as seen by a driver is given by the formula

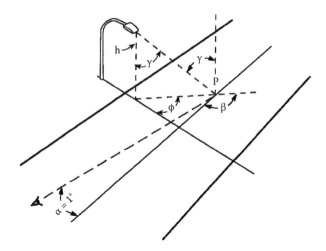

FIGURE 4.4 The luminance coefficient at point P for an observer looking straight down the road axis is dependent on four angles. These are: α = angle of observation from the horizontal, β = angle between the vertical planes of incidence and observation, γ = angle of incidence from the upward vertical, and φ = angle between the transverse axis across the road and the line joining the downward vertical at the luminaire and point P (from IESNA 2000).

$$L_n = q.E_n$$

where L_n is the road surface luminance at point P produced by the nth luminaire (cd/m²), E_n is the illuminance at point P produced by the nth luminaire (lx), and q is the luminance coefficient of the road surface at point P (cd/lm).

The road surface luminance depends on the position and luminous intensity distribution of the luminaire, the road surface reflection properties, and the geometry of the observer and the luminaire relative to the point under consideration. In principle, five angles are needed to define the relevant geometry. Figure 4.4 shows four of them. They are α, the angle of observation from the horizontal; β, the angle between the vertical planes of incidence and observation; γ, the angle of incidence from the upward vertical at point P; and φ, the angle between the transverse axis across the road and the line joining the intersection of the downward vertical through the luminaire with the road and point P. A fifth angle not shown in Figure 4.4 but which has an influence on the luminance at P, is the angle between the vertical plane of observation and the road axis. This angle is ignored because the reflection properties of most road surfaces are almost completely isotropic and the range of possible angles is limited.

To simplify matters further, one of the angles in Figure 4.4 is fixed. The angle to be fixed is α, the angle of observation from the horizontal. This angle has a limited range in practice. For a driver's eye height ranging from 1–3 m, which covers both sports cars and heavy trucks, α ranges from 0.35 to 2.86 degrees. Road surface luminance coefficients within this range show little variation (Moon and Hunt 1938; De Boer et al. 1952). The angle α is conventionally fixed at about 1 degree.

With these assumptions, it is possible to describe the reflection properties of a point on a road surface by a two-dimensional array of luminance coefficients, the dimensions of the array being β, the angle between the vertical plane of incidence and the vertical plane of observation, and γ, the angle of incidence from the upward vertical at point P. However, such a table is not convenient for use in calculation because the fundamental photometric data available for road lighting luminaires consist of a luminous intensity distribution that will be expressed in terms of angles φ and γ. This can be allowed for by replacing the illuminance in the formula for road surface luminance by the luminous intensity and using the inverse square law. The result is an expression for the luminance of a point on the road surface due to a single luminaire of the form

$$L = (q(\beta,\gamma) \cdot I(\varphi,\gamma) / h^2) \cos^3\gamma$$

where L is the road surface luminance at point P produced by the nth luminaire (cd/m^2), $I(\varphi,\gamma)$ is the luminous intensity of the nth luminaire towards point P (cd), $q(\beta,\gamma)$ is the luminance coefficient of the road surface at point P (cd/lm), h is the mounting height of the nth luminaire (m), and γ is the angle of incidence at point P of light from the nth luminaire (degrees).

The term $q(\beta,\gamma) \cdot \cos^3\gamma$ is called the reduced luminance coefficient (r) and is the metric conventionally used in what are called the r-tables that characterize the reflection properties of road materials. The two dimensions of the r-table are the angle β, the angle between the vertical plane of incidence and the vertical plane of observation, and the tangent of the angle γ, the angle of incidence from the upward vertical at point P (see Figure 4.4). Each cell in an r-table contains a value for the reduced luminance coefficient multiplied by 10,000.

With an r-table matched to the road material and the luminous intensity distribution for the luminaire, the luminance produced by a single luminaire at any point on the road surface as seen from a specified position can be calculated. This process can then be repeated for adjacent luminaires and the contributions from all luminaires summed to get the luminance at that point for the whole lighting installation. This process can then be repeated over an array of points on the road so as to get the luminance metrics used in road lighting standards (BSI 2003a). In practice, there is a limit to how large an area of road surface should be included in the array. Calculation is conventionally limited to the road surface between two adjacent luminaires on the same side of the road. By convention, in the UK the observer is placed 60 m in front of the first transverse row of calculation points and 1.5 m above the road. In the US, the corresponding assumptions are 83 m and 1.45 m. As regards the other dimension needed to define the observation position, for the average luminance and overall luminance uniformity metrics, the observation position is taken to be one quarter of the road width from the near side of the road.

While the above process is possible in principle, it is rarely used in practice. This is because, strictly, every piece of road has a unique r-table and that itself will change over space and time as different parts of the road wear differently. This implies that before designing a road lighting installation, measurements should be made of the reflection properties of samples taken from the road to be lit. Such measurements are difficult and time consuming. In consequence, a road reflection classification system has been developed by which many different road surfaces are approximated by a single r-table.

The first step in building a classification system is to identify some descriptive parameters for the measured r-table. Several different attempts have been made to do this (de Boer and Westermann 1964a,b; Roch and Smiatek 1972; Range 1972; Massart 1973; and Erbay 1974). After consideration, the CIE decided that most r-tables could be described by three parameters, one concerned with lightness and two concerned with specularity (CIE 1976).

The parameter adopted for lightness is the average luminance coefficient, Q_0, which is the solid angle weighted average of the reduced luminance coefficients in the r-table. The solid angle weighting ensures that the large reduced luminance coefficient values, corresponding to large γ angles, do not have an overwhelming influence on the value of Q_0. The average luminance coefficient, Q_0, can be calculated from the r-table using a weighting factor procedure developed by Sorensen (1974). The average luminance coefficient, Q_0, has been shown to be highly correlated to the average luminance produced on the road surface (Bodmann and Schmidt 1989).

As for the parameters relating to specularity, there are two, defined as ratios,

$$S1 = r\,(0,\,2)\,/\,r\,(0,\,0)$$

where $r\,(0,\,2)$ is the reduced reflection coefficient for $\beta = 0$ degrees, and $\tan\gamma = 2$, $r\,(0,\,0)$ is the reduced reflection coefficient for $\beta = 0$ degrees and $\tan\gamma = 0$.

$$S2 = Q_0\,/\,r\,(0,\,0)$$

where Q_0 is the average luminance coefficient, and $r\,(0,\,0)$ is the reduced reflection coefficient for $\beta = 0$ degrees and $\tan\gamma = 0$.

Frederiksen and Sorensen (1976) have shown that changes in the two parameters of specularity, S1 and S2, have somewhat different effects on the luminance patterns produced. Specifically, increases in S1 lead to rapid decreases in overall luminance uniformity ratio but have little effect on longitudinal luminance uniformity ratio. Increases in S2 lead to increases in both overall and longitudinal luminance uniformity ratios. As might be expected, there is evidence that the use of these two parameters of specularity together give more accurate predictions than either one alone (Frederiksen and Sorensen 1976). However, data provided by Sorensen (1975) and Frederiksen and Sorensen (1976) from numerous measurements taken on road samples in Europe show that S1 and S2 are highly correlated ($r = 0.93$). As a result, S2 has been dropped from the CIE classification system, leaving just Q_0 as a parameter for the diffuse reflectance and S1 as a parameter for the specular reflectance.

Having identified two descriptive parameters that can be used to characterise any r-table, the next step in developing a classification system is to decide on how many classes to use and where the boundaries should be. In 1976, the CIE recommended the use of two different four-class classification systems, the R system and the N system, the latter being recommended for countries where artificial brighteners are used in pavement materials to give very diffusely reflecting surfaces (CIE 1976). The boundaries of classes in both the R and N systems are determined by the value of S1. Table 4.14 shows the boundary values for the four classes in each system. Different countries opted to use either the R or the N system, the US choosing the R classification system (see Table 4.3).

However, there were doubts about the value of having four classes. Figure 4.5 shows the relationship between the descriptive parameters, Q_0 and S1, for the 286 measured road surfaces reported in Sorensen (1975). It is apparent that there is little change in Q_0 for classes R2, R3, and R4. A similar relationship between Q_0 and S1 is shown in van Bommel and de Boer (1980). Further, Burghout (1979) demonstrated that combining classes R2, R3, and R4 would make little difference to the predicted luminance patterns. In this work, the average luminance, overall luminance uniformity ratio, and longitudinal luminance uniformity ratio were calculated for 413 road surfaces lit by two different lighting systems using five different luminaires at three different spacings, from the measured r-tables for each road surface. These calculations showed that the average road surface luminances for class R1 are markedly different from those of classes R2, R3, and R4, which show little difference. There are clear differences between the predicted mean overall luminance uniformity ratios for the R2, R3, and R4 classes and smaller differences for the mean longitudinal luminance uniformity ratios. Burghout (1979) recommended the use of a two-class classification system for road surfaces. As a result of this work, CIE (1984) recommended the adoption of a two-class classification system, called the C system. The boundary values of S1 and the standard values of S1 and Q_0 for each of the two classes in the C classification system are given in Table 4.14. The category C2 of the C classification system has been adopted in the UK as the representative British road surface.

Given that a representative road surface r-table is used in the calculation of luminance rather than an r-table of the actual road surface it is interesting to consider the

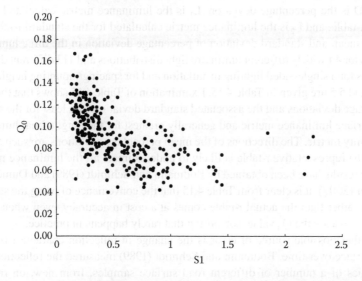

FIGURE 4.5   Relationship between descriptive parameters Q_0 and S1 for 286 measured road surfaces (after Sorensen 1975).

TABLE 4.14

The Boundaries and Standard Values for the Classes in the R, N, and C Road Reflection Classification Systems

System	Class	S1 Boundaries	Standard S1	Standard Q_0
R	R1	S1< 0.42	0.25	0.10
	R2	0.42 < S1 < 0.85	0.58	0.07
	R3	0.85 < S1 < 1.35	1.11	0.07
	R4	1.35 < S1	1.55	0.08
N	N1	S1 < 0.28	0.18	0.10
	N2	0.28 < S1 < 0.60	0.41	0.07
	N3	0.60 < S1 < 1.30	0.88	0.07
	N4	1.30 < S1	1.55	0.08
C	C1	S1 < 0.40	0.24	0.10
	C2	S1 > 0.40	0.97	0.07

Source: CIE (1999).

magnitude of the consequent differences. CIE (1984) reports the errors in terms of the mean and standard deviation of the percentage deviation between the luminance metric calculated using the standard r-table and the actual r-table, after re-scaling the standard r-table for the actual Q_0 value. The expression for percentage difference for the luminance metric is

$$D = ((L_1 - L_2) / L_1) \times 100$$

where D is the percentage deviation, L_1 is the luminance metric calculated for the actual r-table, and L_2 is the luminance metric calculated for the standard r-table.

The mean and standard deviation of percentage deviation in the three luminance metrics for 44 widely different luminaire light distributions and 113 different dry road surfaces for a single-sided lighting installation and for spacing/mounting height ratios of 4.0 and 5.5 are given in Table 4.15. Examination of Table 4.15 shows that the mean percentage deviations and the associated standard deviations are least for the average road surface luminance metric and generally greatest for the longitudinal luminance uniformity metric. The directions of the mean percentage deviations are such that the use of the representative r-table is likely to underestimate all the luminance metrics. Similar results have been obtained by Bodmann and Schmidt (1989) and Dumont and Paumier (2007). It is clear from Table 4.15 that the convenience of using the standard r-tables rather than the actual r-table comes at a cost in accuracy, even when care is taken to re-scale the Q_0 value, something that rarely happens in practice.

Another possible source of error is the change in reflection characteristics of a road surface over time. Bodmann and Schmidt (1989) measured the reflection characteristics of a number of different road surface samples, from new, on roads in Germany over a period of approximately three years. Most of the change in reflection properties occurred over the first six months of exposure. Dumont and Paumier

TABLE 4.15

Mean Percentage Deviation and the Associated Standard Deviation (in Brackets) for Calculations of Average Road Surface Luminance, Overall Luminance Uniformity Ratio and Longitudinal Luminance Uniformity Ratio using Actual and Standard r-Tables after Rescaling the Q_0 Value of the Standard r-Table

Class	Spacing/ Mounting Height Ratio	Average Road Surface Luminance	Overall Luminance Uniformity Ratio	Longitudinal Luminance Uniformity Ratio
C1	4.0	5.3% (3.6%)	6.6% (4.4%)	10.0% (8.1%)
C1	5.5	5.3% (3.6%)	9.0% (6.2%)	13.4% (9.5%)
C2	4.0	7.0% (5.0%)	10.8% (8.4%)	9.9% (8.0%)
C2	5.5	7.0% (5.1%)	11.0% (8.7%)	12.9% (9.8%)

From CIE (1984).

(2007) measured the change in reflection properties in France over three years. They found that some road materials became less specular and more diffuse in reflection over time but others showed the reverse trend. Another cause of variability is the difference in reflection properties across the road. An observation of any well-travelled lane will reveal the existence of five distinct areas, two polished wheel tracks, a darker oil track in the centre of the lane and two unworn tracks at the edges of the lane. Each of these areas will have a different balance of reflection properties although they are all represented by a single r-table.

In summary, despite the precision of the standards used for road lighting, the assumptions made in the design process mean that the reality is likely to be much more variable from site to site (Hargroves 1981). This variability could be identified and then reduced if on-site measurements were routinely taken (ILE 2007). Unfortunately, such measurements are not routinely taken. Rather, the usual approach is to over-design the installation so as to exceed the relevant metric. Using a higher wattage light source and hence a higher light source luminous flux, without changing the luminaire luminous intensity distribution, will result in a higher average road surface illuminance and luminance without changing the uniformity. Reducing the spacing between luminaires without changing the light output of the luminaire or the luminous intensity distribution will increase the average road surface illuminance and luminance and increase uniformity. Increasing luminaire mounting height without changing luminaire light output or luminous intensity distribution will decrease average road surface illuminance and luminance but improve uniformity. Increasing the luminaire overhang without changing luminaire light output or luminous intensity distribution will increase average road surface illuminance and luminance but the effect on uniformity is unpredictable. Another possibility is to change the layout of the luminaires along the road. By changing from a staggered layout to an opposite layout and doubling the spacing between luminaires without changing the light source or luminaire, the uniformity can be improved without changing the average road surface illuminance and luminance. By changing from a staggered layout to a centre-mounted layout and doubling the spacing between luminaires without changing the light source or luminaire, the uniformity will be degraded with a small change in the average road surface illuminance and luminance.

4.6 SPECTRAL EFFECTS

One choice in the design process that has not been considered so far is the spectral power distribution of the light source used. The perceived colour of road lighting varies greatly from the monochromatic yellow of low-pressure sodium, through the orange of high-pressure sodium, to the white of metal halide light sources. Other colours produced by such relics of the past as incandescent or mercury vapour light sources can sometimes be seen. Several studies have been made of the effectiveness of these light sources for making largely achromatic objects on the carriageway visible, without any clear conclusions, suggesting that any effects are small (Eastman and McNelis 1963; de Boer 1974; Buck et al. 1975). One common feature of these evaluations is that all the measurements were taken fixating the object, i.e., the retinal image fell on the fovea. More recent measurements of the effect of light spectrum on the detection of off-axis targets suggest that there may be a significant effect of light colour relevant to road lighting. Specifically, He et al. (1997) carried out a laboratory experiment in which high-pressure sodium and metal halide light sources were compared for their effects on the reaction time to the onset of a 2-degree-diameter disc with the centre either on axis or 15 degrees off-axis, for a range of photopic luminances from 0.003 cd/m^2 to 10 cd/m^2. The luminance contrast of the disc against the background was constant at 0.7. The same light source was used to produce both the background luminance and the stimulus so there was no colour difference between the stimulus and its background. Figure 4.6 shows the median reaction time to the onset of the stimulus, on-axis and off-axis, for a range of photopic luminances, for two practiced subjects. From Figure 4.6, it is evident that reaction time increases as photopic luminance decreases from the photopic to the mesopic state, for both on-axis and off-axis detection. There is no difference between the two light sources in the change of reaction time with luminance for on-axis detection, but for off-axis detection, the reaction times for the two light sources begin to diverge as vision enters the mesopic region. Specifically, the reaction time is shorter for the metal halide light source at the same photopic luminance, and the magnitude of the divergence between the two sources, increases as the photopic luminance decreases. These findings can be explained by the structure of the human visual system. The fovea, which is what is used for on-axis vision, contains only cone photoreceptors, so its spectral sensitivity does not change as adaptation luminance decreases until the scotopic state is reached, at which point the fovea is effectively blind. The rest of the retina contains both cone and rod photoreceptors. In the photopic state the cones are dominant but as the mesopic state is reached the rods begin to have an impact on spectral sensitivity until in the scotopic state, the rods are completely dominant. Given the different balances between rod and cone photoreceptors in different parts of the retina and under different amounts of light, it should not be surprising that the metal halide light source produces shorter reaction times for off-axis detection than

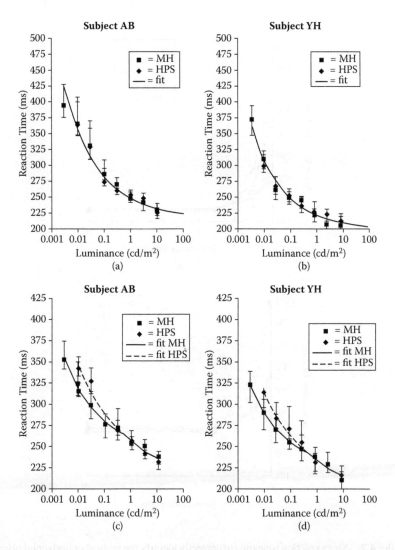

FIGURE 4.6 Median reaction times (ms), and the associated interquartile ranges, to the onset of a 2 degree, high-contrast target seen either (a and b) on-axis or (c and d) 15 degrees off-axis, and illuminated using either high-pressure sodium (HPS) or metal halide (MH) light sources, for photopic luminances in the range 0.003 to 10 cd/m² (after He et al. 1997).

the high-pressure sodium in the mesopic range because it is better matched to the rod spectral sensitivity. It is also evident why there is no difference between the two light sources for on-axis reaction times.

Lewis (1999) has obtained similar results. Figure 4.7 shows the mean reaction time to correctly identify the vertical or horizontal orientation of a large, achromatic, high-contrast, 14 degree by 10 degree grating, where the grating was lit by one of five different light sources: low-pressure sodium, high-pressure sodium, mercury vapour, incandescent, and metal halide, plotted against the photopic luminance. As

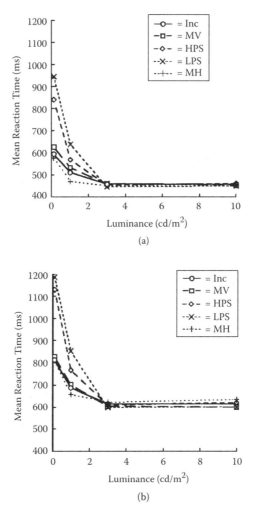

FIGURE 4.7 Mean reaction time (ms) to correctly identify the vertical or horizontal orientation of a grating (a) and the direction a pedestrian (b) located adjacent to a roadway is facing, plotted against the photopic luminance (cd/m²) produced by five different light sources (Inc = incandescent, MV = mercury vapour, HPS = high-pressure sodium, LPS = low-pressure sodium, MH = metal halide) (after Lewis 1999).

long as the visual system is in the photopic range, i.e., at 3 and 10 cd/m², there is no difference between the different light sources provided they produce the same photopic luminance. However, when the visual system is in the mesopic state, i.e., at 1 and 0.1 cd/m², the different light sources produce different reaction times, the light sources that better stimulate the rod photoreceptors (incandescent, mercury vapour, and metal halide) giving shorter reaction times than the light sources that stimulate the rod photoreceptors less (low- and high-pressure sodium).

Such measurements of the time to detect the onset of abstract targets under different light sources may seem irrelevant to the task of driving but, in fact, driving often requires the visual system to extract information from the peripheral visual field. Lewis (1999) verified that the spectral power distribution of a light source does have an effect on the time taken to extract information of relevance to driving, by repeating the experiment described above but replacing the gratings with a transparency of a female pedestrian standing at the right side of a roadway in the presence of trees and a wooden fence. In one transparency the woman was facing towards the road, in the other she was facing away from the road. The subject's task was to identify which way the woman was facing. Figure 4.7 also shows the mean reaction times for this task, under the different light sources, over a range of photopic luminances. Again, there is no difference between the light sources as long as the visual system is in the photopic state but once it reaches the mesopic state, the light sources that more effectively stimulate the rod photoreceptors show faster reaction times.

Another approach to evaluating the effect of light spectrum in mesopic conditions measured the probability of detecting the presence of a target off-axis. Bullough and Rea (2000) used a simple driving simulator based on a projected image of a road, controlled through computer software. The subject could control the speed and direction of the vehicle along the road through a steering wheel and accelerator. The computer monitored the time taken to complete the course and the number of crashes occurring. Filters were applied to the projected image of the course to simulate the spectrum of both high-pressure sodium and metal halide light sources and more extreme red and blue light for a range of luminances. Interestingly, there was no effect of light spectrum on the time taken to complete the course, i.e., on driving speed, but there was a marked effect on the ability to detect the presence of a target near the edge of the roadway. The light spectra that stimulate the rod photoreceptors more (blue and metal halide) led to a greater probability of detection than light spectra that did not stimulate the rod photoreceptors so effectively (red and high-pressure sodium).

Of course, the consequences of making an error while driving a simulator is not quite the same as driving a real vehicle on the road and so may lead subjects to pay less attention to what they are doing. Fortunately, Akashi and Rea (2002) have extended this work into the field, having people drive a car along a short road while measuring their reaction time to the onset of targets 15 and 23 degrees off-axis. The lighting of the road and the area around it was provided by either high-pressure sodium or metal halide road lighting, adjusted to give the same amount and distribution of light on the road, and seen with and without the vehicle's halogen headlamps. There was a statistically significant difference between the high-pressure sodium and metal halide lighting conditions. Specifically, the mean reaction time to the onset of the targets was shorter for the metal halide road lighting than for the high-pressure sodium road lighting at both angular eccentricities (Figure 4.8). Using the same experimental site and equipment, Akashi and Rea (2002) also examined the effect of disability glare, provided by halogen headlamps from a stationary car in the adjacent lane, on the ability of a stationery driver to detect off-axis targets at 15

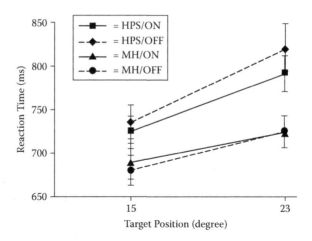

FIGURE 4.8   Mean reaction times (ms) (and the associated standard errors of the mean) to the onset of a target at 15 degrees and 23 degrees off-axis while driving, with high-pressure sodium (HPS) and metal halide (MH) road lighting, and with halogen headlamps turned on and off. The road lighting using the two light sources was adjusted to give similar illuminances and light distributions. The rectangular target subtended 3.97×10^{-4} steradians for the $15°$ off-axis position and 3.60×10^{-4} steradians for $23°$ off-axis position. Both targets had a luminance contrast against the background of 2.77 (after Akashi and Rea 2002).

degrees and 23 degrees when the road lighting was provided by metal halide and high-pressure sodium lighting. Again, the mean reaction times to the onset of the targets were longer for the high-pressure sodium road lighting than for the metal halide road lighting. As might be expected, the mean reaction times were also longer when the headlamps in the opposing vehicle were switched on.

Given the results discussed above there can be little doubt that light spectrum is a factor in determining off-axis visual performance but how important is it? It could be argued that the increase in reaction times in the luminance range of interest is small and would make little difference to traffic safety. For example, an increase of 100 ms in reaction time would mean a vehicle moving at 80 km/h (50 mph) would travel only 2.2 m further because of the longer reaction time. Fortunately for those committed to the importance of light source spectrum, there is some evidence that longer reaction times are associated with more missed events. Rea et al. (1997) measured observers' responses to a change of high-contrast character on a changeable message sign located 15 degrees off-axis. The setting was a roadway lit by either high-pressure sodium or metal halide light sources so as to give an average road surface luminance of 0.2 cd/m². An effective average road surface luminance of 0.02 cd/m² was achieved by asking the observers to wear glasses with a transmittance of 0.1. Figure 4.9 shows the measured reaction times to the changes that were detected and the percentage of off-axis signals missed, plotted against the photopic luminance of the road surface. It is evident that the high-pressure sodium lighting leads to longer reaction times and more missed signals than metal halide lighting, differences that increase as road surface luminance decreases further into the

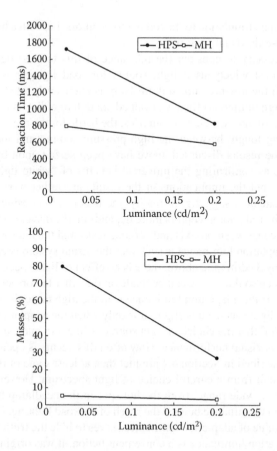

FIGURE 4.9 Mean reaction times (ms) and percentage of misses for changes in a single high-contrast character in a message sign located 15 degrees off-axis for observers looking down a road lit to the same average luminance (cd/m²) by either high-pressure sodium (HPS) or metal halide (MH) lighting (after Rea et al. 1997).

mesopic. Similar increases in missed off-axis changes under high-pressure sodium relative to metal halide illumination have been obtained in other experiments (Bullough and Rea 2000; Lingard and Rea 2002). While, the importance of increases in reaction time of the order shown are debatable, there can be little doubt about the importance of missing off-axis changes altogether.

A number of other studies have been made of the effect of light spectrum on visual performance in mesopic conditions (IESNA 2006). These studies have produced a consistent pattern in which tasks done on-axis, such as measurements of visual acuity (Eloholma et al. 1999) and the visibility distance of small targets (Janoff and Havard 1997) show no effect of light spectrum at the same photopic luminance in the mesopic range, while tasks requiring off-axis activity, such as measurements of effective field size (Lin et al. 2004) and identifying the direction of movement of an off-axis target while driving (Akashi et al. 2007), do. The effect is that light sources that provide greater stimulation to the rod photoreceptors, i.e, with a higher S/P ratio, ensure better off-axis visual performance.

It is now necessary to consider the relevance of this to road lighting practice. Currently, the most widely used light source for road lighting is high-pressure sodium although the low-pressure sodium lamp is extensively used in some countries where its high luminous efficacy is valued enough to offset its complete lack of colour rendering. In most developed countries, the battle for the road lighting market is presently being fought between the high-pressure sodium and the metal halide light sources. The results discussed above have been seized upon by advocates of the metal halide as confirming the universal benefits of "white light" as opposed to "orange light" but the implications of the results are rather more complex than such a simple statement suggests. In reality, the benefit of choosing a light source that stimulates the rod photoreceptors more depends on the driver's adaptation luminance, the balance between on-axis and off-axis tasks and the nature of those tasks. Provided the adaptation luminance is such that the visual system is operating in the photopic state, say 3 cd/m^2 and above, there is no effect of light spectrum on off-axis visual performance so the choice can be made on the other factors such as luminous efficacy and life. If the adaptation luminance is in the high mesopic, say about 1 cd/m^2, the effect of light spectrum is slight. It is only when the adaptation luminance is well below 1 cd/m^2 that the choice of light source is likely to make a significant difference to off-axis visual performance. How often this occurs is open to question. If the standards described in Section 4.4 are met then at least some of the road classes are likely to benefit from a careful choice of light spectrum. However, an average luminance masks a wide range, from the hot spot in the headlamp beam on the lit road to the ambient luminance beyond the reach of the road lighting. Further, discussion of a single value of adaptation luminance serves to hide the truth. The fact is the concept of adaptation luminance is a convenient fiction. It was originally developed to describe the effects of luminance on basic visual functions. Its use for this purpose was not unreasonable as such measurements are usually made on a uniform luminance field but where the visual field has a wide range of luminances, the adaptation of different parts of the retina will be different, depending on where and for how long the eye is fixated. If the driver has one dominant fixation point, such as might be the case of a driver approaching a tunnel entrance, then the average luminance within about 20 degrees of the fixation point is a reasonable estimate of the adaptation luminance (Adrian 1976). If the observer has many fixation points, i.e., the observer is rapidly moving his eyes around a lot, then the concept of adaptation luminance is meaningless and specific luminances need to be considered (EPRI 2005). However, Mortimer and Jorgeson (1974) found that when driving at night, eye fixations tended to be confined to the lit area, an observation also made by Stahl (2004), at least for straight roads (see Figure 3.6). This suggests that when driving at night on a lit road the average road surface luminance seen by the driver can be used as a measure of adaptation luminance. If this is so, then the recommendations in Tables 4.4 and 4.8 imply that the part of the retina receiving light from the part of the road lit by road lighting alone will be operating in the mesopic state and that local and link roads will benefit most from choosing a light source that provides greater stimulation to the rod photoreceptors.

Of course, all these luminances are photopic luminances, calculated using the CIE Standard Photopic Observer. It might be thought that the use of the CIE Standard Photopic Observer for the measurement of light when the visual system is operating in the mesopic state is a fundamental problem. There is no doubt that light sources that more effectively stimulate the rod photoreceptors enhance the performance of off-axis detection tasks when the visual system is operating in the mesopic state but at what luminance the mesopic state begins is the subject of controversy. A unified model of photopic, mesopic, and scotopic photometry based on reaction times has mesopic vision starting at 0.6 cd/m^2 (Rea et al. 2004) while a model of mesopic effects, based on the performance of tasks claimed to be important to driving, shows mesopic vision having an impact up to 10 cd/m^2 (Elohoma and Halonen 2006; Goodman et al. 2007). Most road lighting for traffic routes is designed to produce average road surface luminances between these two limits. Fortunately, comparisons of the predictions of the two models show only small differences (Rea and Bullough 2007) (Figure 4.10), which suggest that either could be used to evaluate the role of spectral power distribution in road lighting. Compared to the uncertainties discussed in Section 4.4, such small differences seem unimportant.

Given that road lighting does produce conditions in which the visual system is operating in the mesopic state while driving at night, it is also necessary to consider the nature of the driver's task and the balance between on-axis and off-axis tasks. The nature of the driver's task can vary widely, both in the stimuli presented to the driver and the information that needs to be extracted from them. This is important because the magnitude of any spectral effect on off-axis visual performance will depend on the exact task (IESNA 2006). For stimuli close to threshold, the spectral effects can be large but for stimuli well above threshold the spectral effects may be

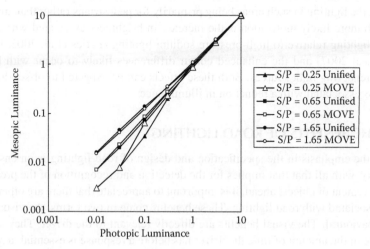

FIGURE 4.10 Calculated mesopic luminances plotted against photopic luminance for three different scotopic/photopic (S/P) ratios, using the unified system of photometry of Rea et al. (2004) and the MOVE model of Eloholma and Halonen (2006). The three S/P ratios approximate to low-pressure sodium (S/P = 0.25), high-pressure sodium (S/P = 0.65), and metal halide (S/P = 1.65) light sources, all of which are used for road lighting.

insignificant. What can be said is that using a light source that better stimulates the rod photoreceptors at a given photopic luminance will not make off-axis visual performance worse and may make it better, except where accurate colour discrimination is required and in this case, a worsening of performance is possible.

As for the balance between on-axis and off-axis tasks, this is important because it is sometimes argued that the metal halide light source can be used at a lower road surface luminance than the high-pressure sodium light source, without penalty. Certainly, the results discussed above suggest that, in mesopic conditions, the same off-axis visual performance can be achieved at a lower photopic luminance with a metal halide light source. However, for an on-axis target, a lower photopic luminance will produce worse visual performance for both light sources. Thus, it is only if off-axis detection is assumed to be the only important task in driving that a reduction in road surface luminance for the metal halide light source can be justified. There cannot be many drivers who would be willing to deny the importance of both on-axis and off-axis vision for driving. Given that both on- and off-axis vision are important to drivers, a responsible approach to introducing the effect of light spectrum into road lighting practice would be to use light sources with high S/P ratios (see Table 2.3) without reducing recommended road surface luminances expressed in photopic measures.

At the moment the only recognition of the role of light spectrum in road lighting standards is the statement in the UK standards that where light sources with a CIE general colour rendering index of 60 or greater are used in residential roads, the recommended average illuminance for those areas can be reduced by one class (see Table 4.13) (BSI 2003b). The motivation for this recommendation is not stated, but it probably has little to do with reaction times and other measures of visual performance, the lighting in such areas being primarily for pedestrians rather than drivers. A much more likely motivation is the increase in brightness associated with metal halide lighting relative to high-pressure sodium lighting (Fotios et al. 2005; Fotios and Cheal 2007) and the enhanced colour differences likely to occur with better colour rendering light sources. Both these effects can be expected to offset the perceptions associated with a reduction in illuminance.

4.7 THE BENEFITS OF ROAD LIGHTING

While the emphasis in the specification and design of road lighting is on ensuring visibility, with all that that implies for the detection and recognition of the presence and movement of objects ahead, it is important to appreciate that there are other benefits associated with road lighting. These benefits come in two forms, the visual and the behavioural. The visual benefits are directly relevant to the driver. They are an increase in the amount of time the driver has before a response is essential, a reduction in the amount of discomfort and disability glare produced by opposing vehicles' headlamps, and guidance on the direction of the road far ahead. The behavioural effects are both direct and indirect. The direct behavioural effect is on the speed at which drivers choose to travel. The indirect effects of road lighting are evident on the level of crime occurring and the possibility of economic development.

Anyone who has made the transition from an unlit to a lit section of road while driving using low-beam headlamps will be aware of the immediate sense of relaxation that results. The reason for this relaxation is the greater distances over which objects on and near the road can be seen and hence the longer times available for selecting an appropriate response. This benefit of road lighting will be most evident on high-speed traffic routes where the amount of additional information revealed by the road lighting is likely to be modest and, without road lighting, the response times are short. Where the amount of additional information revealed by the road lighting is large, as may be the case in urban areas, the sense of relaxation may be less because of the additional information that has to be dealt with.

Road lighting itself will produce some disability and discomfort glare but, given that the standards discussed in Section 4.4 are met, the amount of glare produced by road lighting will be much less than that produced by the headlamps of approaching vehicles. The formulae for disability and discomfort glare produced by vehicle headlamps are discussed in Section 6.6.2. For discomfort glare from headlamps, road lighting will tend to increase the adaptation luminance with the result that discomfort glare is reduced. For disability glare from headlamps, road lighting will not change the equivalent veiling luminance but the impact of the equivalent veiling luminance on luminance contrast will be diminished as the luminance of the background, which is usually the road surface, is increased. Thus, road lighting will always tend to diminish both disability and discomfort glare from headlamps, an achievement that makes driving more comfortable.

As for guidance, the view of road lighting luminaires stretching away into the distance provides easily understandable clues to the run of the road far ahead, further than is possible with retroreflective road markings. Such guidance is most obvious when the road lighting is in a central twin or single-sided layout. Double, staggered, or mixed luminaire layouts can be more difficult to interpret.

The behavioural effects of road lighting can be positive or negative. A negative effect is based on observations of risk compensation in human behaviour. The idea behind risk compensation is that drivers will adjust their behaviour to a constant level of risk. This implies that if road lighting is installed, visibility of hazards is improved so the risk to drivers is reduced so drivers will increase their speed. Assum et al. (1999) found that drivers in Norway did indeed increase their speed at night after road lighting was installed, by about 5 percent on straight roads and about 0.7 percent on curves. Presumably, the reason for the difference between straight and curved roads is that road lighting is more effective in enabling the driver to see further ahead on straight roads than on curved roads. Despite this finding, the results from the meta-analysis of road lighting as an accident countermeasure conducted by Elvik (1995) suggest that such increases in speed are not enough to nullify the value of road lighting but it is undeniable that they may go some way to diminish its impact.

A possible positive effect of road lighting in urban and suburban areas through behaviour is its use as a crime prevention measure. For this to occur, the road lighting has to be designed not just to light the road but also the surroundings. Even then, there can be no guarantees. This is because lighting, per se, does not have a direct effect on the level of crime. Rather, lighting can affect crime by two indirect mechanisms (Boyce 2003). The first is the obvious one of facilitating surveillance by people on the street after dark, by the community in general, and by the authorities. If such increased surveillance is perceived by criminals as increasing the effort and risk and decreasing the reward for a criminal activity, then the incidence of crime is likely to be reduced. Where increased surveillance is perceived by the criminally inclined not to matter, then better lighting will not be effective. The second indirect mechanism by which an investment in better lighting might affect the level of crime is by enhancing community confidence and hence increasing the degree of informal social control. This mechanism can be effective both day and night but is subject to many influences other than lighting.

Finally, urban lighting, which includes the lighting of roads and pedestrian areas, has a role to play in economic development. The purposes of such lighting are to attract people into the area and to make their experience safe and enjoyable. Given this purpose, lighting has to be both bright and interesting (Hargroves 2001). Brightness is essentially a means to an end, namely providing enough light so that visitors can see well enough to feel safe in the space (Boyce et al. 2000). For a given illuminance, brightness can be enhanced by using a nominally white light source with good colour rendering properties. Interest can be produced by many different means ranging from lighting a historic area using "historic" luminaires through floodlighting iconic buildings to filling the surroundings with continuously changing advertising, as in Times Square, New York. Most large-scale developments involve a lighting master plan intended to ensure that the development is seen as coherent rather than muddled. Such plans cover many types of lighting but road lighting is usually one of them.

4.8 SUMMARY

Road lighting, as opposed to urban street lighting, first appeared widely in the 1930s. Incandescent, mercury vapour, fluorescent, low-pressure sodium, high-pressure sodium, and metal halide light sources have all been used for road lighting at different times. The luminaires used for road lighting are classified according to their luminous intensity distribution. There is a large number of luminous intensity distributions available, some symmetrical and some asymmetrical. Different luminous intensity distributions are necessary because roads of different width and layout require different light distributions if the light is to be directed onto the road surface and not wasted. Luminous intensity distribution is also important for limiting light pollution and controlling glare to drivers.

Road lighting luminaires are typically mounted on columns beside the road. Today, columns are mostly made of steel or aluminium. The layout of the luminaires for single carriageway roads can be single-sided, staggered, or opposite. The layout of luminaires for dual carriageways and motorways is central twin, central twin and opposite, or catenary arranged along the axis of the road. These layouts apply to continuous linear stretches of roads not to sharp bends or conflict areas such as crossroads and roundabouts. Layouts for these areas are customized. It is worth noting that lighting columns represent a hazard for the driver. The magnitude of the hazard can be reduced by setting the columns back from the road or shielding them with a crash barrier or designing them to bend or break away on impact.

In the past, most road lighting was controlled by time switches to be on from half an hour after sunset to half an hour before sunrise. Today, the most common control system is the photoelectric cell, this being used to detect the amount of daylight available and thus to ensure that the road lighting is only used when necessary. In the future, light source dimming, mains signaling, and wireless communications are likely to be used to make controls much more sophisticated.

Given the need to make a choice between a wide variety of light sources, luminaire types, column heights, road layouts, and control systems it is necessary for the designer of road lighting to have a target to aim at. The primary purpose of road lighting is to make objects on the road visible without causing discomfort to the driver. Road lighting makes objects on the road visible by producing a difference between the luminance of the object and the luminance of its immediate background, which is usually the road surface or its edges. Two single-number metrics have been developed to quantify the visibility provided by road lighting, revealing power, and small target visibility. Revealing power is the average percentage of pedestrian clothing reflectances that are above threshold at a matrix of points spread across and along the road. Small target visibility is defined as the weighted average of the visibility levels of a number of flat, 18 cm side, square targets, all with a diffuse reflectance of 0.5, mounted vertically in a matrix across and along the road. The visibility level of the target is defined as the ratio of the actual luminance difference of the target to the luminance difference threshold of the target; the higher is this ratio, the more visible is the target.

Neither revealing power nor small target visibility is much used because the visual task of driving is multifaceted and both metrics address only one facet, on-axis detection. The driver certainly has to be able to detect small targets on-axis, but also needs to be able to detect objects on the road off-axis, to detect movement at the edges of the roadway, to judge the relative speeds and direction of movement of other vehicles, and to identify many different signs and signals. Further, the revealing power and small target visibility metrics are only really applicable in light traffic where the driver can see a considerable distance up the road. In heavy traffic, detection of small obstacles is not an issue, much more importance being attached to the relative movements of nearby vehicles. This is probably why most current road lighting standards are expressed in conventional photometric terms.

Road lighting standards vary in detail from country to country but they do have some common features. One is the division of the road network into different classes, according to the type of users and the road geometry. Different lighting metrics are used to form standards for the different classes and these metrics are then varied to match the subdivisions of each class. Another common feature is the aspects of road lighting dealt with. Most lighting standards contain metrics for the amount of light on the road, the distribution of light on the road, and the extent of disability glare produced by the road lighting. In the UK, the amount of light and its distribution across and along the road for traffic routes is specified in terms of maintained levels of average road surface luminance, overall luminance uniformity, and longitudinal luminance uniformity while the amount of disability glare is restricted by setting a maximum threshold increment, this being the percentage increase in luminance difference threshold that occurs because of disability glare. For conflict areas, such as crossroads and roundabouts, and residential areas, the amount and distribution of light is specified in terms of road surface illuminance. The change from a luminance to an illuminance metric is rationalized as being due to the fact that for traffic routes there are a limited number of directions of view but in conflict areas and residential streets there are multiple directions of view.

In the US, standards are available based on three different metrics, illuminance, luminance, and small target visibility supplemented by luminance criteria. Again, these metrics are used to specify different maintained average and uniformity values for different traffic, conflict, and residential areas. Disability glare is taken into account for all three forms of the recommendations. For the illuminance and luminance recommendations, an explicit metric is used, this being the maximum veiling luminance ratio, which is defined as the ratio of the veiling luminance to the average road surface luminance. For the small target visibility criteria, disability glare from the road lighting luminaires is implicit as it is included in the calculation of visibility level.

A comparison of the road lighting recommendations used in the UK and US shows they are similar in concept but differ in detail. The UK recommendations should lead to higher-level, more uniform lighting on roads with faster traffic than the US recommendations. However, the allocation of the highest average road surface luminance to major roads in high pedestrian conflict areas in the US rather than high-speed traffic routes as is the case in the UK is consistent with putting most light where it will have the most beneficial effect in reducing fatalities and personal injury accidents.

Setting standards is one thing, but the reality may different. The extent to which road lighting standards are achieved in practice depends on the assumptions that have to be made. Standards based on illuminance are the easiest to achieve in practice because they require no assumptions about the reflection characteristics of the road surface. Standards based on luminance do. In principle, the reflection properties of a specific road surface can be measured but in practice they rarely are. Rather, the detailed reflection properties of a small number of what are claimed to be representative road surfaces are used in the calculation of road lighting installations. Measurements of the difference between the calculated luminance metrics using the representative reflection characteristics and the actual reflection characteristics for the road suggest that the use of the representative reflection characteristics is likely to underestimate all the luminance metrics.

It should also be noted that the reflection characteristics of a road surface change over time due to wear. An observation of any well-travelled road lane will reveal the existence of five distinct areas, two polished wheel tracks, a darker oil track in the centre of the lane, and two unworn tracks at the edges of the lane. Each of these areas will have different reflection properties although they are all represented by a single r-table. It should be apparent that, despite the precision of the standards used for road lighting, the assumptions made in the design process mean that the reality is likely to be much more variable from site to site.

Another choice in the design process that needs to be considered is the spectral power distribution of the light source used. The perceived colour of road lighting varies greatly from the monochromatic yellow of low-pressure sodium lighting, through the orange of high-pressure sodium lighting, to the white of metal halide lighting.

Measurements in the laboratory and on the road have shown that as the visual system enters the mesopic state, light sources that provide more stimulation to the rod photoreceptors produce faster reaction times and fewer missed signals for off-axis events at the same photopic luminance. For these findings to be meaningful for road lighting, it is necessary for the visual system to be operating in the mesopic state under road lighting, for off-axis tasks to matter to the driver, and for the magnitude of the effects to be significant. If the current road lighting standards are followed, then when fixating the part of the road lit by road lighting alone, part of the retina will be in the mesopic state.

Given that mesopic vision is at work it is necessary to consider the nature of the driver's task and the balance between on-axis and off-axis tasks. The nature of the driver's task can vary widely, both in the stimuli presented to the driver and the information that needs to be extracted from them. This is important because the magnitude of any spectral effect on off-axis visual performance will depend on the exact task. For stimuli close to threshold, the spectral effects can be large but for stimuli well above threshold the spectral effects may be insignificant. What can be said is that using a light source that better stimulates the rod photoreceptors will not make off-axis visual performance worse and may make it better, except where accurate colour discrimination is required. As for the balance between on-axis and off-axis tasks, this is important because it is sometimes argued that metal halide lighting can be used without penalty at a lower road surface luminance than high-pressure sodium lighting. However, for an on-axis target, a lower photopic luminance will produce worse visual performance for both light sources. Given that both on- and off-axis vision are important to drivers a responsible approach to introducing the effect of light spectrum into road lighting practice is to use light sources with high S/P ratios without reducing recommended road surface luminances, which are expressed in photopic measures.

While the emphasis in the specification and design of road lighting is undoubtedly on its impact on visibility, there are other benefits associated with road lighting. These benefits come in two forms, the visual and the behavioural. The visual benefits are directly relevant to the driver. They are an increase in the amount of time the driver has before a response is essential, a reduction in the amount of discomfort and disability glare produced by opposing vehicles' headlamps, and guidance on the direction of the road far ahead. The behavioural effects are both direct and indirect. The direct behavioural effect is on the speed at which drivers choose to travel. The indirect effects of road lighting are evident on the level of crime occurring and the possibility of economic development.

Clearly, road lighting is not as simple as it might seem. The design of road lighting involves many assumptions that influence the photometric conditions achieved. Even when these assumptions are reasonable, doubts must remain as to the suitability of the photometric conditions for driving. The fundamental problem is that our understanding of how lighting affects the driver's ability to extract information of different types of on- and off-axis is limited. Until this deficiency is rectified, road lighting will continue to be based more on experience than science.

5 Markings, Signs, and Traffic Signals

5.1 INTRODUCTION

One of the earliest signals associated with the use of motor vehicles was introduced by the British parliament in the form of the 1865 Locomotives on Highways Act. This called for each self-propelled, steam-powered vehicle to have a crew of three: one driver, one stoker, and a man to walk 60 yards in front of the vehicle carrying a red flag to warn of its approach, the purpose being to avoid frightening the horses. Today, horse-drawn transport and steam-powered vehicles have almost vanished from the developed world, but signals have proliferated dramatically. Today, drivers are faced with a plethora of markings, signs, and signals designed to inform and regulate their behaviour, some fixed, some changeable, some unlit, some lit, but all needing to be seen by day and night. The form and location of markings, signs, and signals are strictly controlled so as to ensure consistency across road networks, although different countries have different rules (FHWA 2003; DfT 2005a). The factors considered in designing markings, signs, and signals are the distances from which they need to be visible; their shapes and colours, shape and colour being used as cues to meaning as well as being important for visibility; the advantages and disadvantages of pictograms rather than text in a specific language for signs; and the need for some means to attract attention to the sign or signal. The aspect of design that will be considered here is the use of light, either as an inherent part of the marking, sign, or signal or as a means to make the marking, sign, or signal conspicuous, visible, and legible at night.

5.2 FIXED ROAD MARKINGS

Some of the simplest means used to guide drivers are markings on the road. Such markings usually consist of different patterns of lines on the road surface, although sometimes solid shapes, text, or pictograms are used and occasionally, the road surface itself is the sign, the colour of the surface indicating a restricted class of use. Road markings are used to indicate lane boundaries, bends in the road, areas where overtaking is prohibited or limited to a specific lane, edges of junctions, parking restrictions, vehicles allowed to use a lane, mini-roundabouts, etc. (Figure 5.1).

The lighting variables important to road markings are colour and luminance. Colour is important because it often carries part of the message. For example, in the UK, road markings that restrict parking in some way are yellow while markings that indicate the road ahead are white. There is rarely a problem in discriminating

(a)

(b)

FIGURE 5.1 Two examples of road marking: (a) a mini-roundabout without road lighting and (b) lane markings with road lighting.

colours by day but at night there can be. Under low-pressure sodium road lighting, which is widely used in the UK, yellow and white road markings will be difficult to discriminate, except in the area illuminated by the vehicle's headlamps where small amounts of light will make colour discrimination possible (Boynton and Purl 1989). In practice, the inability to separate yellow and white lane markings at night is not much of a problem because the different locations of the markings related to parking and driving make the meaning obvious. However, where the colour of the whole lane is used to indicate a restricted use, e.g., buses and taxis only, there can be a problem depending on the colouring material used. Often, this is a dark red, with the result that when low-pressure sodium road lighting is used, the lane marking is indistinguishable from adjacent conventional asphalt lanes. This is why such lanes usually have text or pictograms indicating their use set out in white retroreflective paint at regular intervals.

The main role of road markings is to provide visual guidance and lane definition. Drivers need both long-range guidance (more than five seconds preview time) and short-range guidance (less than three seconds preview time) (Rumar and Marsh 1998). Long-range guidance is accessed intermittently and consciously, using foveal vision. Short-range guidance is accessed continuously and unconsciously, using peripheral vision. Road markings can provide both short- and long-range guidance. Road markings usually consist of a paint or thermoplastic material containing spherical retroreflective beads (see Section 2.8). The paint or thermoplastic material is a high-reflectance, diffuse reflector, which ensures the mark will be seen in positive luminance contrast against the low-reflectance road surface during daylight. At night, the luminance of the white paint has two components, the diffuse reflected component mainly from any road lighting and the retroreflected component from the vehicle headlamps (CIE 1999). Where there is no road lighting, the luminance depends almost entirely on the retroreflective materials. The greatly enhanced reflection of these materials in the direction of the vehicle means that the luminance of the markings will be much greater than the luminance of the adjacent road surface, the resulting luminance contrast making the markings visible at a distance (Figure 5.2). The main limitation of such markings is that they tend to lose their retroreflective properties with wear and they tend to disappear when the road is covered with water, the water surface forming a specular reflector above the markings that reflects the

FIGURE 5.2   Painted retroreflective road markings and road studs lit by high-beam headlamps alone.

grazing-incident light from the vehicle's headlamps along the road away from the driver before it reaches the retroreflectors (see Section 11.3). As a result, visual guidance is much reduced at the time when it is most required (Rumar and Marsh 1998).

Another means used to provide visual guidance, particularly where snow is common, is to place vertical posts of about 1 m height, at regular intervals along the edges of the road. These posts often have small panels of retroreflective material fixed near the top. Such posts define the edges of the road well in rain, snow, and fog but do not give any guidance as to the lanes. A mechanism used to provide both lane definition and guidance as to the run of the road ahead, in rain and fog but not in snow, is the individual retroreflector, originally known as a "cat's eye," but now commonly called a road stud (Figure 5.2). The original "cat's eye" consisted of a cast-iron frame containing a rubber housing, itself containing four refractive/reflective retroreflectors, two facing each way down the road. A series of such assemblies were set into the centre of the road at approximately 10 m intervals to provide visual guidance. The rubber housing deflected when a vehicle's wheel passed over it, the deflection acting to clean the front surfaces of the retroreflectors when wet. There are now a number of simpler road studs available designed to withstand the impact of wheels without deflection (Figure 5.3). All road studs place the retroreflectors high enough above the road surface to stand above water on the road, although this can make them prone to damage by snowploughs. Road studs can be fitted with filters so that colour can be used to carry a message. For example, road studs acting as lane dividers are usually white, while those acting as road edge markers are conventionally red on the nearside and orange on the offside of the road. Where the edge of a road can be crossed, as at a slip road off a major road, the colour of the road studs changes from red to green.

Road studs depend for their visibility on light from a vehicle's headlamps. This inevitably limits the distance over which guidance is delivered to less than 100 m. An alternative now available is a photoelectric-powered road stud containing an LED. Such a stud is self-luminous. By installing studs of this type at regular intervals along a road, visual guidance is available over much longer distances, typically

FIGURE 5.3   A retroreflective road stud fitted with a red filter to mark the edge of the road.

up to 1000 m, and around curves in the road. Anecdotal reports claim that such installations have had dramatic effects on the number of accidents occurring on unlit roads subject to mist and fog.

The effects of road marking on drivers' behaviour are mixed. Adding lines marking the edges of a road where previously there had been no marking results in increased driving speeds with the lateral position of the vehicle being closer to the edge of the road (Rumar and Marsh 1998; Davidse et al. 2004). When edge lines are added to an existing centreline, there is no overall change in speed, but when a centreline marking is replaced with edge lines, there tends to be a decrease in speed (Davidse et al. 2004). The rationale for such changes in behaviour lies in the driver's confidence about what lies ahead. Providing edge markings, or a centreline, on a previously unmarked road will increase the amount of visual guidance and confidence in where the road goes, hence the increase in speed. Adding edge markings to a road with a central line marking adds little to visual guidance, so a change in speed is unlikely. Removing central marking and replacing it with edge markings may have the effect of making the road appear narrower, hence the reduction in speed. These examples serve to make a basic point that providing better visual guidance to the driver at night may not result in safer driving. There are two opposing views on the value of road marking. One view holds that better visual guidance leads to smoother and safer driving. The other is that better visual guidance leads to overconfidence in where to go without consideration of how to get there (see Section 11.5). The problem this conflict exposes is that while some visual guidance is certainly necessary and road markings are a convenient way to provide it, markings only address one part of the driver's task. An overemphasis on visual guidance and a neglect of the other aspects of the driver's task may diminish traffic safety rather than improve it.

5.3 FIXED SIGNS

Another common feature of roads are fixed signs mounted beside or over the road giving information on directions, lane changes, speed limits, etc. The size, shape, colour, and content of such signs have been extensively studied (Forbes 1972; Mace et al. 1986). The first question of interest here is whether or not such signs should be illuminated and, if so, how? The decision on whether or not to illuminate a sign depends first and foremost on the distance at which the sign needs to be detected and the distance at which it needs to be legible. These distances depend on the speed and density of traffic approaching the sign and whether the driver has to carry out some maneuvre in response to the sign. High speeds, dense traffic, and the need for a maneuvre all increase the distances at which the sign needs to be detected and legible. Other factors to be considered are the complexity and brightness of the background against which the sign has to be seen, the location of the sign relative to the driver, and the size of the sign. The more complex the background, the brighter the ambient light level, the further the sign is from the edge of or above the road, and the larger the sign, the more likely it is that individual lighting should be provided. Individual sign lighting is necessary because the alternative sources of light, road lighting, and headlamps, are inadequate. Road lighting and headlamps are designed to illuminate the road surface, not signs remote from it. Attempts to use road lighting or headlamps for sign lighting risk producing increased glare to drivers.

Table 5.1 lists some recommendations from the United States for the maintained average illuminance on externally illuminated signs for different ambient conditions. The maximum illuminance uniformity ratio (maximum/minimum) associated with these recommended maintained illuminances is 6:1. Where external lighting is provided for individual signs, the luminaire is usually positioned close to and either above or below the sign (Figure 5.4). Luminaires mounted above the sign will collect less dirt on the emitting face and will produce less light pollution but may cause glare

---

TABLE 5.1

Recommended Maintained Average Illuminances (lx) for Externally Illuminated Road Signs and Maintained Average Luminances (cd/m²) for White Translucent Elements in an Internally Illuminated Road Sign for Different Ambient Light Levels

Ambient Light Level	Maintained Average Illuminance on the Sign (lx)	Maintained Average Luminance of White Elements (cd/m²)
Low — Rural areas without road lighting	140	20
Medium — Suburban areas with small commercial developments and road lighting	280	40
High — Urban areas with road lighting and complex bright surroundings	560	80
From IESNA (2001).		

FIGURE 5.4 Two small, externally illuminated road signs on a single post. The light is provided by compact fluorescent light sources.

to drivers approaching from the back of the sign, may cast shadows on the sign in daylight, and may be more difficult to reach for maintenance. Luminaires mounted below the sign will collect more dirt and will cause more light pollution but will not cast any shadows on the sign. Regardless of mounting position, all luminaires need to be positioned so that they do not obstruct the view of the sign from any meaningful direction and so as to ensure that any specular reflections of the luminaire do not occur in the direction of approaching traffic. Further, the light source must be chosen to accurately reveal the colours of the sign, colour often being an important element in identifying the sign before any printing is readable. This implies that the best light sources for lighting signs are presently metal halide and compact fluorescent, although white LEDs would also seem to be an attractive option.

An alternative approach to external lighting of signs is the internally illuminated sign. Such signs consist of an inter-reflecting box containing a light source, with the front face of the box providing the information. Both the reflection and transmission properties of the front face are important because the sign has to be legible by both day and night and should look the same under both conditions. It is the reflection properties that dominate the appearance of the sign by day and its transmission properties that dominate by night. The great advantage of the internally illuminated sign is that, compared with external sign lighting, it produces much less light pollution. The risk with internally illuminated signs is that, at night, the luminance is so high that the sign itself becomes a glare source. Table 5.1 gives some recommended maintained average luminances for white translucent material (reflectance = 0.45), at night, for three different ambient light levels. The luminance uniformity ratio (maximum/minimum) should not exceed 6:1.

A special case of the internally illuminated sign is the bollard placed in the middle of the road to indicate an obstruction such as a pedestrian island (Figure 5.5). An interesting feature of such bollards is that the light source and associated electrics

FIGURE 5.5   An internally illuminated bollard placed to warn of a pedestrian island in the middle of the road. The light source and associated electrics are placed below ground level.

are placed below road level. The whole of the bollard above road level is made of translucent plastic, all of which emits light in all directions. Vehicles colliding with the bollard destroy the plastic upper part usually without damaging the below-road part, making replacement relatively inexpensive.

Where neither external nor internal sign lighting is provided, the luminance of the sign at night is dependent on the illumination provided by the headlamps of approaching vehicles, the effectiveness of the retroreflective treatment of the material from which the sign is constructed, and the angular separation of the driver from the headlamps (Sivak and Olson 1985). Olson et al. (1989) examined how the detection distances for differently coloured retroreflective sign materials varied with the effectiveness of the retroreflective material expressed as the specific intensity/unit area of material. Specific intensity is the luminous intensity emitted by the retroreflector per unit of illuminance received at the retroreflector. The distances were obtained from observers driving along unlit roads at night using headlamps alone. There were two linear relationships between the logarithm of the specific intensity/unit area and detection distance, one for yellow, white, blue, and green materials and one for red and orange materials. For all colours, the higher the specific intensity/unit area for the material, the greater the detection distance (Figure 5.6).

As for the role of headlamps, the two questions of interest are, can the colour of the sign be correctly recognized and is there sufficient light to read the sign? McColgan et al. (2002) examined the effect of different light sources designed to be used in headlamps on the ability to correctly identify the colour of road sign materials at high (63 cd/m^2) and low (12 cd/m^2) sign luminances, as well as preference. They showed that tungsten halogen, tungsten halogen with a neodymium coating, tungsten halogen with a blue coating, and xenon (HID) light sources all enabled very accurate identification of the sign colours specified in the US Manual on Uniform Traffic Control Devices (FHWA 2003), at both high and low sign luminances. They also showed that

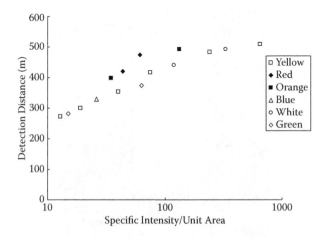

FIGURE 5.6 Mean detection distances (m) for road sign materials of various colours plotted against the efficiency of the retroreflective material, measured as the specific intensity/unit area of the material (cd/lx/m^2) (after Olson et al. 1989).

of the four light sources examined, the tungsten halogen with a neodymium coating was the one most frequently preferred for illuminating the sign materials.

As for the angular separation of the driver from the headlamps, this matters because the retroreflective materials used in signs reflect the incident light back along its own path, i.e., light received at a sign from a headlamp will be reflected back to the headlamp. Of course such material is not perfect so there will always be some spread in the reflected light distribution. The position of the driver relative to the headlamps is not usually a problem for cars but for large trucks it can be. Sivak et al. (1993) have shown that the luminance of retroreflective signs can be much less for truck drivers than for car drivers, and that such reduced sign luminances will seriously reduce the detection distances of signs. It is also important to note that the efficiency of retroreflectors is much reduced when water and dirt cover the sign, suggesting that regular cleaning is required, although this itself will ultimately diminish the retroreflector efficiency through wear.

Another area of concern for signs, both lit and unlit, is the background against which the sign is seen. The background can be important for two different reasons. The first is the presence of a very bright light source close to the sign. Such a source can produce enough disability glare to make the sign invisible. The classic example of this is a sign with the setting sun immediately adjacent to it. This problem is usually solved by surrounding the sign with a low-reflectance screen that cuts off the view of the sun within a few degrees of the sign. The second reason is where the background against which the sign is seen is visually complex so that the sign is just one sign amongst many. This often occurs in city centres where there are a multitude of advertising signs of high luminance to compete with the road sign. Schwab and Mace (1987) examined the detection and legibility distances for signs seen against backgrounds of different complexity. They found that the more complex the background, the shorter was the detection distance, but there was little effect on legibility distance. This is not surprising because legibility is primarily dependent on the details within the sign when the sign is fixated, while a sign is usually first detected off-axis. The effectiveness of off-axis detection during visual search will be influenced by the presence of competing visual information. Schwab and Mace suggest that in the presence of a complex background either the luminance of the sign or its size should be increased, or an advanced warning sign should be used. An alternative would be to separate the sign from the background by using a large, low-reflectance surround. It would also be possible to attract attention to the sign by means of a simple strobe light.

5.4 CHANGEABLE ROAD MARKINGS

A recent development has been the introduction of changeable road studs. These are powered LED road studs that can change their state on instruction from a control unit. The change can be one of luminous intensity, colour, or fluctuation. Installations of these studs along roads have been used to change lane usage patterns and to warn of hazards ahead, the former by changing colour and the latter by changing from continuous to flashing light output. Glare at night can be avoided by reducing the luminous intensity of the stud from that necessary to maintain visibility by day.

Essentially, installations of changeable road studs form part of what is called an intelligent road system, the aim of this being to maximize the throughput of the road network while maintaining safe travel. Changeable road studs are expensive to install because they require connection to an electricity network, although developments in photoelectric technology and wireless communication may ultimately make this unnecessary. Changeable road studs tend to be used mainly at choke points in the road network, i.e., in areas of very high traffic density.

5.5 CHANGEABLE MESSAGE SIGNS

Another element in the intelligent road system that is found with increasing frequency is the changeable message sign (CIE 1994b). These signs are used to provide information about temporary road conditions, such as the presence of road works, variable speed limits, and traffic congestion. Changeable message signs use a series of pixels to display a text message or pictogram. The pixels may be light reflecting or light emitting. Light reflecting pixels are usually of low reflectance on one side and high reflectance on the other, electromechanical means being used to flip each pixel as required. During the day, the reflectances are enough to ensure visibility of the message. At night, some illumination of the pixels is required, illumination that sometimes uses ultraviolet radiation to excite phosphors in the high-reflectance material. For the self-luminous pixels, the light sources are usually either miniature incandescent lamps or LEDs. The legibility and readability of such signs depends on many factors, including the pixel shape, letter width/height ratio, font, and letter separation. Collins and Hall (1992) showed that words with a regular pixel, a letter width/height approaching unity, an upper case font, and a letter separation of two pixels are most legible. Collins and Hall worked with light reflecting signs and so had no need to consider the luminance of the display. Padmos et al. (1988) carried out field evaluations of self-luminous message signs mounted above the road so that the immediate background was the sky. Figure 5.7 shows the mean message luminances set for three different formats of the number 5, for two different visibility criteria, plotted against the horizon luminance. The number 5 was viewed from 100 m. The message luminance is given by the equation

$$L_{mes} = 10^6 \cdot I_{px}/ d^2$$

where L_{mes} is the message luminance (cd/m^2), I_{px} is the pixel luminous intensity (cd), and d is the distance between pixels (mm).

Figure 5.7 shows that the visibility of the message varies with the horizon luminance, the higher the horizon luminance the higher the message luminance required for the message to be visible. By using other visibility criteria, Padmos et al. (1988) were able to show that the message luminances necessary for a rating of "optimum" on a bright day would be rated as "glaring" at night. This finding implies that some degree of luminance control is necessary to ensure comfortable and effective viewing of the message by day and night. Padmos et al. (1988) suggest that a sufficiently legible but not too bright message can be obtained by a two-step message luminance,

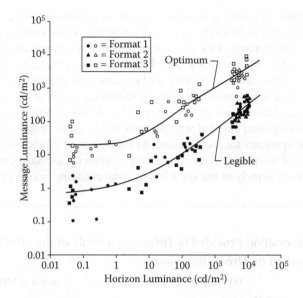

FIGURE 5.7 Message luminances (cd/m²) set by individual subjects for the number 5 presented in three different formats on a self-luminous message sign, seen from 100 m, plotted against horizon luminance (cd/m²). The luminances were set to match two different visibility criteria at different times of day and night and hence for different horizon luminances. The criterion "optimum" is based on the perception that the display is conspicuous but not glaring. The criterion "legible" is based on the perception that the display is just recognizable. The three formats used different numbers of pixels to form the number 5. Specifically, format 1 = 23 pixels, format 2 = 50 pixels, format 3 = 141 pixels (after Padmos et al. 1988).

4000 cd/m² by day and 100 cd/m² by night, although three steps (4000, 400, and 40 cd/m²) would be better.

A changeable message sign is only as useful as the message it displays. Messages fall into three classes, those that are valuable, those that are wrong, and those that are already obvious to the driver. A valuable message is one that gives information that is not otherwise available to the driver, such as an accident causing congestion some miles ahead that can be avoided by taking a different road or a change of speed limit made at times of heavy congestion with the intention of maintaining steady traffic flow. Help in avoiding long delays and the evident benefit to the driver of being in heavy traffic moving smoothly rather than in a stop and start manner will ensure that such messages are appreciated and obeyed. But, if the message is wrong, e.g., a message about the presence of fog being displayed when visibility is clear, or is a statement of the obvious, e.g., a message telling drivers to slow down because of heavy spray when the windscreen wipers have already been set to maximum speed, the message will be ignored because drivers will already have taken what they consider to be the appropriate action.

Where the message is relevant as regards event, location, and timing, changeable message signs can have a beneficial effect on traffic safety. Alm and Nilsson (2000) conducted an experiment looking at the effect of different message content on drivers' behaviour. In a driving simulator, drivers were faced with three incidents, a queue of cars moving at 30 km/h, road works requiring a lane change, and an accident requiring a lane change. Five levels of information where provided at a distance of 1000 m from the incident (Table 5.2). Figure 5.8 shows the mean speed plotted against distance from the slow-moving traffic queue. It is evident that all forms of message result in a slower approach to the slow-moving traffic queue than when no warning is given. Indeed, one of the drivers who did not receive a warning failed to slow soon enough and collided with the back of the queue. Figure 5.9 shows the mean speed of approach for the accident. Again the speed of approach is reduced when any form of warning is given. Interestingly, when the message contains a rec-ommended action, namely to use the left lane, lane changing occurs earlier but the

TABLE 5.2

Levels of Information Provided to Drivers in a Study of the Effect of Message Content on Driver Behaviour

Message Level	Information Provided	Example of Message
0	None	—
1	Warning (flashing red light)	Warning
2	Warning, nature of incident	Warning, congestion
3	Warning, nature of incident, distance to incident	Warning, road works, 1 km ahead
4	Warning, nature of incident, distance to incident, recommended action	Warning, accident, 1 km ahead, use left lane

From Alm and Nilsson (2000).

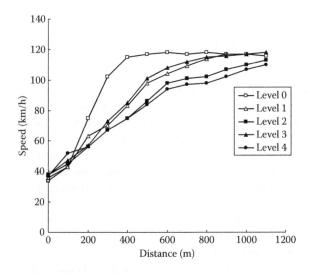

FIGURE 5.8 Mean speed (km/h) plotted against distance (m) from a slow-moving queue of traffic for the five levels of message content listed in Table 5.2 (after Alm and Nilsson 2000).

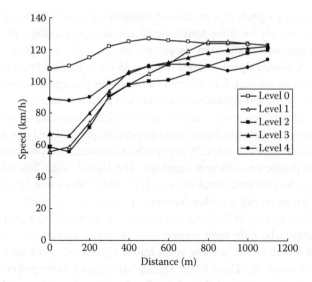

FIGURE 5.9 Mean speed (km/h) plotted against distance (m) from an accident requiring a change of lane for the five levels of message content listed in Table 5.2 (after Alm and Nilsson 2000).

speed past the accident is faster than when less information is given. There can be little doubt that messages that are correct in describing event, location, and timing are helpful to drivers. The only downside is the possibility that a message containing a lot of information may focus the attention of the driver on that incident to the exclusion of others.

5.6 TRAFFIC SIGNALS

A ubiquitous feature of roads in urban and suburban areas is the traffic signal. Traffic signals are placed at intersections to identify priorities for both vehicular and pedestrian traffic. The photometric and colourimetric characteristics of traffic signals are closely regulated in terms of their luminous intensity distributions and colour, the latter because the meaning of the signal is given by its colour (ITE 1985; CIE 1994a; ITE 2005; European Committee for Standardization 2006). The recommendations are consensus decisions made by a committee, but those decisions are based, at least in part, on studies of the reaction time to the onset of the signals and the number of signals that are not detected under different conditions. Bullough et al. (2000) have reported an extensive series of measurements of reaction time and missed signals using a tracking task requiring continuous fixation and simulated traffic signals occurring a few degrees from the fixation point, the traffic signals being provided by both incandescent and LED light sources. Reaction times for all three signal colours tended to become shorter as signal luminance increased until a minimum was reached. However, small changes in reaction time are of little significance for traffic signals because of the delays built in to the sequencing of the signals. Much more important are signals that are missed altogether. Figure 5.10 shows the percentage of missed signals for three traffic signal colours, over a range of signal luminances, seen against a 5000 cd/m² large-area background, i.e., against a simulated daytime sky. A missed signal was one that was lit for more than one second without a response from the subject. It is evident that increasing the signal luminance reduces the percentage of missed signals until a minimum level is reached.

The practical implication of Figure 5.10 is that different traffic signal colours need to have different luminances to achieve the same percentage of missed signals, unless the luminances are so high that virtually no signals are missed. This observation explains differences in practice in different countries. The United States has different luminous intensities for different signal colours (ITE 1985, 2005) while Japan and Europe recommend the same but a higher luminous intensity for all traffic signal colours (National Police Agency 1986; European Committee for Standardization 2006).

Until recently, the only light source used in traffic signals was the incandescent lamp. However, in the last few years traffic signals using coloured LEDs have become widely available. These traffic signals have a much lower power demand and a longer maintenance interval than those fitted with an incandescent light source. Measurements have shown that there is no difference in the percentage of missed signals between LED and incandescent traffic signals of the same luminous intensity and the same nominal colour (Bullough et al. 2000). However, these measurements

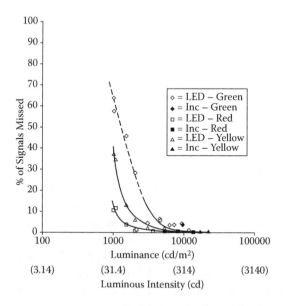

FIGURE 5.10 Percentage of signals missed for each signal colour plotted against signal luminance (cd/m²). The signals were provided by either LED or filtered incandescent light sources. To be counted as a missed signal, the signal had to have been on for one second without a response from the subject. The second horizontal axis is the luminous intensity (cd) corresponding to the signal luminance (cd/m²) for a 200 mm diameter signal (after Bullough et al. 2000).

were taken against a uniform background. Where the background to the traffic signal is complex and filled with other lights there may be an advantage for LED traffic signals in that the colour produced tends to be more saturated than when an incandescent lamp and filter are used to produce the required colour. Saturation matters because, for the same hue, more saturated colours are perceived as brighter than less saturated colours at the same luminance (Ayama and Ikeda 1998). This difference in perception is associated with the stronger signal produced through the colour channels of the human visual system by more saturated light. What this means in practice is that, at the same luminance, traffic signals using LEDs will be seen as brighter than those using an incandescent light source and hence will be more conspicuous (Bullough et al. 2007a). As for the ability to correctly identify the signal colour, Boyce et al. (2000) examined this question for both incandescent and LED signals in daytime, both light sources producing signals with chromaticity coordinates inside the Institute of Transportation Engineers colour boundaries for signal lights (ITE 1985). They found that errors in colour identification were rare, but when they did occur, were most likely to occur for off-axis viewing, at signal luminances below about 8000 cd/m^2, for the incandescent light source, and the yellow signal, the yellow signal being identified as red. Taken together, these results suggest that, visually, the LED light source has a slight perceptual advantage over the incandescent light source for use in traffic signals.

One general conclusion from the results discussed above is that the higher the luminous intensity of the signal, the fewer the number of missed signals and the more likely it is that the signal colour will be correctly identified. This suggests that the higher the luminous intensity, the better the signal, but there is a limit to how far the luminous intensity of a signal can be taken. A traffic signal has to be seen both day and night. A higher luminous intensity is of value during the day because it will tend to increase the conspicuity of the signal but by night a high luminous intensity can become a source of discomfort and even disability glare. Bullough et al. (2001a) measured the percentage of people considering traffic signals of different luminances uncomfortable when viewing them directly (Figure 5.11). Such data can be used to set desirable traffic signal maximum luminances at night, which might be lower than the maximum allowed by day.

5.7 SUMMARY

Road markings, road signs, and traffic signals are designed to inform and regulate drivers' behaviour. Some markings, signs, and signals are fixed, some are changeable, some are unlit, some are lit, but all need to be seen by day and night. The aspect of the design of markings, signs, and signals that is considered here is the use of light, either as an inherent part of the marking, sign, or signal or as a means to make the marking, sign, or signal conspicuous, visible, and legible at night.

Some of the simplest means used to guide drivers are markings on the road. Road markings usually consist of white paint containing some retroreflective materials. The white paint is a diffuse reflector, which ensures the mark will be seen in positive contrast against the road surface during daylight. At night, the luminance of the white paint has two components, the diffuse reflected component mainly from any

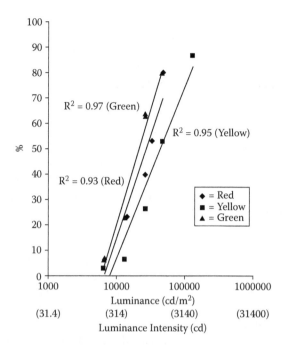

FIGURE 5.11 Percentage of subjects considering a signal uncomfortable for the three signal colours seen in darkness, plotted against signal luminance (cd/m²). The signal simulates a 200 mm diameter signal seen from a distance of 20 m. The second horizontal axis is the luminous intensity (cd) corresponding to the signal luminance (cd/m²) for a 200 mm diameter signal (after Bullough et al. 2001a).

road lighting and the retroreflected component from the vehicle headlamps. Where there is no road lighting, the luminance depends almost entirely on the retroreflective materials. The retroreflective materials produce a greatly enhanced reflection in the direction of the vehicle, which means that the luminance of the markings will be much higher than the luminance of the adjacent road surface, the resulting positive luminance contrast making the markings visible at a distance. The main limitation of such markings is that they tend to lose their retroreflective properties with wear and they tend to disappear when the road is covered with water. As a result, visual guidance is much reduced at the time when it is most required.

One way to overcome this problem is to use road studs. These devices consist of an assembly containing one or two large retroreflectors. The assembly is usually high enough to stand above any water on the road. Road studs depend for their visibility on light from a vehicle's headlamps. This inevitably limits the distance over which guidance is delivered. An alternative now available is a self-luminous, solar-powered road stud containing a white LED. Studs of this type can provide visual guidance over a much longer distance than is possible using light from headlamps alone.

Another common feature of roads is fixed signs mounted beside or over the road giving information on directions, lane changes, speed limits, etc. The decision on whether or not to illuminate a sign depends on the distance at which the sign needs to be detected. High speed, dense traffic, and the call for a maneuvre all increase the distance at which the sign needs to be detected and hence the need for lighting of the sign. To be sure of being detected, the sign also needs to be conspicuous. The minimum sign luminance necessary to ensure conspicuity is influenced by the complexity of whatever is around the sign, the sign colour, and the driver's age. Individual sign lighting is most required where detection distances are large, the background is complex, and the amount of light reaching the sign from the headlamps of approaching vehicles is small.

Where individual lighting is provided for a sign, there are two options, external or internal lighting. For external lighting, the luminaire is positioned close to and either above or below the sign. For internally illuminated signs, both the reflection and transmission properties of the front face are important because the sign has to be legible by both day and night and should look the same under both conditions. It is the reflection properties that dominate the appearance of the sign by day and its transmission properties that dominate them by night. The great advantage of the internally illuminated sign is that, compared with external sign lighting, it produces much less light pollution.

Where individual sign lighting is not provided, the illumination of the sign is dependent on the headlamps of approaching vehicles. The need to control the light distribution from headlamps in order to control glare to other vehicles limits the amount of light that can be expected to reach a sign. To increase the amount of light from headlamps reflected towards the driver, signs contain retroreflective materials.

Road markings and signs are usually fixed but there are changeable versions of both. Changeable road markings are powered road studs that can change their colour or state on instruction from a control unit. Changeable message signs are used to provide information about temporary road conditions, such as the presence of road works, variable speed limits and traffic congestion. A changeable message is only as useful as the message it displays. Messages fall into three classes, those that are valuable, those that are wrong, and those that are already obvious to the driver. Only messages in the first class are likely to be obeyed. To contribute to traffic safety, the message has to be correct about the event, its location, and its timing.

Another ubiquitous feature of roads in urban and suburban areas is the traffic signal. Traffic signals are placed at intersections to identify priorities for both vehicular and pedestrian traffic. The photometric and colourimetric characteristics are closely regulated in terms of their luminous intensity and colour, the latter because the meaning of the signal is given by its colour. Both incandescent and LED light sources can be used for traffic signals. The most important response measure to a traffic signal is missing the signal altogether. Different traffic signal colours need to have different luminances to achieve the same percentage of missed signals, unless the luminances are so high that virtually no signals are missed. This would seem to suggest that all that is necessary to ensure an adequate signal is to use a high luminance, but this may make the signal itself a glare source at night.

Markings, signs, and signals can be considered as a balancing act on a number of levels. There is a need to balance the level of information so that enough information is provided to be useful but not so much that the driver is overwhelmed. There is a need to balance conspicuity and visibility against visual comfort and light pollution, by day and night. Lighting and retroreflection are the means used to make the information provided by markings, signs, and traffic signals accessible to the driver but care is necessary to achieve the proper balance.

6 Vehicle Forward Lighting

6.1 INTRODUCTION

The exterior lighting of vehicles can be conveniently divided into two types, forward lighting and signal lighting. Forward lighting is lighting designed to enable the driver to see after dark. Signal lighting is lighting designed to indicate the presence of or give information about the movement of a vehicle, by night and day. Forward lighting includes headlamps and fog lamps. Often, these are of such a high luminous intensity that they cause glare to approaching vehicles. Signal lighting includes front, side, and rear position lamps; turn lamps; stop lamps; rear fog lamps; reversing lamps; daytime running lamps; hazard flashers; and license plate lamps. These are usually small in size and of limited luminous intensity. Forward lighting is there to see by. Signal lighting is there to be seen. One exception to this crude classification is reversing lamps. Reversing lamps provide both visibility to the rear and a signal to others around the vehicle. This chapter is concerned with the forward lighting fitted to all vehicles, i.e., headlamps. Fog lamps are dealt with in Section 11.5. Forward lighting is now an essential part of vehicle design, serving functional, aesthetic, and marketing demands. This chapter is concerned only with the functional aspects.

6.2 THE TECHNOLOGY OF VEHICLE FORWARD LIGHTING

The technology of vehicle forward lighting involves light sources, optical systems, and structure (Wordenweber et al. 2007).

6.2.1 Light Sources

To be suitable for vehicle forward lighting, light sources have to have sufficient light output to meet the legal requirements that specify the minimum luminous intensities of headlamps. They also have to meet the customer's expectations about the amount of light immediately available on switch-on. Further, they have to meet the manufacturer's need to operate in a wide range of climates as well as to withstand frequent vibration and to last as long as the vehicle, assuming common patterns of use. For the first motor vehicles, the light source used in headlamps was a flame produced by burning acetylene gas generated by dripping water onto calcium carbide. It was only in the 1910s that electric lighting began to be widely used and not until the 1950s that vehicle headlamps became the subject of quantitative legal requirements. The first electric light source used in vehicle headlamps was the basic incandescent. This served until the 1960s when the tungsten halogen light source was introduced but that, in turn, was challenged by the arrival of the xenon discharge (HID) light source in the 1990s. Today, another light source, the light emitting diode (LED) is poised to make an entrance. These light sources differ in their light spectrum, luminous efficacy, and life (see Section 2.5). All are used for the conventional lighting of buildings but when used in vehicles they are modified to be able to withstand the rigours of the vehicle environment that include extremes of heat, cold, moisture, and vibration. The main modification for the incandescent and tungsten halogen light sources is to increase the strength of the filament assembly. For the HID light source used in headlamps, the main modification is the addition of xenon to what is basically a metal halide light source to provide a significant amount of light available immediately on switch-on and to make the run-up time to full light output much shorter. These changes in dominant light source technology have been positively correlated with an increase in light output. Halogen headlamps emit more light than incandescent headlamps, and HID headlamps produce more light than halogen headlamps. Headlamps in modern vehicles use either tungsten halogen or xenon discharge (HID) light sources.

6.2.2 OPTICAL CONTROL

To provide the required light distribution, the light source is housed in a structure that itself is either mounted on the vehicle or integrated into the vehicle body. At least two headlamp light distributions are required by law, high beam for use when there is no other vehicle on the road ahead, either approaching or leading, and low beam for use when there is another vehicle approaching or leading (see Section 6.3). In some countries, high beam is referred to as main beam or driving beam while low beam is also called dipped beam, meeting beam, or passing beam.

There are three systems of optical control used in headlamps to produce the specified luminous intensity distribution, based on reflection, projection, and multiple light sources. For the reflector system, the light distribution is determined by the position of the light source relative to the reflector, the shape of the reflector, and any optical patterning of the front cover glass. The oldest and simplest headlamp consists of a parabolic reflector with the filament of the light source at the focus and having a shading cap over the light source. The presence of the shading cap means that all the light emitted from the headlamp comes from the reflector. The parabolic reflector produces a parallel beam of light so the desired light distribution is achieved through cylindrical or prism lenses built into the cover glass. At the moment, designers prefer smooth clear cover glasses, so in many vehicles, the parabolic reflector has been replaced by smooth, segmented or faceted reflectors designed to produce the desired light distribution without any lenses on the cover glass (Figure 6.1a).

Headlamps using projection have at least three components: a light source, a near ellipsoidal reflector, and a condensing lens. Because the reflector is nearly ellipsoidal, it has two foci. The light source is placed at one focus so the reflector produces an image of the source at the second focus. The light distribution after the second focus is strongly divergent so a condensing lens is used to collimate the beam (Figure 6.1b). Projector headlamps typically use HID light sources.

Headlamps using multiple light sources to form the beam are associated with the LED light source. Unlike tungsten halogen or xenon discharge (HID) light sources,

(a)

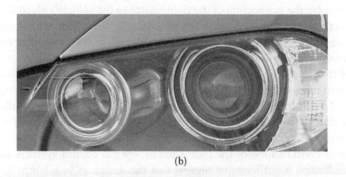

(b)

FIGURE 6.1 Two systems of headlamp optical control: (a) a reflector headlamp on a Jeep Patriot and (b) projector headlamps on a BMW X5.

a single LED does not presently have a high enough light output for a headlamp. Therefore, the LED headlamps that are being developed all have multiple LEDs. The desired light distribution can be achieved by means of reflectors or lenses together with switching or dimming of individual LEDs.

Various methods have been used to provide the two luminous intensity distributions required, high beam and low beam. The simplest is to provide two different headlamps, one for high beam and one for low beam, and to switch between them, thereby creating a four- rather than a two-headlamp vehicle. The same or different optical control methods can be used in the high-beam and low-beam headlamps. Rumar (2000) concludes that four-headlamp systems are better than two-headlamp systems in all respects except cost and size. If a two-headlamp system is required, and they often are, some means has to be found to get two different light distributions from the same headlamp. For reflector headlamps, this can be done by using a single light source and moving the reflector to change the light distribution. An alternative approach suitable for the tungsten halogen light source is to keep the reflector fixed but to use two filaments in the same headlamp. By placing the two filaments at different positions relative to the reflector, two different light distributions can be achieved. For projector headlamps, a moveable baffle within the headlamp can be used to cut off part of the high beam when low beam is required. The change from high to low beam is made by raising the baffle across the second focal plane. For multiple-source headlamps, the different light distributions can be achieved by switching or dimming different combinations of light sources.

6.2.3 HEADLAMP STRUCTURE

Headlamp structure can be considered in four areas: mechanical, electrical, environmental, and supplementary. The mechanical elements are concerned with ensuring that the light source and optical control elements are consistently related to each other and the whole is mounted securely on or in the vehicle body. It is also necessary to provide some means of adjusting the aiming of the headlamp. The electrical elements involve the connection of the light source to the vehicle's electrical system, either directly or through any necessary control gear, and the means to change the light source in the event of premature failure. The environmental elements require careful consideration of the materials used and the mechanical structure so that the sealing of the headlamp is sufficient to prevent water access yet there is adequate ventilation to prevent fogging. The supplementary aspects refer to the systems used for automatic washing of the cover glass, automatic leveling of the headlamp, and automatic change from high beam to low beam and vice versa. Details of these matters are given in Wordenweber et al. (2007).

This brief consideration of the light sources, optical control systems, and structure of headlamps is sufficient to suggest that the design of headlamps is complex and that headlamps can take many different forms. However, all have to meet the relevant regulations.

6.3 THE REGULATION OF VEHICLE FORWARD LIGHTING

Headlamp luminous intensities, light distribution, and placement on the vehicle are closely regulated. The purpose of these regulations is to bring some order to the potential conflict between drivers caused by the fact that headlamps increase the visual capabilities of the driver sitting behind them but simultaneously decrease the visual capabilities of the driver facing them. These regulations are implemented through adherence to them by vehicle manufacturers and by annual government inspections of used vehicles, designed to ensure continued roadworthiness. These regulations can take different forms in different countries, but the vast majority follows either the recommendations of the US Federal Motor Vehicle Safety Standard or the Economic Commission for Europe (ECE). Regardless of which set of standards is followed, there are two features they have in common. The first is that the headlamps fitted to a vehicle have to produce two different luminous intensity distributions, here called high beam and low beam. High-beam headlamps are for use when there is no other vehicle on the road ahead so there is no need to limit glare. Low-beam headlamps are for use when there is an approaching vehicle or a vehicle immediately in front and it is necessary to limit the headlamp luminous intensity so as not to compromise the vision of the other driver. The second is that the colour appearance of the light emitted by headlamps must be white, defined as emitting light with chromaticity coordinates that fall within a specified region of the CIE 1931 chromaticity diagram (see Figure 2.3 for an example). Insisting on a similar colour appearance for all headlamps is an example of using colour as a means to signal the direction of movement of the vehicle. This principle also explains why reversing lights, which are mounted on the rear of the vehicle, are also white, reversing lights only being lit when the vehicle is moving backwards.

6.3.1 FORWARD LIGHTING IN NORTH AMERICA

In the United States, the design, performance, and installation of all motor vehicle lighting equipment are regulated by the Federal Motor Vehicle Safety Standard 108, which incorporates the technical standards of the Society of Automotive Engineers (SAE 2001). Both Canada and Mexico, which have land borders with the United States, generally follow the same standard, although Canada also requires the installation and use of daytime running lights (see Section 7.14). The required photometric properties of headlamps are given as minimum and maximum luminous intensity values for specific directions relative to the beam axis. Interestingly, there are very few points where both minimum and maximum luminous intensities apply. For most points, the limit is either a minimum or a maximum but not both. The points where a minimum luminous intensity is required are those associated with providing good vision for the driver. The points where a maximum is set are those most likely to cause glare to an approaching driver. The outcome of having different low- and high-beam luminous intensity distributions is to produce different illuminances at different locations along the road ahead. Figure 6.2 shows contours of the median vertical illuminance for pairs of headlamps used on the twenty best-selling passenger vehicles in the United States for the 2000 model year, measured at road level for both low and high beams (Schoettle et al. 2002). Other requirements relate to the positioning of the headlamps. Headlamps should be mounted at least 559 mm above the road surface and as far apart as practicable. Left and right headlamps should be equidistant from the vehicle centreline and at the same height above the road surface.

6.3.2 FORWARD LIGHTING ELSEWHERE

The headlamp regulations used in virtually all other industrialized nations follow the ECE recommendations. The photometric requirements are specified as the minimum and maximum illuminances at different positions on a vertical plane positioned 25 m from the headlamp, normal to the beam axis. Figure 6.3 shows the contours of median vertical illuminance for pairs of headlamps used on the twenty best-selling passenger vehicles in Europe for the 1999 model year, measured at road level for both low and high beams (Schoettle et al. 2002).

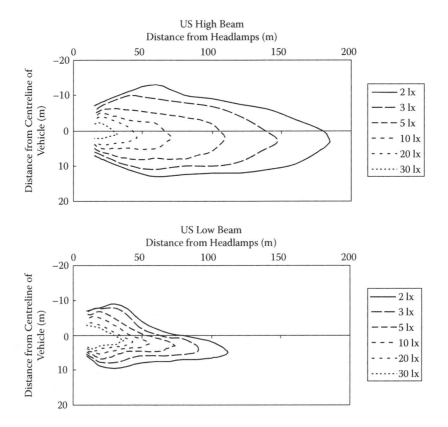

FIGURE 6.2 Contours of median vertical illuminances (lx) produced by pairs of halogen headlamps operating on high beam and low beam for the twenty best-selling passenger vehicles in the US in the 2000 model year. The vertical illuminances were calculated at road level. Vehicles in the United States are driven on the right (after Schoettle et al. 2002).

The ECE regulations require that the headlamps producing the low beam should be mounted at least 500 mm and no more than 1200 mm above the road. A static headlamp leveling device is compulsory for all headlamps on vehicles in Europe. Where a xenon discharge (HID) light source is used, the leveling device has to be automatic.

6.3.3 SIMILARITIES AND DIFFERENCES

A comparison of Figures 6.2 and 6.3 reveals some similarities and some differences between the North American and ECE standards, at least as regards vertical illuminance distributions. Specifically, the European high beam is narrower and provides more light further down the road than does the American high beam. This is consistent with the fact that the luminous intensity values for high beams differ as regards the maximum luminous intensity allowed. In the American regulations the maximum luminous intensity is 75,000 cd while in the ECE recommendations the maximum is 140,000 cd. Rumar (2000) examined this difference on the basis

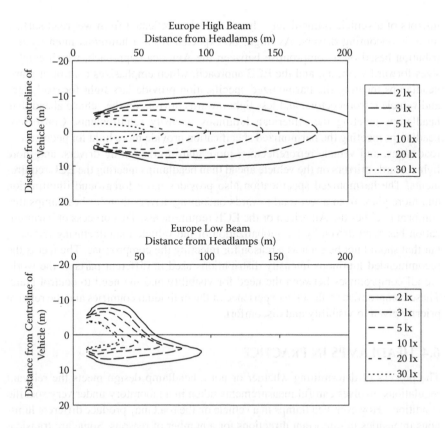

FIGURE 6.3 Contours of median vertical illuminances (lx) produced by pairs of halogen headlamps operating on high beam and low beam for the twenty best-selling passenger vehicles in Europe in the 1999 model year. The vertical illuminances were calculated at road level. Vehicles in most countries of Europe are driven on the right (after Schoettle et al. 2002).

of visibility and glare and concluded that the US maximum should be increased to the ECE level. For low beams, both American and ECE regulations are similar in that they show an emphasis to the near side of the road and a limitation in the vertical illuminance produced towards vehicles in the opposing lane. However, the ECE low beam has a sharper cut-off than the North American low beam, indicating the greater emphasis given to controlling disability glare. Conversely, the American low beam provides more light down the road and more on the edge of the road, indicating a greater emphasis on visibility.

For several years, there have been moves to resolve this difference in low-beam luminous intensity distributions by the production of a harmonized vehicle headlamp specification (SAE 1995; GTB 1999). Sivak et al. (2001) have carried out an evaluation of the proposed harmonized low-beam luminous intensity distribution in terms of the change in luminous intensities in directions where significant elements in the road environment are to be found, e.g., pedestrians at the edge of the road, road edge markings, road signs, glare towards oncoming drivers, glare towards the mirrors of a vehicle immediately ahead, and glare reflected from wet road surface towards oncoming drivers. As might be expected from a luminous intensity distribution based on a compromise between the American approach, which emphasizes forward visibility, and the ECE approach, which emphasizes control of glare, headlamps meeting the harmonized specification provide less light for road signs and vehicle retroreflectors, and less light towards mirrors on the vehicle ahead than headlamps meeting the American luminous intensity requirements. Conversely, headlamps meeting the harmonized specification provide more light for pedestrians, road signs, and vehicle retroreflectors, more glare to oncoming drivers, and more light towards mirrors on the vehicle ahead than headlamps meeting the ECE requirements. The harmonized specification also provides more foreground illumination and more glare from a wet road towards oncoming drivers than for headlamps that conform to either the American or the ECE requirements. This process of harmonization has been driven by the globalization of the vehicle manufacturing industry, but that should not be seen as a reason for rejecting the compromise. The fact is the recommended luminous intensity distributions used in different parts of the world are all compromises between the need for visibility and the need to control glare. These compromises reflect the experience of the individual countries and the relative priority given to visibility and discomfort.

6.4 HEADLAMPS IN PRACTICE

The process of determining whether or not a headlamp design meets the relevant regulations involves careful measurements taken in a laboratory under very specific conditions. However, headlamps in a vehicle on the road may produce different luminous intensities in important directions for a number of reasons. Some are transient and inherent in the road layout or the nature of the vehicle. An example of the former is the reduction in illumination of the road ahead and the increase in glare to opposing drivers that occur when breasting a hill. An example of the latter is the reduction in the illumination of the road ahead and the increase in glare to opposing drivers produced by motorcycles when cornering to the right on right-hand-drive roads due to the tilting of the machine (Konyukhov et al. 2006). Others are permanent and occur because the vehicle is not level, or the headlamp is incorrectly aimed, or the headlamp is dirty. Yerrel (1971) reported a set of roadside measurements of headlamp luminous intensities in Europe and found a very large range of luminous intensities for the same direction despite a common standard. Alferdinck and Padmos (1988) found similar results from roadside measurements in The Netherlands. They also examined the importance of aiming, dirt, and lamp age on the luminous intensity in a series of laboratory measurements. Figure 6.4 shows the cumulative frequency distributions of luminous intensity in a direction important for forward visibility and in a direction important for glare to an oncoming driver, for fifty cars taken from a parking lot. The luminous intensity measurements of the headlamps, as found, but taken in the laboratory, agreed with measurements taken at the roadside. From Figure 6.4 it can be seen that the headlamps, as found, tend to produce less forward visibility and more glare than new headlamps. The forward visibility is most improved by correcting the aiming. Cleaning the headlamps and operating them

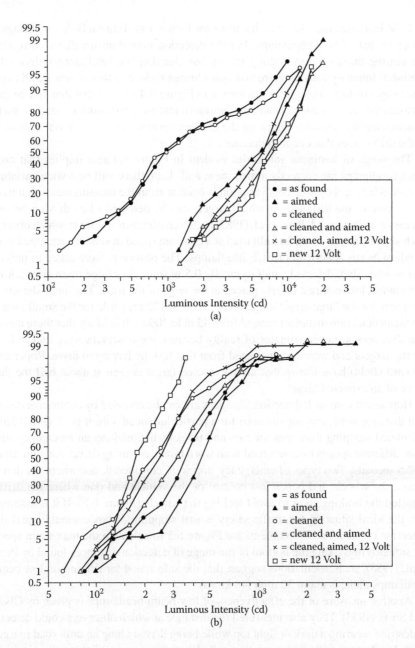

FIGURE 6.4 Cumulative frequency distributions of luminous intensities (cd) in directions (a) important for visibility of the near side of the road and (b) important for glare to oncoming drivers, for headlamps on fifty cars as found; aimed; cleaned; cleaned and aimed; cleaned, aimed, and operated at 12 V; and for new headlamps (after Alferdinck and Padmos 1988).

at 12 V increases the luminous intensity for forward visibility a little and brings it closer to that of new headlamps. For the direction important for glare, correcting the aiming makes things slightly worse, but cleaning the headlamps reduces the luminous intensity causing glare and again brings it close to that of new headlamps. The ranges of luminous intensities shown in Figure 6.4 suggest that fine differences between the recommended headlamp luminous intensity distributions used in North America and by countries following the ECE recommendations are trivial compared to the differences that occur in practice.

The range of luminous intensities evident in Figure 6.4 also implies that even when headlamps are correctly aimed, new, and clean, there will be a wide variation in how effective they are. This variation is evident from the measurements that have been made of the distances at which targets can be detected when driving on low beams on an unlit road. Perel et al. (1983) reviewed nineteen studies in which observers had been driven along an unlit road at a constant speed in vehicles equipped with standard North American or ECE headlamps. The observers were asked to press a button when they detected small (typically 0.5 m square) or large (man sized), low-reflectance, low-contrast targets placed at the edge of the road. The mean detection distances for the large target ranged from 51 m to 122 m, while for the small target, the mean detection distances ranged from 45 m to 100 m. It is likely that these detection distances are overestimates of reality because the observers were told to look for the targets and were not distracted from that task by having to drive. Roper and Howard (1938) have shown that an unexpected target is seen at about half the distance of an expected target.

How significant such detection distances are can be revealed by comparing detection distance with stopping distance for a particular speed. Olson et al. (1984) have calculated stopping distances for cars and trucks when making an emergency stop from different speeds on a wet road with worn tyres, assuming driver reaction times of 2.5 seconds. Two types of emergency stop were considered, one where the driver locked the wheels and hence lost control of the vehicle and one where the driver adjusted the braking so as to avoid locking the wheels (Figure 6.5). If it is assumed that the ideal situation for traffic safety is that stopping distance should equal the detection distance, it is possible to use Figure 6.5 to calculate the maximum speed for safe driving. Using the bottom of the range of detection distances found by Perel et al. (1983), such calculations suggest that the safe speed for driving on low beam headlamps alone is about 30 mph (48 km/h).

Another measure of the effectiveness of low-beam headlamps is given by Olson and Sivak (1983). They also measured the distance at which observers could detect a pedestrian wearing a dark or light top while being driven along an unlit road in a car using low-beam headlamps, as well as calculating the stopping distance for a vehicle moving at 55 mph (89 km/h). The percentage of trials in which the detection distance was less than the stopping distance ranged from 3 percent, which was for young observers with the pedestrian wearing the light top, to 83 percent for old observers with the pedestrian wearing the dark top. Similar estimates of the percentages of people able to stop within the distance at which they can detect low-contrast targets have been made by Yerrel (1976).

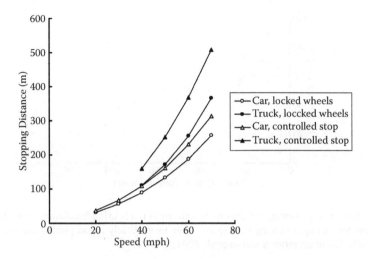

FIGURE 6.5   Stopping distances (m) for cars and trucks with worn tyres making an emergency stop on a wet road surface, with and without locked wheels, plotted against speed (mph), and assuming a driver reaction time of 2.5 seconds (after Olson et al. 1984).

Taken together, these comparisons of detection distances and stopping distances go some way to explain the increase in pedestrian deaths around the change in daylight saving time (see Section 1.5). They also indicate that driving on low beams on unlit roads other than slowly is very much an act of faith. Of course, this might not matter so much if drivers used high beams whenever possible, i.e., whenever there was no vehicle approaching and no vehicle immediately ahead. Unfortunately, field measurements have shown that this is not what happens. Sullivan et al. (2004a) observed drivers' use of low and high beams on unlit rural roads. Figure 6.6 shows the percentage of drivers using high beams when there was no approaching vehicle and none immediately ahead plotted against traffic density. As might be expected, the percentage of vehicles using high beams decreases with increased traffic density, but it is not until traffic density falls below about fifty vehicles/hour that high beams are used by more than about 50 percent of drivers.

These results imply two basic truths. First, that low beams alone do not provide adequate visibility of the road ahead, except at low speeds. Second, that high beams, which do increase visibility, are not used as frequently as they could be. Why high beams are not used as frequently as they could be has been the subject of some theorizing. One hypothesis is that drivers drive faster than they should because they do not recognize the extent to which their vision has been degraded at low light levels because the degradation is not uniform across all visual capabilities (Leibowitz and Owens 1977). Specifically, the ability to steer the vehicle using peripheral vision is maintained at low light levels but the ability to recognize small, low-contrast objects is not (Owens and Tyrrell 1999). Further, the widespread use of high-contrast signs and markings may give drivers an unrealistic estimate of their ability to detect and recognize detail. The problem with this selective degradation hypothesis is that it is difficult to believe there are many drivers who are not aware that they can see

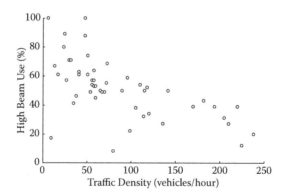

FIGURE 6.6 The percentage of drivers on unlit rural roads using headlamps on high beam when there was no approaching vehicle and none immediately ahead plotted against traffic density (vehicles/hour) (after Sullivan et al. 2004a).

much better in all respects with high beams than low beams. Certainly, professional drivers of heavy trucks seem to be so aware. An early study of high beam use (Hare and Hemion 1968) showed that where there were no opposing vehicles or vehicles immediately ahead, only 25 percent of car drivers used their headlamps on high beam. Under the same conditions, 67 percent of truck drivers used their headlamps on high beam.

An alternative explanation for the failure to use high beams as frequently as possible is based on a combination of probability and inconvenience. The idea is simply that given the low probability of there being a pedestrian on an unlit rural road at night, the increase in visibility that is achieved by changing from low beam to high beam is not worth the trouble, particularly if there is a high probability of having to revert to low beam shortly after. There has been no study of this probability/inconvenience hypothesis. One way to do this would be to examine the pattern of use of headlamp beams on unlit rural roads where there was and was not a high probability of there being large unlit obstacles on the road, such as deer.

The selective degradation hypothesis for the under-use of high beams treats drivers as rational beings and excuses their behaviour. The probability/inconvenience hypothesis treats them as emotional humans and condemns the implicit gamble. While the argument between these two hypotheses is interesting, it may soon be moot because the technology to change between low and high beams automatically is already available (Wordenweber et al. 2007). With this technology, the default state is high beam. Sensors detect the presence of approaching vehicles or vehicles immediately ahead and change to low beam, reverting to high beam as soon as the approaching vehicle has passed or the vehicle ahead has moved on.

6.5 HEADLAMPS AND LIGHT SPECTRUM

The most dramatic change in headlamps over the last decade has been the increasing use of xenon discharge (HID) light sources in vehicles, identifiable by their blue-white colour appearance. HID headlamps differ from conventional halogen headlamps in several respects but the three that are important for visibility are the amount of light produced, the luminous intensity distribution, and the spectral power distribution of the light emitted. HID headlamps typically produce two to three times more luminous flux than halogen headlamps. The recommended minimum and maximum luminous intensities used in regulations apply regardless of the light source used, a fact that raises the question of how the additional luminous flux produced by a HID light source should be distributed. The maximum luminous intensities specified in regulations are mainly restricted to parts of the beam that cause glare to opposing drivers or drivers immediately ahead. In other parts of the beam the regulations specify minimum values but not maxima. Consequently, the additional luminous flux produced by the HID light source tends to be directed to the parts of the beam where no maximum is specified. Figure 6.7 shows contours for the detection of a square target of 40 cm side and of reflectance 0.1 by drivers using either HID headlamps or halogen headlamps, on an ECE low-beam setting (Rosenhahn and Hamm 2001). Clearly, the HID headlamps conforming to the same regulations allow objects to be detected at greater distances and over a wider range of angles than the halogen headlamps. It is also worth noting that the locations where there is close agreement in detection distances for the two headlamp types are the locations where the maximum luminous intensities are specified in regulations.

Van Derlofske et al. (2001) report another way of quantifying the benefits of HID over halogen headlamps. They measured reaction times to the onset of a change in reflectance of targets at various angles off-axis when illuminated by an HID headlamp set and two halogen headlamp sets, all conforming to ECE regulations and used on low beam. Figure 6.8 shows the geometry of the experiment as set out on an unused and unlit asphalt runway.

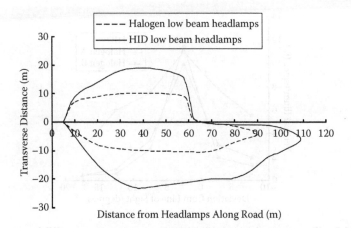

FIGURE 6.7 Contours for the distances (m) at which a square target of 40 cm side, with a reflectance of 0.1, is detected by drivers using either HID headlamps or halogen headlamps, both on low beam (after Rosenhahn and Hamm 2001).

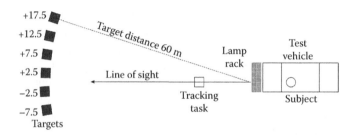

FIGURE 6.8   Geometry of the experiment conducted by Van Derlofske et al. (2001). The subject sat in the test vehicle and did the continuous tracking task. The headlamps were mounted on the lamp rack at the front of the vehicle. The flip dot targets were positioned on an arc 60 m away from the headlamps spaced at 5-degree intervals. (Reprinted with permission from SAE Paper 2001-01-0298, © 2001 SAE International.)

The targets were placed on an arc of radius 60 m from the headlamps. Each target consisted of a 178 mm square grid of 12.7 mm diameter flip dots, each dot being a disc painted black on one side and white on the other. By applying current to the target, the dots are flipped over within 20 ms, thereby changing the target from a black square to a grey square, grey because each dot is surrounded by a black frame and at 60 m, the dots and the frame merge and together appear grey with an average reflectance of 0.4. The luminance of the target varies with position because the illuminance on the target also varies with position. Figure 6.9 shows the illuminances on each target produced by the three headlamp sets.

The subjects performed a continuous tracking task, designed to maintain fixation directly ahead, and released a press switch as soon as they detected the change in reflectance of any of the targets. The reaction time to the onset of the target, i.e., the change from black to grey was measured. Any response longer than one second was

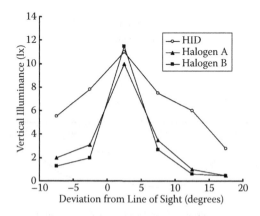

FIGURE 6.9   Illuminance (lx) on the flip dot target produced by one HID headlamp set and two different halogen headlamp sets, all conforming to ECE regulations and used on low beam, plotted against deviation from the line of sight in degrees (after Van Derlofske et al. 2001).

taken as a miss although each such miss was included in the data from which mean reaction time was calculated at an assumed reaction time of 1000 ms. Figures 6.10 and 6.11 show the mean reaction times to the onset of the target and the percentage of missed signals respectively, plotted against deviation from the line of sight, for the three headlamp sets. An examination of Figures 6.10 and 6.11 shows there is little difference between the three headlamp sets for less than 7.5 degrees deviation, but beyond this, the HID headlamps give statistically significantly lower values of mean reaction time and percentage of missed signals than either of the halogen headlamp sets.

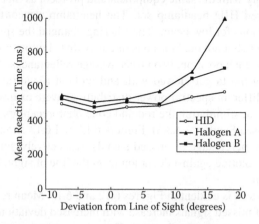

FIGURE 6.10 Mean reaction times (ms) to the onset of the targets for one HID headlamp set and two different halogen headlamp sets, all conforming to ECE regulations and used on low beam, plotted against deviation from the line of sight in degrees (after Van Derlofske et al. 2001).

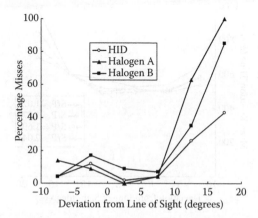

FIGURE 6.11 Percentage of missed signals for one HID headlamp set and two different halogen headlamp sets, all conforming to ECE regulations and used on low beam, plotted against deviation from the line of sight in degrees (after Van Derlofske et al. 2001).

This difference between headlamp sets is what might be expected from the illuminances they produce on the targets (Figure 6.9). However, the HID and halogen headlamp sets differ in spectral power distribution as well as illuminance and there is evidence that at low light levels light sources that provide greater stimulation to the rod photoreceptors allow faster reaction times off-axis than light sources that do not (see Section 4.6). This implies that HID headlamps should allow faster reaction times for off-axis detection than halogen headlamps, even when both provide the same photopic illuminance. Van Derlofske and Bullough (2003) have examined this possibility with the same equipment and protocol as that described above but using a filtered HID headlamp set. The headlamp set conformed to North American regulations for low beam. The filtering changed the spectral power distribution of the light but not the luminous intensity distribution or the illuminances on the targets. For each position, two target average reflectances, 0.4 and 0.2, were created by having the target viewed with and without a neutral density filter in front of it. Four different spectral power distributions were examined, the relative efficiency of each at stimulating the rod and cone photoreceptors being quantified by the S/P ratio (see Section 2.4.3.3). Figures 6.12 and 6.13 show the mean reaction times and the percentage of missed signals, respectively, for the 0.2 average reflectance target, plotted against deviation from the line of sight, for each spectral power distribution.

A common pattern is evident in Figures 6.12 and 6.13. Mean reaction times and the percentages of missed signals increase with increased deviation from the line of sight, for all spectral power distributions. These increases are certainly due to the decrease in illuminance on the target as deviation from the line of sight increases (Figure 6.9). However, at the extreme deviations, where the illuminances on the targets are least, an effect of spectral power distribution is evident. Specifically, the mean reaction time and the percentage of missed signals both decrease as the S/P

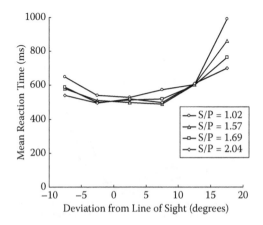

FIGURE 6.12 Mean reaction times (ms) to the onset of the low reflectance (0.2) target for four different light spectra specified by the scotopic/photopic (S/P) ratio, plotted against deviation from the line of sight in degrees (after Van Derlofske and Bullough 2003).

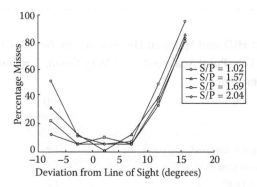

FIGURE 6.13 Percentage of missed signals for the low reflectance (0.2) target for four different light spectra specified by the scotopic/photopic (S/P) ratio, plotted against deviation from the line of sight in degrees (after Van Derlofske and Bullough 2003).

ratio of the light source increases. This is what would be expected from what is known about how the spectral sensitivity of off-axis vision changes in the mesopic state (see Section 3.3.4) and indicates that HID headlamps have an additional advantage for off-axis visibility over and above the greater illuminances produced. Although real, this advantage is small relative to the impact of the greater illuminances. This became apparent when the data for the 0.4 average reflectance targets were examined. For these data, the increases in mean reaction times and percentages of missed signals with increasing deviation from the line of sight were present and of similar magnitude to those obtained for the 0.2 average reflectance targets, but there was no statistically significant effect of S/P ratio.

There can be little doubt that HID headlamps provide better visibility over a larger area than do halogen headlamps, mainly because of the higher illuminances produced over a larger area. This suggests that HID headlamps should be preferred over halogen headlamps. Sivak et al. (2002) asked drivers who had driven the same make of vehicle equipped with HID and halogen headlamps through the same residential neighbourhood to rate the headlamps for twelve different attributes. Table 6.1 gives the mean ratings for the two headlamp types. Given that the possible answers range from 1 = very good to 5 = very poor, it is apparent that both HID and halogen headlamps are good for all the attributes, but HID headlamps are more often very good, although these differences only appeared after the drivers' attention had been drawn specifically to the headlamps rather than the car in general.

6.6 GLARE FROM HEADLAMPS

6.6.1 Forms of Glare

Glare occurs because while the human visual system can operate over about twelve log units of luminance in total, it can only operate over about three log units simultaneously. Any luminance more than about two log units above the average luminance of the scene will be considered glaringly bright. Vos (1999) has classified glare into

TABLE 6.1

Mean Ratings of HID and Halogen Headlamps for Twelve Different Attributes on a Five-Point Scale with 1 = Very Good, 2 = Good, 3 = Fair, 4 = Poor, 5 = Very Poor

Attribute	HID Headlamps	Halogen Headlamps
Overall	1.1	2.5
Making objects in general stand out*	1.2	2.6
Helping to see colours in general	1.8	2.2
Making stop signs stand out*	1.2	2.1
Helping to see the red colour of stop signs	1.6	2.0
Performance on straight sections of road	1.4	2.0
Performance on curves	1.4	2.0
Performance on the foreground	1.6	2.2
Performance in the distance, straight ahead*	1.2	2.4
Performance to the sides*	1.5	2.6
Evenness of light distribution*	1.4	2.9
General appearance of colours*	1.6	2.5

* Statistically significant difference
From Sivak et al. (2002).

eight different forms. Of these eight, only four are experienced while driving: saturation glare, adaptation glare, disability glare, and discomfort glare.

Saturation glare occurs when a large part of the visual field is bright for a long time. The behavioural response is to shield the eyes by wearing low transmittance glasses. Such behaviour is common when driving in very sunny climates during the day.

Adaptation glare occurs when the visual system is exposed to a sudden, large increase in luminance of the whole visual field, e.g., on exiting a long road tunnel during daytime. The perception of glare is due to the visual system being misadapted. On exiting the tunnel the visual system is adapted to the low light level of the tunnel but is exposed to the brightness of sunlight. Adaptation glare is temporary in that the processes of visual adaptation will soon adjust the visual sensitivity to match the new conditions (see Section 3.3.3).

Disability glare, as its name implies, disables the visual system to some extent. This disabling is caused by light scattered in the eye (Vos 1984). The scattered light forms a luminous veil over the retinal image of adjacent parts of the scene, thereby reducing the luminance contrasts of the image on those parts on the retina. Disability glare is experienced most frequently on the roads at night when facing the headlamps of an approaching vehicle.

Discomfort glare is said to be occurring when people complain about visual discomfort in the presence of bright light sources. There is no known cause for discomfort glare, although suggestions have been made ranging from fluctuations in pupil size (Fry and King 1975) through distraction (Lynes 1977) to muscle tension around the eye (Murray et al. 2002). This distinction between disability and discomfort glare does not mean that disability glare does not cause visual discomfort. Headlamps at night can certainly be both visually disabling and visually uncomfortable. In essence, these two forms of glare, disability glare and discomfort glare, are two different outcomes of the same stimulus pattern, namely a wide variation of luminance across the visual field.

6.6.2 THE QUANTIFICATION OF GLARE

Disability glare can be quantified by comparing the visibility of an object seen in the presence of the glare source with the visibility of the same object seen through a uniform luminous veil. When the visibilities are the same, the luminance of the veil is a measure of the amount of disability glare produced by the glare source, and is called the equivalent veiling luminance. The CIE has developed a disability glare formula suitable for use with vehicle headlamps. This is more elaborate than that used for road lighting (see Section 4.3) because headlamps are usually seen much closer to the line of sight. The CIE formula applies at all angles from the line of sight in the range 0.1 to 30 degrees and to either young or old people (CIE 2002). This equation takes the form

$$L_v = \Sigma(10E_n/\Theta_n^3 + (1 + (A/62.5)^4)\ 5\ E/\Theta_n^2)$$

where L_v is the equivalent veiling luminance (cd/m^2), E_n is the illuminance (lx) at the observer's eyes from the nth glare source, θ is the angle of the nth glare source from the line of sight (degrees), and A is the age of the observer (years). The effect of the equivalent veiling luminance on the luminance contrast of an object can be estimated by adding it to the luminance of both the object and the immediate background.

The only photometric quantity relevant to equivalent veiling luminance is the illuminance from the glare source received at the eye. There is little evidence for other aspects of exposure, such as the illuminated area of the headlamp and the light spectra influencing disability glare. Van Derlofske et al. (2004) examined the impact of different illuminances at the eye on the ability to detect off-axis targets using the same equipment and protocol as that described earlier (see Figure 6.8), but using another HID headlamp set positioned 50 m ahead and 5 degrees to the left of the subject's line of sight. The HID headlight set was tilted slightly to produce three different illuminances at the subject's eyes, 0.2, 1.0, and 5.0 lx. Each flip dot target was presented with and without a neutral density filter placed in front, the result being that the average reflectance of the target was, when presented, either 0.4 or 0.2. Again, the subjects performed a continuous tracking task to control fixation and released a button when they detected a change in one of the targets. Any change that was not detected within one second was counted as a missed target and included in the data used to calculate mean reaction time as a reaction time of 1000 ms. Figures 6.14 and 6.15 show the mean reaction times and percentage of missed signals plotted against the target position relative to the line of sight for the three illuminances at the eye and the two target average reflectances. Also shown are the predicted percentages of missed targets when no glare source is present,

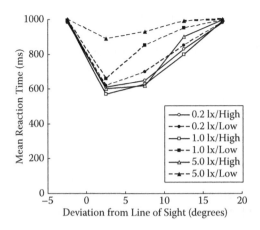

FIGURE 6.14 Mean reaction times (ms) to the onset of the high (0.4) and low (0.2) reflectance targets for three different levels of disability glare specified by the illuminance (lx) received at the eye, plotted against deviation from the line of sight in degrees (after Van Derlofske et al. 2004).

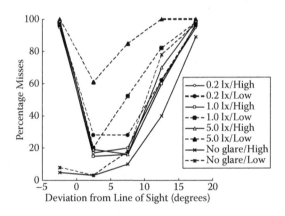

FIGURE 6.15 Percentage of missed high (0.4) and low (0.2) reflectance targets for three different levels of disability glare specified by the illuminance (lx) received at the eye, plotted against deviation from the line of sight in degrees. Also shown are the percentages of missed targets in the absence of glare predicted by the model of Bullough (2002b) (after Van Derlofske et al. 2004).

based on the model of Bullough (2002b). The first point to note about Figures 6.14 and 6.15 is that the mean reaction times for targets of both reflectances positioned at –2.5 degrees and 17.5 degrees are concentrated at 1000 ms. This is because for these two positions virtually all the targets were missed. The targets at –2.5 degrees were closest to and those at 17.5 degrees were furthest from the glare source. The reason for the missed targets at –2.5 degrees is the reduction in luminance contrast caused by the disability glare produced by the glare source, an observation supported by the

low level of misses predicted for the –2.5 degrees position in the absence of opposing headlamps by Bullough (2002b). Even an illuminance at the eyes as low as 0.2 lx
ensures the target at –2.5 degrees will be missed. This is bad news for any pedestrian
caught behind two opposing vehicles when attempting to cross the road. The reason
for the missed targets at 17.5 degrees is not disability glare but rather the failure of
the subject's headlamps to illuminate the target. For the other positions, 2.5, 7.5, and
12.5 degrees from the line of sight, it is clear that reaction times increase and the
percentage of misses increases with increasing deviation from the line of sight and
that these increases are much greater for the low-average-reflectance than for the
high-average-reflectance targets. The difference between the low- and high-average-
reflectance targets is to be expected because the effect of a given equivalent veiling
luminance on visibility will depend on the luminance contrast of the target. The
low–average-reflectance target will have a lower luminance contrast with its immediate background in the absence of glare, so the addition of the veiling luminance
will take the low-average-reflectance target closer to threshold than it will the high-
average-reflectance target. The effect of the illuminance at the eyes is only evident
at 2.5, 7.5, and 12.5 degrees from the line of sight for the low reflectance ($p = 0.2$)
target. For this target, a glare illuminance of 0.2 lx has hardly any effect on mean
reaction time and the percentage of missed targets, but glare illuminances of 1.0 lx
and 5.0 lx both cause increases in mean reaction times and percentage of targets
missed, the increases for 5.0 lx being much greater than for 1.0 lx.

As for discomfort glare, Schmidt-Clausen and Bindels (1974) have produced an
equation relating the illuminance at the eye to the level of discomfort produced by
headlamps, expressed on the de Boer scale. The equation is

$$W = 5.0 - 2\log(E / 0.003(1+\sqrt{(L/0.04)}) \times \varphi^{0.46})$$

where W is the discomfort glare rating on the de Boer scale, E is the illuminance
at the observer's eyes (lx), L is the adaptation luminance (cd/m^2), and φ is the angle
between line of sight and glare source (min. arc). The de Boer scale is a nine-point
glare scale with five anchor points labeled 1 = unbearable, 3 = disturbing, 5 = just
admissible, 7 = acceptable, 9 = unnoticeable. Note that on this scale, lower values
are more uncomfortable. Conditions producing ratings of 4 or less are usually considered uncomfortable.

Figure 6.16 shows the mean ratings of discomfort glare plotted against the illuminance at the eye from the HID headlamps in the experiment of Van Derlofske et
al. (2004) described above. Also shown are the ratings predicted by the Schmidt-
Clausen and Bindels discomfort glare equation for the same experimental situation.
The predictions of the discomfort glare equation show a broad agreement with the
findings for the low-reflectance target of Van Derlofske et al. (2004). More interesting
is the finding that there is a clear difference between the low- and high-reflectance
targets. This implies that the perception of discomfort glare depends not only on
the stimulus to the visual system produced by the glare source but also on what the
observer is trying to do. This should not be too surprising given that it is well known
that the same photometric stimuli can be considered as comfortable or uncomfortable depending on the task being performed (Sivak et al. 1991), the context in which

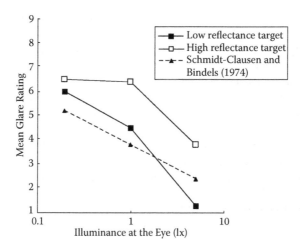

FIGURE 6.16 Mean discomfort glare (de Boer) ratings collected when the subject was attempting to detect the onset of low (0.2) and high (0.4) reflectance off-axis targets, plotted against the illuminance (lx) received at the eye. Also shown are the predicted glare ratings derived from the discomfort glare equation of Schmidt-Clausen and Bindels (1974) (after Van Derlofske et al. 2004).

the stimuli are presented (Boyce 2003), and the information contained in the scene (Tuaycharoen and Tregenza 2005).

The discomfort glare equation produced by Schmidt-Clausen and Bindels (1974) involves three components, the illuminance at the eye, the adaptation luminance, and the angle between the glare source and the line of sight. However, there is evidence that other factors have small effects. While the perception of discomfort from headlamps is dominated by the illuminance at the eye (Sivak et al. 1990; Alferdinck 1996; Van Derlofske et al. 2004), light spectrum (Flannagan et al. 1989; Van Derlofske et al. 2004) and headlight size (Sivak et al. 1990; Alferdinck and Varkevissar 1991; Van Derlofske et al. 2004) both have small effects, although the evidence for the latter is mixed. What this means is that for the same illuminance at the eye, light spectra with more energy at the short wavelength end of the visible spectrum produced by smaller size headlamps will tend to cause slightly more discomfort. There is also some evidence that the magnitude and frequency of changes in the illuminance at the eye as the vehicle moves along the road impact the muscle tension around the eye, a condition that has been associated with driver discomfort and fatigue (Murray et al. 2002).

6.6.3 PERFORMANCE IN THE PRESENCE OF GLARE

The most obvious and best understood consequence of exposure to disability glare is a reduction in the luminance contrasts of the scene around the glare source. This leads to a reduction in the ability to detect targets that are close to threshold in the absence of glare (see Figure 6.15). However, the measurements on which this

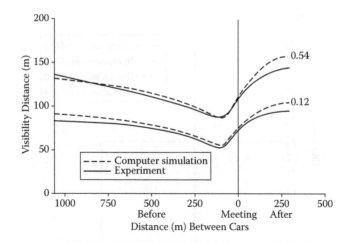

FIGURE 6.17 Visibility distance (m) for targets of reflectance 0.54 and 0.12, plotted against the distance (m) between two vehicles approaching each other, with headlights of equal luminous intensity (after Mortimer and Becker 1973).

conclusion is based were taken in a static situation, but glare is most usually experienced in a dynamic situation as two vehicles approach and pass each other. Mortimer and Becker (1973), using both computer simulation and field measurements, have shown that the distances at which targets of reflectances 0.54 and 0.12 become visible diminish as opposing cars close, and then start to increase rapidly (Figure 6.17). The separation at which the visibility distance is a minimum depends on the relative luminous intensity distribution of the headlamps, the relative positions of the two vehicles, the obstacles to be seen, and the physical characteristics of the obstacle.

Helmers and Rumar (1975) measured visibility distances for flat, dark-gray 1.0 m by 0.4 m rectangles with a reflectance of 0.045. Observers were driven towards a parked car with its headlamps on and asked to indicate when they saw the obstacles. It was found that for the small dark-gray obstacle, a headlamp system with the maximum high-beam luminous intensity gives a visibility distance of about 220 m when no opposing vehicle is present. This is the same as the stopping distance for a vehicle moving at 110 km/h (68 mph) on wet roads (AASHTO 2001). However, when two opposing vehicles have equal luminous intensity headlamps, the visibility distance is reduced to about 60–80 m, which is much less than the stopping distance, and when the opposing vehicle had a luminous intensity about three times more than the observer's vehicle, the visibility distance is reduced to about 40–60 m. Again, it is clear that driving at high speeds against opposing traffic at night approaches an act of faith.

6.6.4 RECOVERY FROM GLARE

The discomfort experienced when exposed to headlamps is replaced by a feeling of relief almost immediately after the other vehicle passes. Likewise, the light scattered in the eye disappears with the glare source, but that does not mean that vision is

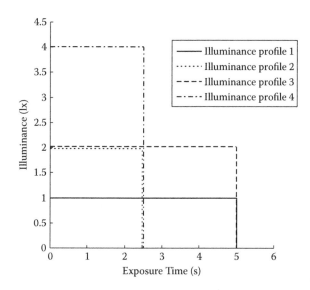

FIGURE 6.18   The four glare stimuli used by Van Derlofske et al. (2005) showing the illu-minance at the eye (lx) and the duration of exposure (s). The effect of these stimuli is to pro-duce three different maximum illuminances and two different light doses.

immediately restored to the state that existed before exposure to glare. The additional light that has reached the retina of the driver from the approaching headlamps will have had an effect on the state of adaptation of the photoreceptors, so immediately after the other vehicle passes, the driver's vision will be misadapted. The process of adjusting adaptation is called recovery from glare.

Van Derlofske et al. (2005) examined what factors determined the time taken to recover from glare. The subject was exposed to four different glare stimuli (Figure 6.18) differing in maximum illuminance and light dose, this latter being the product of illuminance and time of exposure. Specifically, illuminance profiles 1 and 2 had different maximum illuminances but the same light dose. Illuminance profiles 3 and 4 also had different maximum illuminances but the same light dose, although the light dose was twice that of illuminance profiles 1 and 2. Immediately after exposure, the subject was presented with a square target, the contrast of which was a fixed ratio of the individual's threshold contrast, i.e., at a fixed visibility level. The subject's task was to indicate when the target could first be detected. Figure 6.19 shows the mean detection times for different target contrast ratios and for the different glare exposure profiles. From Figure 6.19 it is evident that detec-tion times are shorter for the higher-contrast target ratios and that the detection time is determined by the light dose and not the maximum illuminance. It is inter-esting to note that in the same experiment it was shown that ratings of discomfort on the de Boer scale were more closely related to the maximum illuminance at the eye than the light dose.

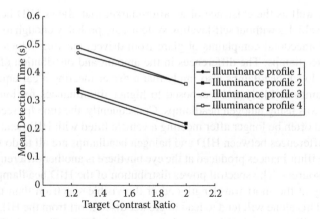

FIGURE 6.19 Mean detection time (s) for targets after exposure to the four glare stimuli shown in Figure 6.18 plotted against target contrast ratio. Target contrast ratio is the ratio of the actual contrast to the threshold contrast without glare (after Van Derlofske et al. 2005).

6.6.5 BEHAVIOUR WHEN EXPOSED TO GLARE

Different illuminances received at the eye can be associated with different behaviours. The range of illuminance received at the eye during normal driving is from 0 to 10 lx (Alferdinck and Varkevisser 1991). Illuminances of 3 lx and more are likely to be considered very uncomfortable (Bullough et al. 2002). Illuminances of the order of 1 to 3 lx are sufficient to cause drivers to request dimming from the approaching vehicle (Rumar 2000). Given that the approaching driver does not respond to a request for dimming, how does the requesting driver respond? Theeuwes and Alferdinck (1996) had people drive over urban, residential, and rural roads at night, with a glare source simulating the headlamps of an approaching vehicle mounted on the bonnet of the car. They found that people drove more slowly when the glare source was on, particularly on dark winding roads where lane keeping was a problem. Older subjects showed the largest speed reduction. The presence of glare also caused the drivers, particularly older drivers, to miss many roadside targets.

6.6.6 HID AND HALOGEN HEADLAMPS

HID headlamps differ from conventional halogen headlamps in two respects relevant to glare; the luminous intensity distribution from the headlamp and hence the illuminance at the eye and the spectral power distribution of the light emitted.

The luminous intensity distributions of HID headlamps typically have a higher maximum luminous intensity than halogen headlamps, as well as putting more light to the sides of the vehicle in areas where the maximum luminous intensity is not controlled by the current regulations (see Figure 6.7). These differences in the amount and distribution of light from HID headlamps, together with the variability introduced by aiming, dirt, and the different geometries that can occur between two approaching vehicles, as well as the existence of an after-market that allows HID headlamps to be fitted to vehicles without self-leveling systems, are probably enough to explain the widespread anecdotal complaints of glare from drivers meeting vehicles equipped with HID headlamps. The differences in the amount and distribution of light from HID and halogen headlamps also imply that a driver meeting a car equipped with HID headlamps is likely to be exposed to higher illuminances for longer than if the vehicle was using halogen headlamps. Consequently, the time for recovery from glare should often be longer after meeting a vehicle fitted with HID headlamps.

These differences between HID and halogen headlamps are all to do with differences in the illuminances produced at the eye but there is another difference between these light sources. The spectral power distribution of the HID headlamp has much more energy at the short wavelength end of the visible spectrum than the halogen headlamp. This alone will tend to lead to greater discomfort from the HID headlamp for the same illuminance at the eye (Bullough et al. 2003). Figure 6.20 shows the mean ratings of discomfort on the de Boer scale for halogen and HID headlamps positioned at 5 and 10 degrees from the line of sight and plotted against the illuminance at the eye (Bullough et al. 2002). As would be expected, both the illuminance at the eye and the deviation from the line of sight show statistically significant effects on the magnitude of discomfort glare, but so does light spectrum.

To evaluate the effect of any specific headlamp light spectrum on discomfort glare, Dee (2003) has proposed a spectral sensitivity curve as follows:

$$V_{dg}(\lambda) = V_{10}(\lambda) + (0.19\ SWC\ (\lambda))$$

where $V_{dg}(\lambda)$ is the discomfort glare spectral sensitivity, $V_{10}(\lambda)$ is the photopic spectral sensitivity for a 10-degree field, and SWC $(\lambda))$ is the short wavelength cone

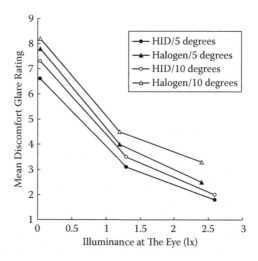

FIGURE 6.20 Mean discomfort glare (de Boer) ratings for exposure to halogen and HID headlamps at 5 and 10 degrees from the line of sight plotted against the illuminances (lx) at the eye (after Bullough et al. 2002).

spectral sensitivity. This discomfort glare spectral sensitivity has been shown to rectify discomfort glare ratings for conditions simulating exposure to headlamps from both white light and monochromatic glare sources (Watkinson 2005).

After the above discussion, there can be little doubt that glare, particularly disability glare, is a major constraint on the visibility available to drivers at night. The detection distances achievable in the presence of opposing vehicles is insufficient unless speeds are dramatically reduced, and everyday experience shows that this rarely happens. There is a desperate need to find a better way to deal with glare than the two-beam approach used at the moment.

6.7 RECENT INNOVATIONS

So far, this chapter has concentrated on the headlamps that are fitted to the vast majority of vehicles on the roads, headlamps that use either halogen or HID light sources, that have two possible states, high and low beam, each producing luminous intensity distributions conforming to either the ECE or North American regulations. However, the last few years have seen the introduction of a number of innovations in headlamps, covering light sources, light spectrum, and luminous intensity distributions.

6.7.1 LIGHT SOURCES

One light source from which much is expected is the light emitting diode (LED), or rather LEDs because multiple LEDs are required in each headlamp to generate the required light output. At the moment, the main attraction of LEDs for headlamps is the freedom they give the designer. The factors limiting their use are cost, regulatory uncertainty, and stability in different climatic conditions, the light output and life of LEDs being sensitive to ambient temperature (Wordenweber et al. 2007). No doubt, as LED technology evolves and the range of uses increases, many of these limitations will be overcome. The most exciting possibility is that an LED headlamp system is developed in which different parts of the headlamp beam are produced by different LEDs. Combining this flexibility with sensors to identify what is ahead of the driver could allow the headlamp beam to be tailored to specific circumstances.

Another innovation in light sources is the mercury-free HID lamp. The reason for eliminating mercury from the HID lamp is nothing to do with the efficiency or effectiveness of the light source and everything to do with the environmental damage caused by mercury when such lamps are discarded. A consequence of removing mercury is a change in the spectral power distribution to give greater emphasis to short wavelengths. Sivak et al. (2006a) have examined the effects of this change in spectrum for discomfort glare and the colour rendering of red retroreflective materials. They concluded that discomfort glare for the mercury-free HID headlamps would be similar to that produced by the bluest of the conventional HID headlamps and that the colour rendering and brightness of red traffic signs would be acceptable.

Yet another innovation involving a light source is a headlamp in which different parts of the beam have different spectral power distributions. Specifically, a projector type headlamp has been designed that uses a dielectric coating to filter and redirect short wavelength light away from an opposing vehicle and towards the nearside edge of the road. One outcome is less light going towards an opposing vehicle and what does go in this direction is spectrally shifted to longer wavelengths resulting in a more yellow appearance. Another outcome is more light going to the nearside edge of the road, this light being shifted towards short wavelengths resulting in a more blue appearance. From Sections 6.5 and 6.6.6, it would be expected that such a spectrally tuned headlamp would result in less glare to opposing drivers and better detection of off-axis targets at the edge of the road. Van Derlofske et al. (2007) confirmed these expectations in a study using the same methodology as described in Section 6.6.2. What they found was that the spectrally tuned headlamps had no effect on disability glare but did reduce discomfort glare by about one unit on the de Boer scale from that produced by standard HID headlamps. Such a difference is comparable with the difference between halogen and HID headlamps producing the same illuminance at the eye. As for off-axis performance, the spectrally tuned headlamp set reduced reaction times to the onset of the targets by about 70 ms relative to the standard HID headlamp set at 12.5 degrees from the line of sight. This improvement is brought about by both the increase in illuminance and the shift in the spectrum.

6.7.2 ADAPTIVE FORWARD LIGHTING

The most dramatic innovation in vehicle lighting in recent years has been the introduction of the adaptive forward lighting systems (AFS). This has occurred in two stages. The first was the introduction of the bending light designed to increase visibility around a curve. Two forms of bending light are permitted. Either the headlamps are swiveled to better illuminate the curve in the road without changing the luminous intensity distribution (dynamic) or the headlamps are fixed but the luminous intensity distributions are changed by switching on additional light sources to increase the illuminance around the curve (static). The swivelling headlamp is the most widely used, now becoming available on many up-market cars. The movement of the headlamp beam is automatic, determined by some combination of signals from sensors providing information on the vehicle's motion. This movement increases visibility distances around the bend, particularly for short-radius bends (Sivak et al. 2004).

The next stage in the development of AFS is presently awaiting regulatory approval but involves systems to produce different luminous intensity distributions for a number of different commonly occurring driving situations. In addition to the usual low and high beams, modified luminous intensity distributions are proposed for use in towns where speeds are low, for use on motorways and divided carriageways where speeds are high and there is a large separation between traffic streams, and for use on wet roads, where the increased specular reflection leads to more glare to opposing drivers (Wordenweber et al. 2007). All these luminous intensity distributions will be usable in static or dynamic bending modes. The transition between these beams is automatic, determined by some combination of signals from sensors giving the vehicle's speed and direction, ambient light level, use of windscreen wipers, and the turning of the steering wheel.

The town beam is wider than the conventional low beam and extends a shorter distance up the road, thereby emphasizing the visibility of pedestrians, road signs, and road markings. In addition, the light output of the headlamp is halved. This town beam is activated when speeds are below 50 km/h (31 mph) or if the road surface luminance is higher than 1 cd/m². The motorway beam is created by moving the low beam up a quarter of a degree to extend the beam further down the road. This beam is activated when the vehicle is moving over 110 km/h (68 mph). The wet road beam involves a reduction in the illumination just in front of the vehicle and increased light to the sides of the vehicle. This beam is activated when either rain is detected on the road or when the windscreen wipers are switched on. As if this were not enough, predictive AFS systems for lighting bends have been suggested. In these, information from a satellite navigation system is used to predict road curvatures immediately ahead with the result that the forward lighting beams can be moved to illuminate a curve before the vehicle enters it. Evaluations of bending lighting, motorway lighting, and town lighting by drivers indicate that bending lighting is considered the most valuable, followed by motorway lighting, with town lighting having the least value (Hamm 2002).

Assuming regulatory approval and given the tendency for expensive options first introduced in up-market vehicles to gradually spread into cheaper vehicles, it seems likely that AFS in various forms will soon become much more widely available. If so, then according to Sullivan and Flannagan (2007) there is a potential to significantly reduce accidents involving pedestrians, particularly on high-speed roads.

6.7.3 NON-VISUAL LIGHTING

Despite the impressive technology and engineering skill involved in developing AFS, the end result is akin to shifting the deckchairs on the Titanic. This is because AFS does not really deal with the fundamental problem of forward lighting, namely the conflict between visibility and glare. Fortunately, there are a number of other potential solutions to the problem of maximizing visibility while minimizing glare (Mace et al. 2001). One solution to the problem of how to ensure visibility without glare to oncoming drivers is to use radiation outside the visible range to illuminate the road ahead. An example of this has been the development of ultraviolet (UV) emitting headlamps. These headlamps emit radiation in the wavelength range 320–380 nm and are designed to supplement conventional low-beam headlamps. When these UV headlamps are combined with fluorescing materials in road markings, the distances at which the road markings can be seen increase dramatically, as do the distances at which pedestrians who are wearing clothing washed in detergents containing a fluorescent whitening agent can be detected (Turner et al. 1998). The UV radiation emitted by such headlamps does not pose a health threat to people exposed to it (Sliney et al. 1995). The combination of UV headlamps with fluorescing materials is considered by some to be a very cost-effective approach to improving the safety of driving at night (Lestina et al. 1999). There are two main problems associated with the introduction of a supplementary lighting system based on UV headlamps and fluorescing materials. The first is technical. It is simply the extent to which materials retain their fluorescing properties over time. The second problem is political. Such a system requires the organizations who manufacture vehicles and who design and construct the road system to work together, something which past history suggests is not easily achieved.

A variation on this approach that does not require action by anyone other than the vehicle manufacturer is to use an infrared night vision system. These come in two forms, active and passive. Active night vision systems use a headlamp emitting infrared radiation in the wavelength range 800–1000 nm coupled to a camera sensitive to these wavelengths, which is linked to a display available to the driver (Holz and Weidel 1998; Wordenweber et al. 2007). Because the wavelengths used are outside the visible range, the infrared headlamp can be kept on high beam even when there are opposing vehicles. This high level of infrared radiation, together with the fact that many materials that have a low reflectance in the visible wavelengths have a much higher reflection in the near infrared, means that active infrared night vision systems are effective at exposing people and animals at much greater distances than would be possible using low beams alone.

An alternative night vision system is passive in that it simply detects infrared radiation emitted by surfaces at different temperatures, usually in the wavelength range 8 to 14 µm. There is no additional radiation emitted from the vehicle. Such a system is effective in detecting warm objects with a distinct temperature difference from the background, such as pedestrians and other animals, but not objects at a similar temperature as the background, such as lane markings.

Tsimhoni et al. (2005) had people perform a simulated steering task while viewing video recordings of a drive along a road where pedestrians were to be seen and the same trip as seen by active and passive night vision systems. The mean distances at which the pedestrians were detected were about three times greater for the passive night vision system than for the active night vision system. Hankey et al. (2005) report measurements of the distances at which drivers were able to detect pedestrians wearing black or white clothing, crossing or standing beside the road, with and without a passive infrared system using a head-up display. The drivers were also asked to detect a pedestrian wearing white clothes standing behind a stationary vehicle with its headlamps on and standing behind a crash barrier on a curve, as well as a piece of tyre tread on the far side edge line. For the pedestrians crossing and standing by the road, the drivers were not aware of their location, but they were for the pedestrian near the glare source. The pedestrian standing behind the crash barrier on the curve was outside the field of view of the passive system and the tyre tread had a similar temperature to the road. Table 6.2 shows the mean detection distances for each detection task, with and without the passive system. In addition, the percentage of times the pedestrian crossing or standing beside the road was detected at less than 150 m are given, as are the percentages of times the pedestrian standing by the glare vehicle, the pedestrian standing behind the crash barrier, or the tyre tread were detected at less than 50 m. Any detection of pedestrians crossing or standing beside the road at less than 150 m was counted as a miss and given a detection distance of zero metres. Similarly, any detection of pedestrians beside the glare vehicle or behind the crash barrier, or the tyre tread that occurred at less than 50 m was given a detection distance of zero metres. An

TABLE 6.2

Mean Detection Distance (m) and Percentage of Misses for Detecting Pedestrians and Tyre Treads at Night Using Headlamps Alone or Headlamps with a Passive Infrared Night Vision System

Target	Headlamps Only		Headlamps and Passive Infrared Night Vision System	
	Mean Distance (m)	Misses (%)	Mean Distance (m)	Misses (%)
Pedestrian in black crossing road	61	31	455	0
Pedestrian in white crossing road	119	3	444	0
Pedestrian in black standing beside road	42	26	414	0
Pedestrian in white standing beside road	137	0	409	0
Pedestrian in white standing near glare vehicle	87	0	379	0
Pedestrian in white standing behind crash barrier on curve	50	12	36	29
Tyre tread on far side edge line	49	6	44	23

From Hankey et al. (2005).

examination of Table 6.2 reveals large, statistically significant increases in mean detection distances as well as zero misses for pedestrians crossing and standing beside the road and standing beside the glare vehicle, when the passive system is operating. When it is not operating, the increased mean detection distances for pedestrians wearing high-reflectance clothing is clear. Another interesting point is that mean detection distance for the pedestrian standing behind the crash barrier on a curve is statistically significantly shorter and there are more misses when the passive system is operating than when headlamps alone are used. This pedestrian is outside the field of view of the passive system, suggesting that when the passive system is used, attention is focused on the area it covers. Nonetheless, the greatly increased detection distances for pedestrians on or close to the road, who are much more at risk than a pedestrian standing behind a crash barrier, are clear.

By now it should be apparent that this is an exciting time to be involved in the design of vehicle forward lighting. After decades of little change, there are now many possibilities for enhancing the ability of the driver to see the road ahead without blinding those approaching. Some of these possibilities are evolutionary in that they involve introducing more beam types and more automation to the existing high-beam/low-beam options. Others are revolutionary in that they offer additional information based on nonvisual parts of the electromagnetic spectrum. Presently, these possibilities are used to supplement human vision, but they may ultimately replace it. It will be interesting to see which of these possibilities thrive and which decline.

6.8 SUMMARY

Forward lighting of vehicles is designed to enable the driver to see after dark. The most common form of forward lighting, headlamps, is fitted to all motor vehicles. Headlamps in modern vehicles use either tungsten halogen or xenon discharge (HID) light sources. At least two headlamp light distributions are required by law, high beam for use when there is no other vehicle on the road ahead, either approaching or leading, and low beam for use when there is another vehicle approaching or leading. High beam is sometimes called main beam or driving beam while low beam is sometimes called dipped beam, meeting beam, or passing beam.

Headlamp light distributions and placement on the vehicle are closely regulated. The purpose of these regulations is to bring some order to the potential conflict between drivers caused by the fact that headlamps increase the visual capabilities of the driver sitting behind them but simultaneously decrease the visual capabilities of the driver facing them. The vast majority of countries follow either the recommendations of the Economic Commission for Europe (ECE) or the US Federal Motor Vehicle Safety Standard. A comparison between these standards shows some similarities and some differences. For high beams, the maximum luminous intensity values allowed are higher in the ECE regulations than in the regulations used in North America. For low beams, both North American and ECE regulations are similar in that they show an emphasis to the near side of the road and a limitation in luminous intensity directed towards vehicles approaching in the opposing lane. However, they differ in the luminous intensity distribution towards the approaching driver. The ECE low beam has a much sharper cut-off than the North American low beam, indicating the greater emphasis given to controlling disability glare.

While the regulations applied to headlamps are exact, headlamps in a vehicle on the road may depart from the regulations for a number of reasons. Some are transient and inherent in the road layout or the nature of the vehicle. Others are permanent and occur because the vehicle is not level, or the headlamp is incorrectly aimed or is dirty. Even when headlamps are correctly aimed, new, and clean, there is still a question about how effective they are, especially on low beam. Comparisons between detection distances using low beam and stopping distances suggest that the maximum safe speed when using low beams is about 30 mph (48 km/h). This might not matter so much if drivers used high beams whenever possible. Unfortunately, they do not.

The most dramatic change in headlamps over the last decade has been the increasing use of xenon discharge (HID) light sources in vehicle headlamps. HID headlamps typically produce two to three times more luminous flux than halogen headlamps and emit light with a spectrum better suited to off-axis detection. Consequently, HID headlamps conforming to the same regulations allow objects to be detected at greater distances and over a wider range of angles than halogen headlamps. This probably explains why drivers who have experienced both types prefer

HID headlamps. However, the same photometric characteristics lead to more glare to the drivers of opposing vehicles.

When driving, glare is commonly experienced in two forms, disability glare and discomfort glare. Disability glare reduces the capabilities of the visual system due to light being scattered in the eye, the important variable being the illuminance received at the eye. Discomfort glare is said to be occurring when people complain about visual discomfort in the presence of bright light sources. The illuminance at the eye is also important in determining discomfort glare, although light spectrum and headlamp size also have small influences.

Drivers most commonly experience glare in a dynamic situation as two vehicles approach and pass each other. In this situation, the visibility distance depends on the relative luminous intensity distribution of the headlamps, the relative positions of the two vehicles, and the physical characteristics of the objects to be seen. When two opposing vehicles have equal luminous intensity headlamps, the visibility distance for low reflectance objects is about 60–80 m. When the opposing vehicle has a luminous intensity about three times more than the observer's vehicle, the visibility distance for low-reflectance objects is reduced to about 40–60 m. Clearly, driving at high speeds against opposing traffic at night approaches an act of faith. Further, it should not be thought that once the opposing vehicles have passed each other, the effects of glare are over. They are not. This is because after exposure to glare the driver's vision will be misadapted. The time taken to recover from exposure to glare depends on the light dose, i.e., the product of illuminance at the eye and the exposure time.

Over the last few years a number of innovations in headlamps have occurred. One such innovation has been the development of the light emitting diode (LED) headlamp. At the moment, the main attraction of LEDs for headlamps is the freedom they give the designer. The factors limiting their use are cost, regulatory uncertainty, and stability in different climatic conditions. No doubt, as LED technology evolves and the range of uses increases, many of these limitations will be overcome. Another innovation has been the introduction of the adaptive forward lighting systems (AFS). This has occurred in two stages. The first was the introduction of the bending light designed to increase visibility around a curve. The second involves the automatic selection of different luminous intensity distributions for different commonly occurring driving situations. In addition to the usual low and high beams, modified luminous intensity distributions are proposed for use in towns where speeds are low, for use on motorways and divided carriageways where speeds are high and there is a large separation between traffic streams, and for use on wet roads, where the increased specular reflection leads to more glare to opposing drivers.

While AFS changes the balance between visibility and glare, it does not really solve the fundamental problem of forward lighting. One approach that does is to use radiation outside the visible range to illuminate the road ahead. Among the systems that have been developed to do this are ultraviolet headlamps, and active and passive infrared imaging systems. All have been shown to extend detection distances.

This is an exciting time to be involved in the design of vehicle forward lighting. After decades of little change, there are now many possibilities for enhancing the ability of the driver to see the road ahead without blinding those approaching. Some of these possibilities are evolutionary in that they involve introducing more beam types and more automation to the use of the existing high-beam/low-beam options. Others are revolutionary in that they offer additional information based on nonvisual parts of the electromagnetic spectrum. Presently, these possibilities are being used to supplement human vision but they may ultimately replace it.

7 Vehicle Signal Lighting

7.1 INTRODUCTION

Signal lighting is lighting designed to indicate the presence of or give information about the movement of a vehicle. Signal lighting includes front position lamps, rear position lamps, side marker lamps, license plate lamps, turn lamps, hazard flashers, stop lamps, rear fog lamps, reversing lamps, and daytime running lamps, as well as retroreflectors. Some signal lamps, such as front and rear position lamps and side marker lamps, are used only at night or in conditions of poor daytime visibility, while others, such as turn lamps and stop lamps, have to be visible at all times, both day and night. To fulfill its function of transmitting information from one vehicle to the driver of another, the signal lamp has to be visible and its meaning has to be clear. It also helps if it is conspicuous.

Today, signal lighting is closely regulated but this has not always been the case (Moore and Rumar 1999). Early motor vehicles had little by way of signal lighting. It was not until the time of the First World War that motor vehicles started to have electric signal lighting, this taking the form of one lamp showing a red light to the rear and a white light to the side to illuminate the license plate. In the 1930s signal lamps that could indicate braking and turning began to appear. These were usually red but sometimes green and yellow. By the 1950s most cars had two front position lamps, two rear position lamps, two stop lamps, and a license plate lamp. Vehicles in the United States also had flashing turn signals but Europe was still using the trafficator, an internally illuminated yellow semaphore that swung out from the side of the vehicle to indicate turning. It was in the 1960s that vehicle signal lighting became regulated, a development brought on by the increased density of traffic and the need to ensure consistency between vehicles so that the meaning of the signal was clear. Since then, the main developments have been the introduction of the centre high-mounted stop lamp, rear fog lamps, reversing lamps, and daytime running lamps, at various times, in various combinations, in different countries.

7.2 THE TECHNOLOGY OF VEHICLE SIGNAL LIGHTING

For many years the technology of vehicle signal lighting hardly changed, consisting of little more than an incandescent light source covered by a clear or coloured lens. However, over the last two decades this situation has been transformed. Today, signal lighting uses a variety of light sources and methods of optical control and is an integral part of the styling of the vehicle (Wordenweber et al. 2007).

7.2.1 LIGHT SOURCES

There are two dominant light sources in the field of vehicle signal lighting, incandescent and LEDs. By far the most widely used light source is the incandescent. This has the advantage of being simple and inexpensive but the disadvantage of requiring regular replacement throughout the life of the vehicle. For example, a double filament incandescent light source for use in a combined stop/rear position lamp has a life of approximately 100 hours for the stop function and 1500 hours for the rear position function. Based on typical patterns of use, after 150,000 km (93,200 miles) the stop lamp will have been used for about 600 hours, which suggests that the light source will have had to have been replaced six times. This problem is overcome by the LED. LEDs have much longer lives than incandescent light sources, so much so that they should not need to be replaced during the life of the vehicle. LEDs have other advantages over the incandescent. To create a coloured signal, the incandescent has to be filtered while the LED, if judiciously chosen, emits light of the desired colour, without filtering. This means that LED signal lamps have smaller power demands than incandescent signal lamps fulfilling the same function. LEDs, being solid-state devices, are also less sensitive to vibration than light sources that rely on a heated filament and, because they are smaller, offer the designer a wider range of possibilities. LEDs are rapidly becoming the light source of choice for vehicle signal lighting.

7.2.2 OPTICAL CONTROL

The regulations applicable to signal lighting specify different luminous intensity distributions for each signal function. To meet these regulations, signal lamps require some form of optical control. There are three such systems used in signal lamps based on reflection, refraction, and total internal reflection (Wordenweber et al. 2007). For the reflector system, the light distribution is determined by the position of the light source relative to the reflector, the shape of the reflector, and any optical patterning of the front cover glass. Where a smooth cover glass is preferred, some combination of smooth, segmented, or faceted reflectors can be designed to produce the desired luminous intensity distribution.

Rear position lamps and turn lamps commonly use refractor optics. In this system, a Fresnel lens cover glass is used. Where a smooth clear cover glass is preferred, a Fresnel lens system behind the cover glass is used. With this approach, no reflector is needed.

Total internal reflection is the physical principle used in light guides. A light guide consists of a transparent material with a refractive index higher than the surroundings, which for motor vehicles is air. The light source is closely coupled to the light guide. Light is transmitted down the light guide by repeated reflections at the surface of the guide. Light is extracted from the light guide by having a prismatic element. Such an element increases the angle of incidence of the light in the guide and thereby defeats total internal reflection. LEDs are the preferred light source for light guides, their low temperature and small size making it easy to couple the LED closely to the light guide. The light distribution is largely determined by the prismatic element. Light guides offer the stylist a great deal of freedom.

7.2.3 STRUCTURE

Although the different types of signal lamps are considered separately, it is necessary to appreciate that they can be packaged individually or combined in a common structure called a cluster. In a cluster there will be multiple light sources for the different signal lamps but even when individual signal lamps are used it is possible to have multiple light sources in a number of different compartments. Until the advent of LEDs, the main reason for using multiple compartments for the same lamp was to increase the visible surface area without having to make the lamp too deep, although there was also a benefit in introducing redundancy in that the signal did not disappear if one light source failed. Where LEDs are used as the light source, it is usually necessary to have multiple LEDs in the same lamp because the light output of one is insufficient. Whether or not such a use constitutes a multicompartment lamp has caused some confusion in the interpretation of regulations.

The structure of signal lamps can be considered in mechanical, electrical, environmental, and colour terms. The mechanical elements involve making sure that the light source and optical control elements are consistently related to each other and the whole is mounted securely on or in the vehicle body. The electrical elements call for the connection of the light source to the vehicle's electrical system, either directly or through any necessary power supply, as well as the means to change the light source when it fails. The environmental elements require careful consideration of the materials used so that the lamp can cope with the extremes of heat, cold, moisture, and vibration. Finally, it is necessary to consider the means used to generate the colour of light. One approach is to use a light source that emits light of the required colour, such as an LED or a painted incandescent. This light source can then be used with a clear cover glass or a coloured cover glass, both having a high spectral transmittance in the relevant wavelengths. The advantage of the coloured cover glass is that the signal looks the expected colour by day, even when the lamp is not lit.

7.3 THE REGULATION OF VEHICLE SIGNAL LIGHTING

The visibility of a signal lamp depends on its luminance, size, shape and colour, the background against which it is seen, and the state of adaptation of the driver. Much effort has been put into measuring the minimum values of some of these variables necessary to make the signal lamp visible under different conditions (Dunbar 1938; de Boer 1951; Moore 1952; Hills 1975a,b; Sivak et al. 1998). Figure 7.1 shows the relationship between luminance and size for red rear position lamps, disc obstacles, and pedestrian dummies to be "just visible," i.e., at threshold, under no road lighting and no glare conditions (Hills 1976). It can be seen that log luminance plotted against log visual area gives a nearly straight line; the smaller the signal area, the greater its luminance has to be before it is just visible. This is just what would be expected from the spatial summation of the visual system for small targets (see Section 3.4.3).

Using data similar to those shown in Figure 7.1, Hills (1976) produced a predictive model of the relationship between luminance increment and area for small objects to be just visible for a wide range of background luminances (Figure 7.2). A small object in this model is one for which spatial summation occurs in the visual

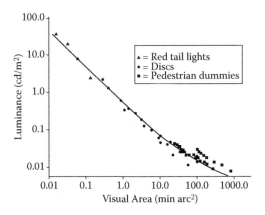

FIGURE 7.1 The relationship between luminance (cd/m²) and visual area (min arc²) for rear position lamps (red tail lights), discs, and pedestrian dummies to be "just visible" under no road lighting/no glare conditions (after Hills 1976).

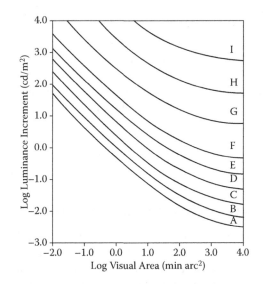

FIGURE 7.2 Relationships between luminance increment (cd/m²) and visual area (min arc²) at different background luminances, for small targets to be just visible. Each curve is for one background luminance as follows: A = 0.01 cd/m², B = 0.1 cd/m², C = 0.32 cd/m², D = 1.0 cd/m², E = 3.2 cd/m², F = 10 cd/m², G= 100 cd/m², H = 1000 cd/m², I = 10,000 cd/m² (after Hills 1976).

system. For foveal vision, spatial summation is complete within a circle of diameter subtending about 6 min arc. For targets that occur 5 degrees off-axis, spatial summation occurs over a circle of diameter about 0.5 degrees (Boff and Lincoln 1988). Given the usual size of signal lamps and the distance from which they need to be seen, spatial summation should apply. The ordinate in Figure 7.2 is the logarithm of the increment of the object luminance necessary for it to be just visible against the background luminance. Different values of background luminance enable the effects of different lighting conditions to be estimated, from starlight, through road lighting, to daylight. Hills (1976) also shows that by using such curves he can plausibly predict the field results of Dunbar (1938) and Moore (1952). Such information forms the background to the regulations governing signal lighting.

The two most widely used sources of standards relevant to vehicle signal lighting are the US Federal Motor Vehicle Safety Standard 108 (FMVSS) and the Economic Commission for Europe (ECE) recommendations, used in the European Union. Other countries follow one or other of these standards, though with modifications to suit local circumstances. These standards are used to form a legal framework for vehicle lighting, but most of the actual work is done by two other bodies, the Society of Automotive Engineers (SAE) for the FMVSS regulations and the Groupe de Travail-Bruxelles (GTB) for the ECE recommendations. The tendency is for the recommendations of the SAE or the GTB eventually to be incorporated into their respective standards. FMVSS regulations are implemented by a self-certification process in which manufacturers have to test any proposed lighting systems to ensure they meet the requirements of the regulations. ECE recommendations are implemented through a type approval process in which manufacturers have to submit products to an independent testing laboratory. The type approval process is usually more stringent than self-certification. Over the last three decades, there has been convergence between the FMVSS and ECE standards, a process called harmonization. For many types of signal lamp, harmonization, although not complete, has resulted in sufficient overlap between standards for a manufacturer to be able to make one signal lamp that can meet both sets of requirements, something that is desirable for a global industry that is conscious of costs.

Both the FMVSS and ECE standards cover the minimum and maximum luminous intensities that should be provided in different directions, the colour of the signal, the lit area of the lamp, the allowed positions of the lamp on the vehicle, and, if flashing is called for, what the frequency of flashing should be. Although the standards cover many of the same topics, they differ in detail. To expound all the similarities and differences would require a tome so what is done here is to summarize what the FMVSS and ECE regulations require for the two most obvious features of a signal lamp, the colour and the luminous intensity. Anyone who wants to know the exact requirements for a specific type of signal lamp should consult current regulations applicable to the country of interest. Table 7.1 sets out the allowed colour, the minimum luminous intensity on the optical axis, and the maximum luminous intensity in any direction for different types of signal lamp fitted to cars, according to the FMVSS and ECE standards. Other regulations apply to commercial vehicles. Where there are no FMVSS regulations, as for rear fog lamps, there may be SAE recommendations. Vehicle signal lamp colours are specified by the chromaticity coordinates of the light emitted. The point represented by these coordinates has to fall within an area set by specific boundaries on the CIE 1931 chromaticity diagram (see Figure 2.3 for an example).

These regulated aspects of vehicle signal lighting are measured when the lamps are new and clean and it is always salutary to be reminded of reality.

TABLE 7.1
Colour, Minimum Luminous Intensity (cd) on the Optical Axis and Maximum Luminous Intensity (cd) in any Direction for Different Types of Signal Lamp on Cars according to the FMVSS and ECE Recommendations (Other Standards Apply to Commercial Vehicles)

Signal Lamp Type	FMVSS 108 Regulations			ECE Regulations		
	Colour	Minimum Luminous Intensity on Optical Axis (cd)	Maximum Luminous Intensity (cd)	Colour	Minimum Luminous Intensity on Optical Axis (cd)	Maximum Luminous Intensity (cd)
Front position	White or yellow	4	125 above and 250 below horizontal	White	4	60 or 100 when built into a headlamp
Rear position	Red	2, 3.5, and 5 for 1, 2, and 3 compartment lamps	18, 20, or 25 for 1, 2, and 3 compartment lamps	Red	4	12
Front side marker	Yellow	0.62	—	Yellow	4	24
Rear side marker	Red	0.25	15	Red, or yellow if in a cluster	2	12
Front turn	Yellow	200 or 500 if less than 100 mm from headlamp	—	Yellow	175 or 400 if less than 20 mm from headlamp	700
Side turn	—	—	—	Yellow	0.6	200
Rear turn	Yellow or red	130, 150, and 175 for 1, 2, and 3 yellow compartment lamps. Same as stop lamp when red	750, 900, and 1050 for 1, 2, and 3 yellow compartment lamps. Same as stop lamp when red	Yellow	50	350
Stop	Red	80, 95, and 110 for 1, 2, and 3 compartment lamps	300, 360, and 420 for 1, 2, and 3 compartment lamps	Red	60	185
CHMSL	Red	25	160	Red	25	80

Rear fog		—	—	Red	150	300
Reversing	White	80 for two reversing lamps, 160 for one	300 for two reversing lamps, 600 for one	White	80	300 and 600 below horizontal
Daytime running	White or yellow	500	3000, or 7000 if high-beam headlamps used	White	400	800

FIGURE 7.3 A front position lamp on a Honda Civic. The front position lamp (upper centre) is integrated into a cluster also containing a headlamp (left) and a front turn signal (right).

Schmidt-Clausen (1985) measured the luminous intensities of rear position lamps and stop lamps of new cars and cars in use. He found that the luminous intensities of rear signal lamps in use were half the minimum values required by regulations. Similarly, Cobb (1990) carried out a roadside survey of vehicle lighting in the UK, including rear signal lighting, and found that dirt typically reduced the luminous intensity of vehicle lighting by 30–50 percent.

7.4 FRONT POSITION LAMPS

In different parts of the world, front position lamps are known as parking lamps, sidelights, position lamps, standing lamps, or city lights. Figure 7.3 shows a front position lamp in a cluster containing a headlamp and a turn lamp as well as the front position lamp. The signal it is intended to convey is that of presence when stopped or parked. In America, front position lamps can be either white or yellow but in countries that follow the ECE regulations, white is the only colour allowed. Front position lamps are required to remain illuminated once the vehicle forward lighting is lit. In many countries it is illegal to drive a vehicle at night using front position lamps alone.

7.5 REAR POSITION LAMPS

Rear position lamps, also known as tail lights or rear lights, are always red in colour. Their function is to indicate presence when moving or stopped. Figure 7.4 shows a pair of individual rear position lamps combined with a stop lamp. The different information conveyed by the rear position lamp and the stop lamp is revealed by brightness, the stop lamp looking much brighter than the rear position lamp. To ensure this perception, regulations specify a much higher minimum luminous intensity for the stop lamp than for the rear position lamp (see Table 7.1). Rear position lamps have to be lit when the front position lamps are lit. Depending on the country, rear position lamps are required, permitted, or forbidden to be lit when daytime running lights are in use.

FIGURE 7.4 A rear lighting assembly on a Lotus Elise consisting of pair of individual rear position/stop lamps (inner pair), a pair of individual turn lamps (outer pair), and a pair of individual retroreflectors (lower pair). Behind the grille are a single reversing lamp and a single rear fog lamp. Above the license plate is the centre high-mounted stop lamp.

7.6 SIDE MARKER LAMPS

The basic function of side marker lamps is to indicate the presence of the vehicle to drivers at oblique angles to the direction of movement, a situation of importance at intersections. In America both side marker lamps and side retroreflectors are required, both being yellow at the front and red at the rear of the vehicle. The side marker lamps are connected so that they are lit whenever the front and rear position lamps are lit. The front yellow side marker lamps may also be wired so that the lamp on the relevant side flashes when a turn signal is activated, thereby adding another function.

In countries following the ECE regulations, side marker lamps are permitted rather than required. If they are installed, the side marker lamps are required to be visible over a wider range of angles than are those in America. Further, the side marker lamps must be continuously lit when the front and rear position lamps are lit, which means they cannot be connected to the turn signal. They must be yellow at the front and red at the rear, unless incorporated into a rear lamp cluster, in which case the lamp can be red or yellow.

7.7 RETROREFLECTORS

Retroreflectors are sometimes referred to as passive lighting, not an unreasonable description when given the facts that such elements reflect incident light predominantly towards the light source (see Section 2.8) and that they are regulated as vehicle lighting devices. The function of retroreflectors is to signal presence, even when there is no power on the vehicle. Both FMVSS and ECE based regulations require vehicles to have red retroreflectors at the rear. American regulations also require yellow front-facing retroreflectors and red side-facing retroreflectors at the rear. Some

FIGURE 7.5   Two rectangular retroreflecting markers on the rear of a truck trailer.

other countries require white front-facing retroreflectors. Pairs of retroreflectors at
the rear are a common feature of heavy truck trailers (Figure 7.5).

Another situation where retroreflectors are sometimes used is on the inside edge
of a door. An open door is a hazard to passing traffic and passing traffic is a hazard
to people exiting the vehicle. The door open warning retroreflector is red in colour
and is mounted so that it is visible to drivers approaching from the rear. Sometimes a
red warning lamp is used instead of a retroreflector. If so, this is activated by opening
the door and extinguished when the door is closed.

7.8 LICENSE PLATE LAMPS

All countries require that the rear license plate should be illuminated by white light
using one or two hidden lamps, after dark. This is for identification purposes not
for signaling presence although the illumination of the license plate does add some
variation in luminance contrast within the vehicle, which should help maintain visibility (see Figure 4.2). The regulations for license plate lighting are complex. In
America, the outcome is specified in terms of illuminance, the minimum illuminance being 8.1 lx. In countries following the ECE regulations, the outcome of lighting the license plate is specified in terms of luminance, the minimum value being
2.5 cd/m^2. Moore and Rumar (1999) have argued that now that license plates are

almost always retroreflective and the level of illumination from the license plate lamps is low, the requirement for license plate lamps should be abolished. However, given that authorities are increasingly using automatic reading of license plates as means of identification for catching toll dodgers and for revenue generation through fines for minor speeding and parking offences, this seems unlikely to happen.

7.9 TURN LAMPS

So far, all the lamps considered are intended to make the vehicle more conspicuous and are permanently lit when the vehicle is moving after dark. Now, attention will be switched to signal lamps with a wider range of messages that are intermittently lit when the message is sent. The first to be considered are turn lamps, also know as direction indicators, flashers, or blinkers. These are lamps mounted at the front and rear corners and on the sides of the vehicle and are used to indicate to other drivers that the vehicle is about to turn or change lane (see Figure 7.4). Regulations for turn lamps specify minimum and maximum luminous intensities, angles from which the signal should be visible, colour, and flash rate, as well as feedback signals for the driver. Minimum luminous intensities on the optical axis of front turn lamps vary according to the separation of the turn lamp from the nearest headlamp. In America the minimum of such luminous intensity is increased from 200 cd to 500 cd if the separation is less than 100 mm, the separation being measured as the distance between the optical centre of the turn lamp to the edge of the light emitting area of the headlamp. In countries following the ECE recommendations, the minimum luminous intensity on the optical axis is increased from 175 cd to 400 cd when the separation of the edge of the turn lamp and the edge of the headlamp is not more than 20 mm. This increase in minimum turn signal luminous intensity when in close proximity to a headlamp is intended to overcome the masking effect of disability glare from the headlamp. Turn lamps flash at a rate of 1 to 2 Hz, with all the turn lamps on one side operating in phase. In countries following the ECE recommendations, front, rear, and side turn lamps are yellow in colour, but the FMVSS regulations also allow the rear turn signal to be red, if desired. The side-mounted turn signal can take several different forms. In countries following the ECE recommendations, it takes the form of a dedicated turn signal. In America, the side marker lamp may also be used as a turn signal by making it flash in phase or anti-phase when the front and rear turn lamps are activated. Both ECE and FMVSS standards require that audiovisual feedback of the operation of turn lamps be given to the driver. This usually takes the form of a flashing light on the instrument panel together with an audible click at the same frequency. Failure of one or more of the turn lamps is indicated by an increased frequency of flashing and clicking.

Current practice in turn lamps has given rise to three questions about their effectiveness. The first question concerns the use of red rather than yellow as a colour for the rear turn lamp in the United States. Current styling tends to group all the rear signal lamps into two clusters, one of each side of the vehicle (Figure 7.6). Stop lamps and rear position lamps are both red in colour. It might be thought that a rear turn lamp of the same colour would be less conspicuous than one of a different colour, specifically, yellow, although the flashing of the signal may be sufficiently potent to

FIGURE 7.6   A rear signal lamp cluster from a Skoda Octavia. The cluster contains a rear position lamp, a stop lamp, a turn lamp, a reversing lamp, and rear fog lamp.

negate the difference in colour, particularly as colour discrimination deteriorates in peripheral vision (Sivak et al. 1999). This may be why Taylor and Ng (1981) were unable to find any significant difference in the prevalence of rear-end collisions in Europe, where rear turn lamps are yellow, and in the United States, where rear turn lamps can be red or yellow.

Another trend in vehicle design is the insertion of side turn lamps in the mirror housings (Figure 7.7). An analysis of the geometry between vehicles when one is about to pass the other in adjacent lanes indicates that a mirror-mounted turn lamp can be seen over a wider range of relative positions by the driver of the overtaking vehicle than conventional side turn lamps because the latter are often obscured by the bodywork of the passing vehicle (Reed and Flannagan 2003). Further, the mirror-mounted turn lamp is closer to the line of sight of the passing driver and hence more likely to be detected, a probability confirmed in a later study (Schumann et al. 2003). This is an important finding for one particular type of accident, lane change/

FIGURE 7.7   A mirror-mounted turn signal on a Honda Civic.

merging. In this, the driver wishing to change lanes is unaware of the vehicle in the adjacent lane because it is in the blind spot between the driver's peripheral vision and what can be seen in the side mirror (Wang and Knipling 1994). As long as the lane changing vehicle is using the turn lamp, a greater ability to see it by the driver in the blind spot may allow that driver to take evasive action before contact. So far, attempts to test this hypothesis using crash data have failed to show a statistically significant effect (Sivak et al. 2006b), but this may be due to a lack of sensitivity caused by a small sample size rather than the absence of a real effect. Time will tell.

Yet another trend in signal lamp design is the use of a clear lens and a coloured light source rather than a coloured lens and a white light source. This difference is trivial in darkness but in daylight it is not because then the greater amount of daylight that is reflected from the inside of the clear lens turn lamp will increase the luminance on which the luminance produced by the operation of the light source is superimposed. The effect of this is to reduce the luminance increment when the turn lamp is lit. In addition, daylight entering and then exiting through a clear lens will desaturate the colour of the turn lamp when lit although the colour difference between the lamp on and off will still be greater for the clear lens than for the coloured lens. Sullivan and Flannagan (2001) measured the reaction time to the onset of turn lamps using coloured and clear lenses in bright sunlight. The turn lamp with the coloured lens had a slightly shorter mean reaction time than those with clear lenses. Sivak et al. (2006c) measured the luminance contrasts for turn lamps on, with and without the sun, for a wide range of commercially available turn lamps. On average, the clear lens turn lamps provided lower luminance contrasts for turn lamps on and off in sunlight than did the coloured lens turn lamps. However, examination of the results shows that this is not inevitable. Clear-lens turn lamps can be designed that are resistant to strong sunlight but it does require an awareness of the potential problem and attention to detail. Sivak et al. (2006c) also point out that the higher luminous intensities required for front turn lamps in close proximity to headlamps means that clear-lens turn lamps of this type will be more resistant to confusion in bright sunlight.

7.10 HAZARD FLASHERS

Since the 1960s, turn lamps have been adapted to give a warning to other drivers that the vehicle is stopped in or near moving traffic, is disabled, or is moving very slowly. This signal is given by making all front, side, and rear turn lamps flash in phase. The photometric conditions produced and the flash rate of the turn lamps are the same as when they are used to signal a turn.

7.11 STOP LAMPS

Stop lamps, also known as brake lights, are mounted at the rear of the vehicle. They are red in colour and are lit when the driver applies pressure to the brake pedal. The function of the stop lamps is to inform drivers behind that the vehicle is decelerating, although it may not stop. For many years, there were only two stop lamps mounted symmetrically on the left and right rear corners of the vehicle. However, over the last 20 years, various countries have adopted a third stop lamp for cars and light trucks. This is called the centre high-mounted stop lamp (CHMSL) or centre brake lamp, third brake lamp, eye-level brake lamp, safety brake lamp, or high-level brake lamp. The CHMSL is mounted on the centreline of the vehicle at a level above the left and right stop lamp although either one lamp with off-centre mounting or two lamps mounted symmetrically about the centreline are allowed if doors at the rear of the vehicle are divided centrally. Different vehicle forms mean that the CHMSL is sometimes mounted at the roofline, sometimes at the top of the boot, and at all points between (see Figure 7.4).

An increase in luminous intensity is used to differentiate the stop lamp from the rear position lamp. Exactly how big a difference is required is open to question. Rockwell and Safford (1968) found that up to a luminous intensity ratio of 5.3 (stop lamp/rear position lamp), reaction time to the onset of the stop lamp was reduced, as was the likelihood of confusion between stop lamp and rear position lamp. In the United States, the minimum luminous intensity ratio (minimum for stop lamps/ maximum for rear position lamps) for cars is in the range 4.4 to 4.7 while in those countries that follow the ECE recommendations the minimum luminous intensity ratio is 5.0.

Visually, the problems associated with detecting the onset of a stop lamp are different by day and night. By day, detection can be difficult because daylight reflected from the stop lamp increases the luminance on which the luminance of the stop lamp is superimposed and hence reduces its luminance increment. In addition, some designers like to use rear lamp clusters that are the same colour as the vehicle bodywork when the lamp is unlit. Such a body-coloured lamp may reduce or enhance the colour difference between the stop lamp and the surrounding area of the vehicle depending on the colour of the vehicle body. Chandra et al. (1992) measured reaction times to the onset of body-coloured stop lamps in simulated sunshine. Provided the vehicle body is not red, the onset of the stop lamp produces a change in both luminance and colour. Four chromatically neutral lamps varying from black to white when off and four body-coloured lamps when off were examined. Figure 7.8 shows the mean reaction times for the stop lamps divided into two groups. One group had equal lightnesses (Munsell Value = 4) but different colours when the stop lamp was off so they had equal luminance shift but different chromaticity shifts as the stop lamp changed from off to on. The other group had similar chromaticity shifts but different luminance shifts when the stop lamp changed from off to on because they were all chromatically neutral, but differed in lightness when the stop lamp was off (Munsell Values of 2, 4, 7, and 9 for black, dark grey, light grey, and white, respectively). The smallest chromaticity shift occurred for the red body-coloured stop lamp. The smallest luminance shift occurred for the white body-coloured stop lamp. An examination of Figure 7.8 shows that increasing the magnitudes of both luminance shifts and chromatic shifts are effective in decreasing reaction time although the effects are small over the range examined. This suggests that the colour appearance of the stop lamp when off could be used to enhance its effectiveness.

Another aspect is the form of the stop lamp. Stop lamps can be part of a rear signal lamp cluster or separate. For both types, the stop lamp can vary in area and that area can have different aspect ratios, particularly CHMSLs. Sayer et al. (1996) found

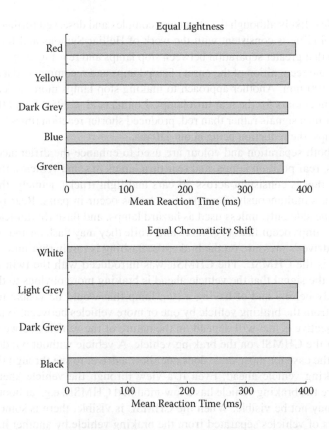

FIGURE 7.8 Mean reaction times (ms) to body-coloured stop lamps sorted into two groups. One group has equal lightnesses (Munsell Value = 4) but different colours when off and hence equal luminance shifts but different chromaticity shifts when the stop lamp comes on. The other has an equal chromaticity shift to red when the stop lamp comes on but varies in luminance shift when changing from off to on because the stop lamps have different lightnesses when off (Munsell Values of 2, 4, 7, and 9 for black, dark grey, light grey, and white respectively) (after Chandra et al. 1992).

that both luminous intensity and aspect ratio influenced reaction time to the onset of a stop lamp. While luminous intensity was the major factor in determining reaction time, the impact of a low luminous intensity was particularly bad when the stop lamp had a high aspect ratio.

By night, the problem facing anyone wishing to detect the onset of a stop lamp is one of confusion because the rear position lamps will be lit, as may rear fog lamps and, on occasion, the rear turn signal. Rear position lamps, rear fog lamps, and stop lamps are all red in colour and rear turn lamps may be in the United States. Further, rear fog lamps and rear turn lamps may have similar luminous intensities to stop lamps. Luoma et al. (2006) examined the influence of having rear stop lamps in a cluster or separate on rear-end crashes at night using data from Florida and North Carolina. They concluded that separate stop lamps tended to make rear-end collisions less likely although the effect was complex and deserved further investigation. This finding is consistent with the work of Helliar-Symons and Irving (1981) who found that greater separation between stop lamps and rear fog lamps resulted in more accurate recognition of the onset of stop lamps and recommended a separation of at least 100 mm. Another approach to making stop lamps more noticeable is to use different colours for the rear turn lamps. Louma et al. (1995a) found that having yellow rear turn signals rather than red, produced shorter reaction times to the onset of stop lamps, the reduction being about 110 ms.

While both separation and colour are used to enhance the difference between stop lamps, rear position lamps, and rear turn lamps in some vehicles, there is one difference that is consistent across all cars and light trucks, namely, the number of lamps lit simultaneously. Rear position lamps occur in pairs. Rear turn lamps occur on one side only, unless used as hazard lamps, and flash rhythmically at 1 to 2 Hz. Stop lamps occur in a set of three and while they may flash on and off according to the driver's pressure on the brake, the flashing is rarely rhythmic. The third stop lamp is the CHMSL. The CHMSL was introduced with the twin objectives of making the signal that the vehicle ahead is braking more obvious to the vehicle immediately behind and of having a stop lamp that would be visible to vehicles separated from the braking vehicle by one or more vehicles between. Whether this second objective is met will depend on the nature of the vehicles between and the position of the CHMSL on the braking vehicle. A vehicle without windows at the rear, as is the case for most vans, does not allow a driver to see through the vehicle to the braking vehicle ahead. Even if a view through the vehicle ahead is possible, where the braking vehicle has a low-mounted CHMSL, e.g., at boot level, the CHMSL may not be visible. When the CHMSL is visible, there is some evidence that drivers of vehicles separated from the braking vehicle by another had shorter reaction times to the application of the brakes than when there was no CHMSL on the braking vehicle (Crosley and Allen 1966). If the use of CHMSLs leads to shorter reaction times to the onset of stop lamps, it seems reasonable to suppose that this would have an effect on the number of rear-end collisions. Rear-end collisions are one of the most frequent types of accidents and although they are rarely fatal they often cause injury, particularly to the neck. Kahane and Hertz 1998) have used accident data to estimate that the widespread use of CHMSLs is responsible for a 4.3 percent reduction in rear-end collisions.

One aspect of stop lamp design that is applicable to stop lamps in all positions, by day and night, is rise time. The rise time of a light source in a stop lamp is determined by the technology used to generate the light and the voltage applied. Incandescent light sources have a slow rise time, of the order of 200 ms, but LEDs and the rarely used neon light source, the latter being a discharge through a low-pressure neon gas, potentially have a very fast rise time, of the order of nanoseconds. The effect of rise time in light output on reaction time to the onset of the stop lamp can be understood by considering the reaction time as the sum of two components, a visual reaction time and a nonvisual reaction time. The visual reaction time is the time it takes for the light received from the stop lamp to be transformed to an electrical signal in the retina and transmitted up the optic nerve to the visual cortex. The nonvisual component includes the time required for information perceived at the visual cortex to be processed and for neural signals to be sent to the muscles that make the response. Differences in rise time of light output for the stop lamp can be expected to influence the visual reaction time but not the nonvisual reaction time. The effect of different rise times in light output can be estimated from the fact that to see the signal a constant level of energy, i.e., luminance integrated over time, is required to reach the retina (Teichner and Krebs 1972). In other words, the visual reaction time is determined by the temporal summation properties of the visual system (see Section 3.4.4). Given a fixed maximum luminous intensity of the stop lamp, the shorter the rise time of the light output, the shorter the reaction time. This suggests that a fast-rise stop lamp, such as those using LEDs or neon discharge, would lead to shorter reaction times to onset and this, in turn, might help diminish the number and severity of rear-end collisions (Sivak and Flannagan 1993).

7.12 REAR FOG LAMPS

In countries that follow the ECE regulations, rear fog lamps are required. Rear fog lamps are the same as rear position lamps in colour but have a much higher luminous intensity (see Table 7.1). Rear fog lamps are not required in the US although they may be used, following the relevant SAE standard, which is the same as for the one-compartment stop lamp. The function of rear fog lamps is, as the name implies, to increase the conspicuity of the vehicle in poor atmospheric conditions. Rear fog lamps are controlled manually by the driver, an action that sometimes leads to complaints of glare from drivers behind when the rear fog lamps are lit in only slightly degraded atmospheric conditions. When used, rear fog lamps can be installed as a single lamp on the driver's side of the vehicle or as a pair mounted symmetrically about the centreline. The case for the single rear fog lamp is that it is easily distinguished from stop lamps while a pair of rear fog lights is not, even when a CHMSL is present (Akerboom et al. 1993). The disadvantage of the single rear fog lamp is that it does not provide a cue to distance for an approaching driver. The need to differentiate rear fog lamps from stop lamps is the reason why the ECE requires a minimum separation, edge to edge, of 100 mm between adjacent stop lamps and rear fog lamps.

7.13 REVERSING LAMPS

The reversing lamp, also known as the back-up lamp, is unusual in rear lighting in that it is white in colour. The reversing lamp has two functions, to illuminate the road behind the vehicle so that the driver can see any obstructions and to alert other drivers and pedestrians that the vehicle is about to reverse. The reversing lamp is automatically lit by placing the vehicle in reverse gear. One or two reversing lamps can be installed on a vehicle, but, in the United States, if only one reversing lamp is used it has to provide twice the minimum luminous intensity.

The problem with reversing lamps is not discriminating them from other rear signals; the different colour is enough to ensure that. The problem is in providing enough light to allow the driver of the reversing vehicle to see clearly. This problem is often exacerbated by the use of low-transmittance glazing in rear windows.

Passenger cars in the United States are required to have glazing with a transmittance of at least 0.70. However, some common vehicle types, such as minivans and sports utility vehicles (SUVs), are classified as light trucks for which there are no limits on transmittance of the rear window. An option available for many minivans and SUVs is privacy glazing for the windows behind the driver. The transmittance of privacy glazing is of the order of 0.18. The installation of privacy glazing has consequences. Freedman et al. (1993) found a decreased probability of detection of children and debris with lower transmittance glazing when preparing to reverse. Sayer et al. (2001) used a US database to examine the impact of lighting conditions, driver age, and vehicle type on backing accidents as a proportion of all accidents. The database, the General Estimates System, contains a nationally representative sample of police-reported accidents. Table 7.2 shows the percentage of reversing accidents and all accidents for cars, minivans, and SUVs involving drivers less than 66 years of age, who had not been drinking. Minivans and SUVs are involved in a higher percentage of backing accidents than would be expected from all accidents. Table 7.3 shows the percentage of reversing accidents and all accidents for minivans and SUVs in different ambient lighting conditions, involving drivers less than 66 years of age who had not been drinking. The percentage of reversing accidents are what would be expected from the overall accident distribution for daylight,

TABLE 7.2

Percentages of Reversing Accidents and Total Accidents for Cars, Minivans, and SUVs for Drivers Less than 66 Years of Age Who Had Not Been Drinking

Vehicle Type	Reversing Accidents (%)	All Accidents (%)
Cars	82.1	88.3
Minivans / SUVs*	17.9	11.7

*Statistically significant differences. From Sayer et al. (2001).

TABLE 7.3

Percentages of Reversing Accidents and Total Accidents for Minivans and SUVs by Ambient Illumination, for Drivers Less than 66 Years of Age Who Had Not Been Drinking

Ambient Illumination	Reversing Accidents (%)	All Accidents (%)
Daylight	66.6	76.1
Dark	7.2	7.6
Dark/Lighted	15.4	11.5
Dawn/Dusk*	9.3	3.7
Unknown	1.5	1.2

*Statistically significant differences. From Sayer et al. (2001).

dark but lighted, and after dark conditions, but for dawn and dusk conditions the percentage of reversing accidents is statistically significantly greater than would be expected. Taken together, these results suggest that in dark conditions, drivers reversing are careful because they are aware that they cannot see well. In daylight, there is no problem in seeing well but in dawn and dusk conditions, and possibly in dark but lighted conditions, drivers of minivans and SUVs overestimate how well they can see when reversing. There are a number of possible solutions to this problem, among them being the banning of low-transmittance glazing, an increase in luminous intensity for reversing lamps where low transmittance glazing is used, and the use of sensors to detect obstacles behind the vehicle when reversing. Such measures would be particularly helpful for older drivers who are over involved in reversing accidents (see Section 12.2).

7.14 DAYTIME RUNNING LAMPS

Daytime running lamps are pairs of lamps positioned at the front of the vehicle, used during the day to increase the conspicuity of the vehicle (Figure 7.9). Some countries require their use, some allow their use, and some prohibit their use. Daytime running lamps may be provided by dedicated lamps or by the use of low beam or high beam headlamps or fog lamps, although exactly which are permitted varies from country to country. ECE regulations call for dedicated daytime running lamps to emit white light with a minimum luminous intensity of 400 cd on axis and no more than 800 cd in any direction. In the United States, the minimum luminous intensity for dedicated daytime running lamps is 500 cd on axis with a maximum luminous intensity in any part of the beam of 3000 cd. However, if a low-beam headlamp is being used as a daytime running lamp there is no maximum luminous intensity set. If a high-beam headlamp is being used, the maximum luminous intensity on axis is 7000 cd and the headlamp must be no higher than 864 mm above the ground.

But why should a vehicle need to have its conspicuity increased in daytime when the visibility of everything on the road should be high? There are two answers to this question. The first is the sad fact that about half of fatal vehicle accidents occur in daytime, so there is certainly a problem (Bergkvist 2001). The second is that the most basic error of drivers is late detection (Rumar 1990). Increasing conspicuity should enable a driver to detect other vehicles earlier. But why does late detection occur so frequently? The answer to this question involves both cognitive and visual factors (Hughes and Cole 1984; Rumar 1990). The cognitive factor is a matter of expectation and hence the allocation of limited attention. If attention is given to the wrong part of the visual world, late detection of an approaching vehicle is likely. Conspicuity is essentially a measure of the ability to attract attention. The visual factor is a matter of a weak stimulus, particularly in the peripheral visual field. A vehicle approaching or being approached head-on at speed causes little change in the retinal image, which may make it difficult to detect until too late. Daytime running lamps are an attempt to attract attention to the other vehicle and hence to get drivers to use foveal vision to examine the other vehicle's movement.

From the above it would seem that daytime running lamps are an obvious means to reduce daytime accidents, but that may not be so. Daytime running lamps have

FIGURE 7.9   A Volvo with daytime running lamps amongst other cars without such lamps.

potential drawbacks as well. Concerns have been expressed about the possibility that daytime running lamps may cause glare, may mask turn signals, and may reduce the conspicuity of vehicles who already use them, such as motorcycles, or who do not have them, such as bicycles (Rumar 2003). The extent to which daytime running lamps might cause glare will depend on the ambient illuminance. Studies of discomfort glare received through rearview mirrors show that for a low ambient illuminance of 700 lx, a luminous intensity of 1000 cd is just permissible, but for a high ambient illuminance of 90,000 lx, a luminous intensity of 5000 cd is acceptable (Kirkpatrick et al. 1987; SAE 1990). The FMVSS regulations applicable to daytime running lamps (see Table 7.1) allow luminous intensities that seem likely to cause discomfort. As for masking of turn signals, SAE (1990) found masking when the daytime running lamps had a luminous intensity of 5000 cd and higher and were observed from a short distance, but at longer distances masking could occur for luminous intensities of 1000 cd, especially if the separation between the turn signal and the daytime running lamp was small, findings that are consistent with what is known about disability glare (see Section 6.6.2). Again, the FMVSS regulations applicable to daytime running lamps allow luminous intensities that are capable of masking turn signals.

In response to these concerns it would seem to be a simple matter to reduce the maximum luminous intensity allowed, but there is a problem with this. It is simply that the higher the ambient illuminance, the greater the luminous intensity required for daytime running lamps to increase the conspicuity of the vehicle (Rumar 2003). This observation suggests that daytime running lamps, as currently regulated, should have a greater effect on conspicuity and hence accidents in countries at high latitudes where the ambient illuminance is low more frequently and the sun is low in the sky for longer (Koornstra 1993). Elvik (1996) has examined this possibility and concluded that it is correct, which may explain why many of the countries that currently require daytime running lamps are at high latitudes.

There is also concern about the relative conspicuity of different vehicle types. This concern arises because if having daytime running lamps makes one vehicle more conspicuous and the available attention is limited, then another vehicle without daytime running lamps should become less conspicuous. Attwood (1979) has shown that it is more difficult to detect a car without daytime running lamps when it is between two cars that have them than when none of the cars have them. The inverse of this situation is a particular concern for motorcyclists who, of all road users, are the most likely to be killed or injured. Even where daytime running lamps are not required, motorcyclists are encouraged to drive with headlamps on during the day to increase their conspicuity, something that is very desirable given the small frontal area of a motorcycle and the consequent difficulty in detecting presence and estimating distance and speed. Wells et al. (2004) have shown that motorcyclists who use headlamps by day have a 27 percent lower risk of being killed or injured than those who do not. With regard to daytime running lamps, the motorcyclists' concern is that if every vehicle were to have daytime running lamps, the conspicuity of motorcycles would be reduced. Whether or not such a reduction would matter is almost certainly related to traffic density and attentional capacity. Where there are only a few vehicles on the road, there should be enough attentional capacity for a driver to examine all of them, starting with those that are using headlamps or daytime running lamps. Where there are many vehicles on the road, there may not be enough attentional capacity to examine all the vehicles. If daytime running lamps are widespread and motorcycles use their headlamps by day, the only advantage motorcycles have is the greater luminous intensity of headlamps over daytime running lamps. This argument implies that motorcyclists' concerns about reduced conspicuity following the widespread introduction of daytime running lamps are justified for heavy traffic. One possibility would be to maintain the conspicuity advantage of motorcycles above cars by making their headlamps flash or pulse by day.

As for other road users without daytime running lamps, such as cyclists, Cobb (1992) examined the conspicuity of bicycles near cars equipped with daytime running lamps of different luminous intensities. He found that daytime running lamps increased the conspicuity of cars but did not reduce the conspicuity of bicycles until the luminous intensity of the daytime running lamps was very high, presumably high enough to produce masking by disability glare. This unexpected finding may simply indicate a halo effect around daytime running lamps. If the daytime running lamps attract attention to the car, and the bicycle is close to the car, the retinal image of the bicycle is closer to the fovea and is more likely to be detected. However, if the bicycle is some way away from the car, directing attention to the car may reduce the chances of its being detected. A similar argument may apply to pedestrians in the road. Daytime running lamps do not make pedestrians more visible to drivers so the presence of daytime running lamps on other vehicles may attract drivers' attention away from pedestrians, particularly in heavy traffic. Fortunately, this disadvantage may be more than offset by making a vehicle with daytime running lamps easier to detect by the pedestrian. Thompson (2003) found that the largest accident reduction associated with the use of daytime running lamps concerned collisions with pedestrians, particularly children.

These observations indicate that introducing a legal requirement for daytime running lamps is a matter of balance between the positive effect of enhanced conspicuity for some and the negative effects of glare and reduced conspicuity for others. A legal requirement for daytime running lamps has been introduced in a number of countries and the consequences studied (Andersson and Nilsson 1981; Elvik 1993; Arora et al. 1994). These studies indicate that daytime running lamps are effective in reducing multivehicle accidents. However, Theeuwes and Riemersma (1995) have criticized the reliability of the statistical method used in these evaluations and argued that the effects found in Sweden are due to an unexplained increase in single-vehicle daytime accidents. Fortunately, Elvik (1996) has carried out a meta-analysis of seventeen such studies and concludes that the beneficial effects of daytime running lamps are robust. He further concludes that the use of daytime running lamps on cars reduces the number of multiparty daytime accidents by about 10–15 percent for vehicles having daytime running lamps and reduces the total number of multiparty daytime accidents by about 3–12 percent, He also states that there is no evidence that the use of daytime running lamps affects any type of accident other than multiparty accidents. Such reductions are somewhat greater than indicated by Farmer and Williams (2002), who found that, in the United States, vehicles with automatic daytime running lamps were involved in 3.2 percent fewer multiparty accidents than those without. Despite this variation, there can be little doubt that the large-scale use of daytime running lamps is advantageous to traffic safety, particularly in high-latitude countries.

7.15 RECENT DEVELOPMENTS

In many ways the development of vehicle signal lighting has been characterized by the piecemeal addition and removal of signals as the need arises. For example, in the UK there was, for a number of years, a requirement for a dim-dip lamp. This was a low-beam headlamp operated at about 20 percent of normal luminous intensity. It was intended for use in towns where road lighting was present and where front position lamps were considered inadequate but low-beam headlamps too glaring. In the United States, sequential turn signals were used for a number of years in some vehicles. In these, the turn signal did not flash as one unit, rather a number of small light sources were energized in sequence so that there appeared to be movement in the direction of the turn or lane change. Both these signal lamps are now prohibited, but initiative has not been completely crushed by bureaucracy. Indeed, there have been a number of suggestions made for improving the effectiveness of vehicle signal lamps. These suggestions have varied from those attempting to improve existing signal lamps by increasing visibility and removing ambiguity, through those that aim to increase the amount and type of information conveyed by the signal lamp to those that combine signal lamps with sensors to make the signal more responsive to prevailing conditions.

In the first group comes the suggestion by Mortimer (1977) that rear signal lamps should be colour coded for function rather than the present situation in America where rear position lamps, stop lamps, and rear turn signals can all be the same colour. What was suggested was that rear position lamps should be greenish-blue, turn signals should be yellow, and stop lamps red. The idea was that by colour coding, the speed of response to the signal would be increased. Another suggestion is to change the location and number of all signal lamps so as to minimize the probability of them being hidden by parts of other vehicles. For heavy trucks this is already common, with many having additional high-level rear position and stop lamps as well as multiple side marker lamps. A similar approach could readily be implemented in small trucks and vans. Even if this were unacceptable, it would be a good idea for the CHMSL in cars really to be mounted high up on the vehicle and not somewhere convenient. Yet another proposal made by several authors is for turn and stop signal lamps to have two levels of light output, one for use by day and one by night (Moore and Rumar 1999). The idea behind this proposal is that the ambient lighting is very different by day and night, yet current turn and stop lamps have a fixed luminous intensity distribution that is inevitably a compromise between providing a high enough luminous intensity for the lamp to be conspicuous by day without causing glare at night. By having different luminous intensities by day and night, the luminous intensity could be increased by day and decreased by night so that conspicuity is increased by day and glare is reduced at night. Finally, Huhn et al. (1997) have suggested that hazard flashers would be more easily discriminated from turn signals by having these two signals flash at different frequencies, unlike the present situation where they flash at the same frequency. Despite the logic of these proposals and the ease with which they could be implemented, most of them have fallen on stony ground.

The second group reflects the desire to provide more information by signal lamps. One example that has already attracted attention is the stop lamp. At the moment, the activation of the stop lamp simply tells drivers behind that the brakes have been applied in the vehicle ahead but nothing about how strongly they have been applied. Horowitz (1994) suggests the use of a combination of flashing and colours to discriminate between deceleration without breaking, sudden accelerator release, antilock braking system activated, and braking at low speeds or stopped. Recently, some manufacturers have released vehicles with stop lamps that have a normal appearance when the brakes are applied normally but change appearance when the driver attempts an emergency stop. The change in appearance involves either an increased brightness, an increase in lit area, or flashing. However, studies of the use of a signal indicating sharp deceleration have given only limited support to its value for enhancing traffic safety (Rutley and Mace 1969; Voevodsky 1974; Mortimer 1981). Another suggestion is to arrange signal lamps so as to make it easier to identify the type of vehicle at night and to estimate the rate of closure. Identifying the type of vehicle is useful because different vehicles have different dynamics. Estimating the rate of closure is valuable for avoiding rear-end collisions. One approach to identifying the type of vehicle is to use retroreflective material to outline the vehicle. Support for this approach comes from the finding that contour lighting of heavy trucks is effective in reducing collisions at night (Schmidt-Clausen and Finsterer 1989). As for estimating the rate of closure at night, the primary cue used is the angular separation of the rear position lamps of the vehicle ahead, so much so that Janssen et al. (1976) suggested that the separation between rear position lamps should be standardized and set as wide as possible. Estimating relative speed is a particular problem for vehicles with a single rear signal lamp or headlamp, a fact that has led to the idea of fitting motorcycles with a specific pattern of multiple daytime running lamps (CIE 1993).

The third group, integration with sensors, is already evident in the latest vehicles. For example, many cars now have a sensor to activate and deactivate headlamps and position lamps according to the ambient illuminance. It is not too difficult to see a similar sensor being used to adjust the luminous intensity of rear fog lamps and of daytime running lamps so as to maintain a constant level of conspicuity in different ambient conditions. Another role for sensors is to ensure that the correct signal is sent every time it is required. This would overcome the problem of drivers turning or changing lanes without signaling or carrying on straight ahead while signaling a turn (Ponziani 2006). It should be possible to develop a system whereby any attempt to change lanes or turn without signaling would trigger the relevant turn signal although this would still give little notice to nearby vehicles. More useful would be a more sensitive system to automatically cancel a turn signal after completion of the maneuvre. Another possibility is a system that provides feedback to the driver in the event of the failure of any signal lamp, something that is presently only available for turn signals. Finally, there is also the possibility of communication between vehicles. For example, it should be possible to use a proximity sensor so that the approach of another vehicle too close behind causes the rear position lamps to be pulsed.

Clearly, there is no shortage of ideas for improving the rather ambiguous and confusing system that currently constitutes vehicle signal lighting (Bullough et al. 2007b). What is required to get some of these proposals implemented is evidence that the perceived problems of current practice are real and not just wishful thinking on the part of researchers, that the proposed changes are effective in changing driver behaviour in a desirable direction, that the new equipment is reliable in use and, when installed on a large scale, that the new equipment does indeed reduce accidents and injuries. This is a rational approach to developing better vehicle signal lighting. But rational development takes time and money. Until these resources are available, it is likely that vehicle signal lighting will continue to develop in an ad hoc manner, driven by whatever the market tells the vehicle manufacturer is most attractive to potential customers.

7.16 SUMMARY

Signal lighting is lighting designed to indicate the presence of or give information about the movement of a vehicle. Signal lighting includes front position lamps, rear position lamps, side marker lamps, license plate lamps, turn lamps, hazard flashers, stop lamps, rear fog lamps, reversing lamps, and daytime running lamps as well as retroreflectors. Some signal lamps are used only at night or in conditions of poor daytime visibility while others have to be visible at all times, both day and night.

For many years, vehicle signal lighting was based on incandescent light sources, but today LEDs are becoming more and more widely used. The optical control of vehicle signal lamps is usually done through reflection, refraction, or total internal reflection, while the structure has to meet mechanical, electrical, environmental, and colour objectives. On top of all this, vehicle signal lighting has become an important part of the styling of the vehicle.

Vehicle signal lighting is closely regulated. The two most widely recognized bases for regulations are the US Federal Motor Vehicle Safety Standard 108 (FMVSS) and the recommendations of the Economic Commission for Europe (ECE) used in the European Union. Other countries follow one or other of these standards though with modifications to suit local circumstances. Both the FMVSS and ECE recommendations cover the allowed colour of the light, the minimum and maximum luminous intensities that should be provided in different directions, the lit area of the lamp, the allowed positions of the lamp on the vehicle, and, if flashing is called for, what the frequency of flashing should be. Although these recommendations cover many of the same topics, they differ in detail.

Vehicle signal lamps can be divided into two classes, those lamps intended to make the vehicle more conspicuous and which are permanently lit when the vehicle is moving after dark and those with a wider range of messages that are intermittently lit when the message is sent. The first group includes front and rear position lamps, side marker lamps, rear fog lamps, and retroreflective markers. The second group includes turn lamps, stop lamps, hazard flashers, reversing lamps, and daytime running lamps. Discriminating the second group from the first is usually done by using a different colour, e.g., reversing lamps, by increasing luminous intensity, e.g., stop lamps, by flashing, e.g., turn lamps, or by changing the number of lamps lit, e.g., hazard flashers.

Despite these differences there is still concern over how rapidly the information presented by a signal lamp can be accessed and how limited the information is, even when it is accessible. There have been a number of suggestions made for improving the effectiveness of vehicle signal lamps. These suggestions have varied from those attempting to improve existing signal lamps by increasing visibility and removing ambiguity, through those that aim to increase the amount and type of information conveyed by the signal lamp to those that combine signal lamps with sensors to make the signal more responsive to prevailing conditions. It would be worthwhile to examine some of these suggestions because past innovations in signal lighting have been beneficial. It has been estimated that the introduction of the centre high-mounted stop lamp has reduced the number of rear-end accidents by about 4 percent. Similarly, it has been estimated that the widespread adoption of daytime running lamps has reduced multiparty accidents by between 3 and 12 percent. The history of vehicle signal lighting is one of ad hoc development. The potential for more information to be made available through a combination of sensors, computers, and LEDs suggests that a more systematic approach to vehicle signal lighting will be possible in the future.

8 Vehicle Interior Lighting

8.1 INTRODUCTION

From the start, it should be made clear that interior lighting, as discussed here, refers to the lighting of the internal compartments of the vehicle and does not include the lighting of the instrument panels and displays. The lighting of instrument panels and displays is a specialist field that is dealt with using different techniques and technologies (Wada et al. 2006; Chang and Wang 2007).

Until recently, interior lighting was the Cinderella of vehicle lighting. This was for a number of reasons. First, there were no detailed legal requirements for interior lighting. Second, the interior lighting was switched off once the vehicle was moving, so it was considered of little practical significance. Third, there were many other aspects of design that had bigger impacts on vehicle reliability and attractiveness to potential purchasers than interior lighting. However, since the 1990s the situation has changed. There are still no detailed legal requirements for interior lighting but as vehicle reliability has improved, competition between vehicle manufacturers has increased the need to find other ways to differentiate one product from another. The number and size of cup holders is one such differentiator, interior lighting is another. Further, great care goes into the selection of the materials used to furnish the vehicle. The appearance of these materials in daylight and in the showroom has always been considered important, but recently the impact of interior lighting on the appearance of these materials when entering the vehicle after dark has begun to be appreciated.

8.2 FUNCTIONS OF INTERIOR LIGHTING

For many years, the only form of interior lighting was the dome lamp, a simple translucent diffuser mounted on the centre of the roof and operated either automatically, by switches on the doors, or manually, by a switch on the lamp itself. Simple dome lamps can still be found on down-market vehicles. Mid-market vehicles tend to have a more sophisticated dome lamp containing several light sources with two distinct light distributions, a diffuse distribution and a focused beam. The diffuse distribution is for ambient illumination of the interior. The focused beam is intended for visual tasks requiring discrimination of detail, such as reading by passengers. Up-market vehicles have multiple interior lighting systems for a range of functions. Table 8.1 lists the functions for which interior lighting is considered desirable and the forms of lighting used to meet these functions. Given the tendency for innovations to first occur in up-market vehicles and then, if found to be attractive to buyers, to percolate down to less expensive vehicles it seems likely that dedicated interior lighting for some of these functions will soon become commonplace.

TABLE 8.1

Functions and Forms of Interior Lighting

Function	Lighting Forms
Entering or exiting the vehicle	Door lock and handle lighting
	Door threshold lighting
	Puddle lighting
	Footwell lighting
	Courtesy mirror lighting
Getting ready to start driving	Marker lighting of all controls
	Ambient lighting
	Storage lighting
	Reading lighting
Driving	Boot lighting
Maintenance	Engine compartment lighting

The factors that are important in selecting the equipment used to provide the various forms of interior lighting are not only light output, power demand, light colour properties, and cost, but also the heat gain and depth of the product. The reasons for the importance of the latter two factors is that interior lighting has to be installed within the body shell so minimizing the space taken up by lighting equipment is important for maximizing interior space. Whatever the lighting equipment needed to fulfill the functions in Table 8.1, it has to be controlled. Two forms of control are used, switching and dimming. Where the lighting is confined to a closed space such as the boot, engine compartment, or glove box, or the lighting is unwanted after a specific event, such as the puddle lighting after the door is closed, switching is used. For all others, dimming is used to reduce the light output slowly, usually to zero.

8.3 TECHNIQUES OF VEHICLE INTERIOR LIGHTING

8.3.1 DOOR LOCK AND HANDLE LIGHTING

The purpose of door lock and handle lighting is simply to show the driver and passengers where the locks and handles are as they approach the vehicle. This can be done very simply by using LEDs to mark the locations or by having a wide beam lamp in the mirror housing. Door lock and handle lighting is usually activated by the remote key system and will be switched off when the door is closed.

8.3.2 DOOR THRESHOLD LIGHTING

Tripping over the threshold of the door will spoil any attempt at elegance. The lighting of the threshold can be done by having a wide-beam incandescent or LED light source mounted in the top or side of the doorframe. Such lamps may also be used for footwell lighting or puddle lighting. An alternative is to use electroluminescent strips or light guides mounted actually on the threshold. Door threshold lighting is switched on by opening the door and switched off when the door is closed.

8.3.3 PUDDLE LIGHTING

The purpose of puddle lighting is to reveal what is on the ground immediately adjacent to the open door. It gets its name from the risk that there might be a deep puddle just outside the door. Puddle lighting can be provided by a wide beam incandescent or LED light source mounted either in the door frame, below the pocket on the door, or in the bottom of the door. Puddle lighting is switched on by the opening of the door and switched off when the door is closed.

8.3.4 FOOTWELL LIGHTING

Footwell lighting has two functions, to reveal the interior of the vehicle so that the whole of the cabin can be seen before entering and to help occupants pick up items that have been dropped at any time. For the former, all the footwells have to be lit, while for the latter, only the footwell of interest has to be illuminated. Footwell lighting can be provided by using incandescent or LED light sources with a diffuse light distribution mounted under the bulkhead for the front footwells and under the seats for the rear footwells. Footwell lighting can be switched on by the operation of the remote key system and dimmed to off with a delay after the closing of the doors. Ad hoc operation of the footwell lighting during driving requires a specific control switch.

8.3.5 MARKER LIGHTING OF ALL CONTROLS

Marker lighting of controls can take two forms, either simple marking that indicates where the control is but relies on some other form of lighting to illuminate how to use the control, or marking and illuminating combined. The marking can be done by internally illuminating the controls. The marking and illuminating can be done by having high-contrast icons on the controls and then lighting the area around it. The controls in question extend beyond the obvious stalks on the steering column and audio controls to such items as window switches, door handles, seat adjustments, and fuel tank cap release. Anyone who has got into an unfamiliar rental car after dark is aware of the value of being able to easily see, identify, and understand all the controls that need to be adjusted before and during a journey. Internal illumination of controls can be done using miniature incandescents or LEDs and light guides to deliver the light where it is needed. Marking and illuminating can be done by miniature incandescents or LEDs mounted close to the control. Marking lighting for all the controls should be activated by the insertion of the key into the ignition system prior to starting the engine.

8.3.6 AMBIENT LIGHTING

Ambient lighting is designed to illuminate the whole of the vehicle cabin. In down-market vehicles it is the only form of interior lighting provided, often in the form of a dome lamp. Usually, ambient lighting is mounted on the centreline of the roof or on the two long edges of the roof and uses either incandescent or LED light sources. The light distribution is diffuse. Usually, ambient lighting is switched on when any door is unlocked and remains on until all the doors are closed, after which there is a slow decline in light output to extinction, typically over a period of 5 to 15 seconds. This allows the people entering the vehicle and particularly the driver to orientate themselves in the vehicle. The ambient lighting is extinguished while driving unless the manual override switch is used. On stopping the vehicle and removing the ignition key, the ambient lighting is switched on and stays on for a few seconds to allow time for the doors to be opened. Opening the doors returns the ambient lighting to full light output where it stays until all the doors are shut, after which there is a slow dimming to extinction.

8.3.7 STORAGE LIGHTING

Most vehicles now contain a number of separate boxes and compartments for storing items, such as glove boxes and parts of the central console. Finding an item in one of these boxes without some form of illumination can be difficult. Glove box lighting is designed to illuminate the inside of the box and is switched on and off when the lid is opened and closed. Storage compartments in the console may use the same system or may use an overhead lamp with a narrow beam and a manual switch. The problem with storage lighting is usually that items inside the box may obstruct the lighting. Where the lighting is built into the box, the solution to this problem is to use several lamps distributed around the box or an area light source. Where the lighting is remote, there is little that can be done.

8.3.8 COURTESY MIRROR LIGHTING

Courtesy mirrors are commonly mounted on the back of sun visors for use by the driver and front seat passenger and, sometimes, in the back of the front seats for use by rear seat passengers. The lighting of these mirrors is usually done by incandescent light sources mounted on either side of the mirror. These are activated by the lifting of the cover over the mirror and extinguished by the closing of the cover. Incandescent lighting is usually used because of its good colour rendering qualities for skin tones. However, such mirror lighting is often unsatisfactory because the small size of the mirror and the position of the lamps close to the mirror mean that the lamps are directly in front of the person looking into the mirror. In this situation, the lighting produces glare to the viewer and does not illuminate the sides of the face well. A better solution would be to light the face of the viewer by more widely spaced lamps, as would be the case in bathroom mirror lighting.

8.3.9 READING LIGHTING

Today, more and more vehicles make provision for the entertainment of passengers during a journey through audio systems and DVD players. One simple approach to ensuring entertainment is to provide a reading lamp for each seat. The essence of the reading lamp is that it should provide a tightly controlled beam of light that produces a sharp-edged area of relatively high illuminance. A tightly controlled beam is necessary to limit the amount of light that will reach the eyes of the driver directly. Some additional light at the driver's eyes is inevitable when reading lamps are used because light will be diffusely reflected from most printed publications, but, unless the interior of the vehicle is finished in very light materials, this should not be enough to cause disturbance to the driver (see Section 8.5). Reading lamps can use tungsten halogen or LED light sources. They are usually controlled by manual switching by the user, but there is also a lot to be said for having a dimming control so that the amount of light can be adjusted by the user.

8.3.10 Boot Lighting

The boot or trunk is often filled with loot or junk. Lighting is necessary to find what is being sought after dark. Like the storage boxes in the interior of a vehicle, the problem facing boot lighting is obstruction. One solution is to use several lamps distributed around the boot rather than just a single lamp. These lamps should have a diffuse light distribution. Another option is to employ an area light source. Either incandescent, LED, or electroluminescent light sources can be used. The boot lighting is switched on and off by the opening and closing of the boot lid. In the United States, it is now a requirement for there to be an emergency release latch that can be operated by someone trapped inside the boot. A simple marking system for this latch, employing a flashing LED, would be of value.

8.3.11 Engine Compartment Lighting

Occasionally, someone may need to access to the engine compartment after dark. Lighting that is switched on and off by the opening and closing of the bonnet is useful. As is the case for the boot, the main problem for lighting is obstruction, most engine compartments being very full. Again, the solution is to use several lamps with a diffuse light distribution rather than a single lamp. It also helps if the underside of the bonnet has a high-reflectance finish.

8.4 INTERIOR LIGHTING RECOMMENDATIONS

While there are no detailed legal requirements for vehicle interior lighting, there are some quantitative recommendations for functional lighting derived from preferences (LRC 1996) or based on current practice (Wordenweber et al. 2007). Table 8.2 is an amalgamation of these recommendations. The mean illuminances are consistent with what is known about the interior lighting of buildings, specifically, the illuminance requirements for safe movement under emergency lighting (Boyce 2003) and for ensuring adequate visual performance for high-contrast printing (Rea and Ouellette 1991). Where ambient lighting is the only form of interior lighting, a reasonable mean illuminance is 20 lx measured on a horizontal plane at the height of the seat cushions. There are no illuminance uniformity criteria for the door lock and handle lighting and for the marking of the controls. This is because such lighting

TABLE 8.2

Vehicle Interior Lighting Recommendations

Function	Mean Illuminance (lx)	Illuminance Uniformity (Maximum: Minimum)	CIE General Colour Rendering Index	Coverage
Ambient lighting only	20	15:1	> 80	Horizontal plane over whole cabin at seat level
Door lock and handle	1	—	—	Local area of lock and door handle
Door threshold	5	15:1	> 50	Whole threshold
Puddle light	10	15:1	> 50	1 m out from door and full width of door
Footwell lighting	10	15:1	> 50	Whole footwell without person present
Marking and illuminating controls	1	—	—	Local area of controls
Storage lighting	20	15:1	> 50	Floor of storage box when empty
Mirror lighting	20	5:1	> 80	Area of face when using mirror
Reading lighting	80	3:1	> 80	300 mm diameter disc on lap of reader
Boot lighting	20	15:1	> 50	Floor of boot when empty
Engine compartment lighting	20	15:1	> 50	Top of engine compartment with bonnet raised

is designed to mark where the lock, handle, and the controls are, not to illuminate these objects. Even where some illumination is provided, the area covered is very small. There are no color rendering index limits for marking the controls, door lock, and handle because saturated colours are often used for these applications. Higher CIE colour rendering index limits are given for applications where the appearance of skin tones and the interior finishes of the vehicle are important, e.g., courtesy mirror lighting, than for applications where exposure is brief, e.g., puddle lighting.

8.5 DISTURBANCE TO THE DRIVER

Interior lighting used when entering or exiting the car or for examining the contents of the boot or engine compartment has no impact on the driver. However, interior lighting used when the vehicle is in motion may be disturbing to the driver because it produces a veil of light over the retinal image of the road ahead thereby reducing the luminance contrasts of the scene. The higher the illuminance received at the driver's eye from the interior lighting, the more likely it is that disturbance will occur. Table 8.3 shows the percentage of drivers disturbed by various forms of interior lighting and the illuminance received at the eye from the interior lighting installed in a Cadillac (LRC 1996). It is clear that the illuminance received at the eye is an important factor in determining the disturbance felt by the driver, but the variation in the percentage disturbance at similar illuminances suggests it is not the whole story. The other factor is most likely the distraction produced when the lamp providing the illuminance is itself visible to the driver, e.g., the glove box lamp when the glove box is open. In passing, it is worth pointing out that a similar pattern of results probably exists for stray light from information and entertainment displays in use during driving.

Table 8.3 suggests that many forms of interior lighting can disturb a significant number of drivers. Further, some of these forms are not directly visible to the driver, being outside the driver's field of view when looking ahead. The reason these forms of interior lighting cause disturbance is that the light they produce is diffusely reflected around the interior of the vehicle thereby increasing the luminances of the dashboard and windscreen. This implies that the reflectances of the interior finishes are important, particularly those of the dashboard and bulkhead. If these reflectances are high, the risk of disturbance to the driver's ability to see down the road at night is higher. To accommodate the different interior finishes used in the same vehicle it would be

---

TABLE 8.3

Percentage of Drivers Disturbed and the Illuminance (lx) Received at the Eye from Various Forms of Interior Lighting in a Left-hand Driver Car

Source	Illuminance at the Eye (lx)	Percentage Disturbed
No interior lighting	0.11	0
Rear footwell lighting	0.11	5
Glove box lighting	0.12	50
Front footwell lighting	0.14	15
Left rear door ambient lighting	0.16	39
Right rear door ambient lighting	0.18	55
Rear map lighting	0.25	50
Right front door ambient lighting	0.30	49
Front map lighting	0.33	40
Dome light ambient lighting	0.60	90
Left front door ambient lighting	1.05	90

From LRC (1996).

a good idea to provide a dimming system for all the interior lighting, similar to that now provided to allow the driver to adjust the brightness of the instrument display.

8.6 DESIGN

So far, this discussion of interior lighting has been confined to functional factors. However, interior lighting has increasingly become one of the means used to give added value to a car and to establish a brand image. This implies that interior lighting has not only to fulfill its functions but also to be a coherent part of the design. The concept behind this approach to vehicle interior lighting is that every lighting installation does two things: it makes whatever it illuminates visible and it sends a message about the people who bought or use the installation. Probably the most obvious example of this approach in buildings is the lighting of shops. In retail premises, the lighting is designed to display the merchandise to advantage and to deliver a message about the nature of the shop, the type of service that can be expected, and the clientele the owners are trying to attract. For example, anyone viewing a shop lit by a uniform array of bare fluorescent light sources is likely to conclude that what is on offer is going to be cheap and of moderate quality, the service will be self-service, and the clientele will be those who are mainly interested in value for money. For the interior lighting of a car, the functional requirements discussed above must be met but consideration also needs to be given to the message sent. It could be argued that meeting the functional requirements alone sends a message, and so it does, but there are other messages that might be sent. For example, it is possible to use lighting to emphasize the style of the interior, the quality of the materials, and the status of the occupants. If any of these possibilities are of interest, consideration has to be given to the lighting of the whole interior space and the people in it rather than just the locations where tasks have to be performed.

There are a number of different technologies available for this purpose. In addition to the incandescent, tungsten halogen, and LED light sources, there is the possibility of using organic LEDs, which are area light sources that can be formed into complex shapes (Kraus et al. 2007). Another possibility is to use electroluminescent foils. Again, these are flat area light sources available in a range of colours. There are also different light distribution systems (Jalink 2002). Conventionally, electricity is distributed to the point where light is required by wire and there it is converted into light by the light source. An alternative is to generate light remotely and distribute it to the desired location using a fibre optic network. The advantage of such remote lighting is that fibre optic cable is very fine and so requires very little depth, the light source being located where space is available. The disadvantage is that any light source failure extinguishes the whole network. Finally, there is the possibility of introducing more individual control. Building lighting already has access to scene-setting control systems, which allow the occupants to choose between a number of pre-set scenes as well as making individual adjustments to the lighting. Similar systems are now becoming available on up-market vehicles.

8.7 SUMMARY

For many decades, interior lighting was the Cinderella of vehicle lighting, but recently this has changed. Over the last decade, it has become increasingly important to differentiate one product from another. Interior lighting is one way to do this. For many years, the only form of interior lighting was the dome light, a simple translucent diffuser mounted in the centre of the roof and operated either automatically, by switches on the doors, or manually, by a switch on the lamp itself. Dome lights can still be found on down-market vehicles. Mid-market vehicles tend to have a combined dome light containing several light sources with two distinct light distributions, a diffuse distribution and a focused beam. The diffuse distribution is for ambient illumination of the interior. The focused beam is intended for visual tasks requiring discrimination of detail, such as reading by passengers. Up-market vehicles have multiple interior lighting systems designed to fulfill a range of functions.

The more sophisticated interior lighting systems provide different lighting for entering and exiting the vehicle, preparing to drive, actually driving, and vehicle maintenance. These systems consist of different combinations of door lock and handle lighting, door threshold lighting, puddle lighting, footwell lighting, marking of controls, storage lighting, courtesy mirror lighting, ambient lighting, reading lighting, boot lighting, and engine compartment lighting. The technologies used to provide this lighting commonly involve the use of incandescent, tungsten halogen, and LED light sources, the light being distributed through a conventional electric network or through a fibre optic network.

There are no detailed legal requirements for interior lighting, but there are lighting recommendations. The main restriction on interior lighting is the need to avoid interfering with the vision of the driver when driving. This can be done by limiting the amount of light reaching the driver's eyes from the interior lighting. Achieving such limits requires attention to the distribution of light within the vehicle, particularly where the interior is finished in high-reflectance materials.

The role of lighting in many buildings today is expanding from the visual performance of tasks alone to the appearance of the space as well. The interior lighting of vehicles is following the same path. This means that interior lighting of vehicles is becoming concerned with revealing the style of the interior, the quality of the materials, and the status of the occupants, as well as fulfilling its functions. Cinderella is going to the ball.

9 The Interaction Between Road and Vehicle Lighting

9.1 INTRODUCTION

Up to this point, road and vehicle lighting have been considered separately. This is a reflection of current practice. Road lighting is designed to promote visibility without reference to vehicle lighting. Vehicle lighting is designed to provide visibility in the absence of road lighting and, with the exception of the town beam proposed as part of the adaptive forward lighting system (see Section 6.7.2), ignores the existence of road lighting. The justification for this apartheid is that some road lighting has to provide visibility for road users who have little by way of lighting, such as cyclists and pedestrians, and vehicle lighting has to be adequate where there is no road lighting. While there is some truth in these explanations, with today's technology there is less and less excuse for failing to consider how current road and vehicle lighting interact and how one or the other might be adjusted to optimize their combined effect.

9.2 EFFECTS ON VISIBILITY

The usual way the interaction between road lighting and vehicle lighting is considered is as the impact of introducing vehicle forward lighting into a road lighting installation (Gallagher et al. 1974; Janoff 1992b; Bacelar 2004). This is reasonable for anyone primarily interested in road lighting but unreasonable from the point of view of the driver. The driver always has vehicle lighting but does not always have road lighting. The approach used here will be to consider the integration of road lighting with existing vehicle lighting.

Both vehicle forward lighting and road lighting are designed to make what is ahead visible to the driver. For objects ahead to be visible they have to have a visual size and a luminance contrast or a colour difference above threshold. Lighting can do little to change visual size, and colour difference is only of importance when luminance contrast is low. Further, many of the light sources used for road lighting have nonexistent or poor colour rendering properties. Therefore, the most fitting way to examine the effect of adding road lighting to existing vehicle forward lighting is to estimate the consequences for luminance contrast. The first step in this process is to look at the illuminances received by a target from both vehicle forward lighting and road lighting at different distances from the vehicle. Figure 9.1 shows the

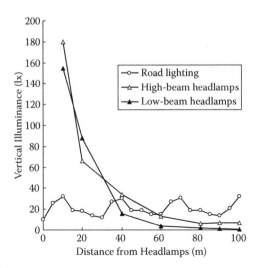

FIGURE 9.1 Vertical illuminance (lx) at road level plotted against the distance from the headlamps (m) for road lighting alone, low-beam headlamps alone, and high-beam headlamps alone. The spacing between the road lighting columns was 30 m (after Bacelar 2004).

illuminances on a square target of side 20 cm with a reflectance of 0.2, placed vertically on the road at different distances from the vehicle and oriented so that the normal to the plane of the target is along the axis of the road (Bacelar 2004). The vehicle forward lighting used was that fitted to a Renault Clio, this being halogen headlamps conforming to the ECE recommendations. The road lighting consisted of a single-sided layout of five luminaires at 30 m spacing. Each luminaire contained a 150 W high-pressure sodium light source and was mounted 8 m above the road surface. The road surface had reflection characteristics $Q_0 = 0.08$ and $S1 = 0.42$, meaning it lies at the boundary of road classes R1 and R2 (see Table 4.14) and is representative of many pavement materials. The resulting photometric characteristics of the road lighting were average road surface luminance = 2.45 cd/m², overall luminance uniformity ratio = 0.6, and longitudinal luminance uniformity ratio = 0.7.

From Figure 9.1 it can be seen that the distances from the vehicle can be divided into three zones. From 10 to 40 m from the vehicle, the illuminance on the vertical target is largely due to the vehicle forward lighting. Between 40 and 60 m, the road lighting and vehicle forward lighting make similar contributions to the vertical illuminance. Beyond 60 m from the vehicle, road lighting makes the major contribution to the illumination of the vertical target, particularly when low-beam headlamps are used. Of course, these boundaries are somewhat moveable, depending on the forms of the vehicle forward lighting and the road lighting. The road lighting used by Bacelar (2004) produces a higher average road surface luminance than is normally recommended (see Section 4.4). For road lighting producing lower average road surface luminances but with the same light distribution, it can be assumed that the boundaries of the three zones will be shifted further away from the vehicle. The same is true for vehicles equipped with HID headlamps. Nonetheless, there will still be three zones, one where vehicle forward lighting is dominant, one where road lighting is dominant, and one where the two forms of lighting are approximately equal.

For the distant zone, where very little light from the vehicle forward lighting reaches the target, the presence of road lighting will usually increase the target's visibility. Visibility is measured as visibility level, this being the ratio of the actual luminance contrast of the target to the threshold luminance contrast of the target. Increasing the adaptation luminance by increasing the road surface luminance using road lighting will increase the contrast sensitivity of the visual system (see Section 3.4.3), thereby reducing the threshold luminance contrast and tending to increase the visibility levels of all targets. While this is generally true, there are some targets for which the visibility level will be reduced. This is because the actual luminance contrast of the target may be reduced by the use of road lighting. One factor that determines whether or not this happens is the relative reflection characteristics of the target and the road surface. Targets that are seen in negative luminance contrast against the road surface, i.e., darker than the road surface, will show an increased luminance contrast when road lighting is introduced unless the luminance of the target is increased proportionally more than the road. Targets that are seen in positive luminance contrast, i.e., brighter than the road, may show a decreased luminance contrast when the road surface luminance is increased, unless the luminance of the target is increased proportionally. Another important factor is the luminance uniformity of the road lighting. Guler and Onaygil (2003) have shown that road lighting with overall and longitudinal luminance uniformity ratios below the minima recommended tend to have larger areas where visibility levels are close to zero.

What this means is that, for the distant zone, introducing road lighting meeting the recommendations will generally increase visibility but may reduce it for specific targets. Within this zone, the range over which targets will remain visible will depend on their visual size. Threshold luminance contrast increases with decreasing visual size (see Figure 3.7) so the visibility level of a target will decrease as the distance between the observer and the target increases until the threshold luminance contrast approaches the actual luminance contrast and the target disappears.

For the near zone, the illuminance on the target is dominated by the vehicle forward lighting, as is the road surface luminance. The increase in adaptation luminance produced by introducing road lighting will again reduce threshold luminance contrasts, although because of the dominance of the vehicle forward lighting, this effect will be small. As for the actual luminance contrast, the impact of introducing road lighting will depend on the relative increases in luminance produced for the target and the road surface. Given that road lighting is designed primarily to light the road surface, it is likely that the increase in road surface luminance will be greater than the luminance of the target. This implies that for targets seen in positive luminance contrast against the road surface when lit by vehicle forward lighting alone, the actual luminance contrast will be reduced. Whether this reduction leads to a decreased visibility level will depend on the extent to which the decrease in actual luminance contrast is compensated by the reduction in threshold luminance contrast. For targets seen in negative luminance contrast against the road surface when lit by vehicle forward lighting alone, the introduction of road lighting will most likely lead to an increase in actual luminance contrast which, together with the reduction in threshold luminance contrast, will always produce an increase in visibility level.

It is in the intermediate zone that things get really interesting. In this zone both road lighting and vehicle forward lighting make similar contributions, although the road lighting emphasizes the horizontal road surface while vehicle forward lighting emphasizes the vertical target. Bacelar (2004) has calculated visibility level from measurements of target and background luminance for the conditions described above and using the model of target visibility developed by Adrian (1989). The target was placed at a constant distance of 40 m from the vehicle for low-beam headlamps and 90 m for high-beam headlamps. Forty metres is assumed to be the stopping distance for inner-city areas where vehicle speeds are of the order of 50 km/h (31 mph) and 90 m is the stopping distance for suburban areas where vehicle speeds are in the range 75 to 110 km/h (47 to 68 mph). The target was moved in 5 m steps along the road between the second and third road lighting columns, successive columns being separated by 30 m. Figure 9.2 shows the variation in visibility level for headlamps alone, road lighting alone, and headlamps and road lighting together. For headlamps alone, the visibility levels are constant because the target is at a constant distance from the vehicle. Visibility levels are lower at 90 m using high-beam headlamps alone than at 40 m using low-beam headlamps alone because of the lower illuminance on and smaller angular size of the target at the greater distance. A lower illu-

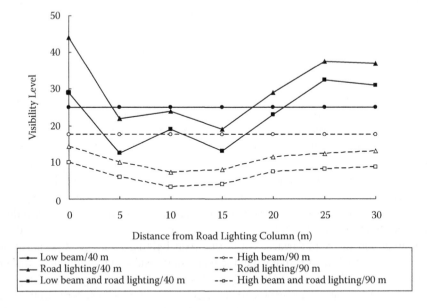

FIGURE 9.2   Visibility levels calculated from luminance measurements taken for a vertical target of reflectance 0.2 at a distance from the driver of 40 m for low-beam headlamps and 90 m for high-beam headlamps, plotted against the position of the target relative to a road lighting column (m). Successive road lighting columns were separated by 30 m (after Bacelar 2004).

minance implies a lower adaptation luminance and consequently a higher threshold luminance contrast, as does the smaller angular size of the target.

For road lighting alone, there is some variation in visibility level because of the variations in illuminances on the road and target at different positions relative to the road lighting luminaires. The visibility levels are lower at 90 m than at 40 m because of the smaller visual size of the target. When the target is between 5 m and 20 m from the column and 40 m from the vehicle, low-beam headlamps and road lighting together produce lower visibility levels than either system alone. When the target is 90 m from the vehicle and high-beam headlamps are used, high-beam headlamps and road lighting together produce lower visibility levels than either system alone, at all positions. A similar pattern of visibility levels for different combinations of vehicle forward lighting and road lighting has been found by others (Guler et al. 2005).

So far, this discussion of visibility has concentrated on the effect of introducing road lighting on the luminance contrast of the target and the adaptation luminance of the visual system, but there is another effect that needs to be considered. This is the effect on disability glare from the road lighting (see Section 4.3). Fortunately, Bacelar (2004) also calculated the visibility level of the target in a fixed position relative to the road lighting and 40 m ahead of the vehicle, which was using low-beam headlamps. Figure 9.3 shows that, for this position, introducing road lighting results in an increase in the visibility level from 25 to 33, suggesting that, with respect to visibility, the changes in road surface luminance and the actual luminance contrast

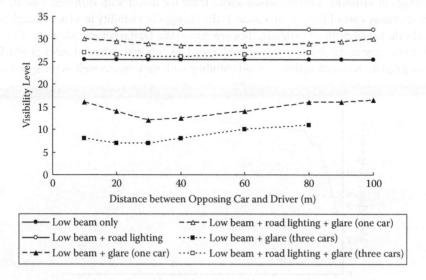

FIGURE 9.3 Visibility levels calculated from luminance measurements taken for a vertical target of reflectance 0.2 at a fixed position relative to the road lighting and 40 m from the driver of a vehicle using low-beam headlamps, with and without road lighting, and with none, one, or three opposing vehicles, plotted against the distance (m) between the opposing vehicles and the driver (after Bacelar 2004).

of the target caused by adding road lighting are more than enough to offset any additional scattered light in the eye.

Of course, this may not always be true. Another interaction of road lighting and vehicle lighting involves the change in the effects of the disability glare caused by the headlamps of opposing vehicles. Bacelar (2004) also reports changes in visibility level for the target positioned at a fixed point 40 m ahead of a vehicle using low-beam headlamps, with and without road lighting, in the presence of one or three opposing vehicles using low-beam headlamps. Figure 9.3 shows the calculated visibility levels, for different distances between the vehicles, with and without the road lighting. It is clear that disability glare from opposing vehicles reduces visibility levels, that three opposing vehicles produce greater reductions in visibility levels than one opposing vehicle, and that the reduction in visibility level caused by disability glare from opposing vehicles is less when road lighting is present.

These results suggest three conclusions. The first is that introducing road lighting is likely to improve the visibility of most targets, particularly when they are in the distant zone. The second is that there can be no guarantees that visibility will improve for all targets. There are some targets for which the combination of light distributions from the vehicle forward lighting and road lighting and the reflection properties of the target and the road surface may lead to reduced visibility. The third is that introducing road lighting alleviates the effects of disability glare from opposing vehicles on visibility.

The question that now arises is what visibility level is enough for driving. If the changes in visibility level discussed above leave the driver with sufficient visibility, these changes are of little significance. If the changes in visibility level are enough to make the target difficult to detect, they are important. Gallagher and Meguire (1975) placed a cone as an obstacle on a lit public road and observed the distances at which unwarned drivers took action to avoid colliding with the cone, as well as the speed of

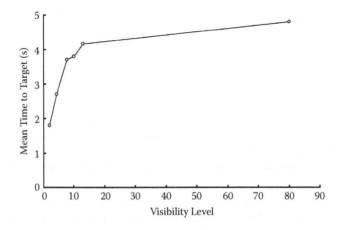

FIGURE 9.4   Mean time to target (s) when the first response to a cone obstacle on the carriageway occurred plotted against the visibility level of the obstacle (after Blackwell and Blackwell 1977).

the vehicle at the time. The combination of distance and speed was used to calculate the time to target. Blackwell and Blackwell (1977) calculated the visibility levels for the cone target from luminance measurements taken at the site of the experiment. Their results showed that time to target increased with increasing visibility level (Figure 9.4), following a law of diminishing returns with little improvement in time to target occurring above a visibility level of 20.

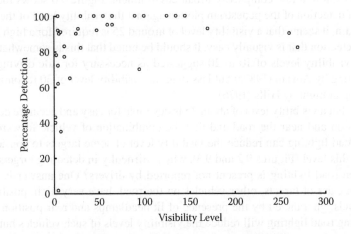

FIGURE 9.5 Percentage detection of a pedestrian at the edge of the road or on the carriageway plotted against the visibility level of the pedestrian for a background with a luminance uniformity (minimum/maximum) around the pedestrian of greater than 0.25 (after Brusque et al. 1999).

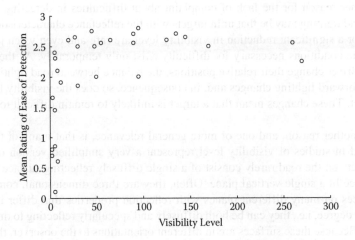

FIGURE 9.6 Mean rating of ease of detection of a pedestrian at the edge of the road or on the carriageway plotted against the visibility level of the pedestrian for a background with a luminance uniformity (minimum/maximum) around the pedestrian of greater than 0.25. The rating scale is 0 = not seen, 1 = difficult task, 2 = average difficulty task, and 3 = easy task (after Brusque et al. 1999).

In another study, Brusque et al. (1999) showed observers a series of street scenes, some of which had a pedestrian at the side of the road or on the carriageway 65 m away. The presentation time was 100 ms. The street scenes varied in the complexity of the background, quantified as the uniformity of background luminance (minimum/maximum). Figure 9.5 shows the percentage of presentations in which the pedestrian was detected and correctly located plotted against the visibility level of the pedestrian in a low-complexity urban environment. Figure 9.6 shows the rating of ease of detection of the pedestrian plotted against the visibility level of the pedestrian. Again, it seems that a visibility level of around 25 is required for a high level of correct detection that is visually easy. It should be noted that this is somewhat higher than the visibility levels of 10 to 20 suggested as necessary for safe driving under road lighting by Adrian (1987a) but less than the visibility level of 30 recommended for driving at night by Hills (1976).

Given that a visibility level of about 25 is desirable for easy and accurate detection of targets on and near the road and that the combination of vehicle forward lighting and road lighting can reduce the visibility level of some targets to considerably less than this level (Figures 9.2 and 9.3), why is difficulty in detecting targets on the road when road lighting is present not reported by drivers? One answer is that an important class of targets, other vehicles on the road, have very high, positive visibility levels, guaranteed by the presence of lit headlamps and rear position lamps. Introducing road lighting will reduce the visibility levels of such vehicles but not by enough to make any difference to the ease with which they can be detected, i.e., the reduced visibility level will still be on the saturated part of the responses shown in Figures 9.4, 9.5, and 9.6. It is targets that rely on reflected light for their visibility, such as pedestrians, unlit vehicles, and cyclists without lights, that are sensitive to the introduction of road lighting.

Another reason for the lack of complaint about difficulties in detecting targets under road lighting may be that unlit targets with the reflectance characteristics necessary for a significant reduction in visibility level are rare and even when they do occur, the conditions necessary for difficulty exist only temporarily. As the target and the driver change their relative positions, the balance between road lighting and vehicle forward lighting changes and, in consequence, so does the visibility level of the target. These changes mean that a target is unlikely to remain difficult to detect for long.

Yet another reason, and one of more general relevance, is that many of the targets used in studies of visibility level represent a very simplified version of reality. Targets on the road rarely consist of a single diffusely reflecting surface of one reflectance in a single vertical plane. Often, they are three dimensional, composed of surfaces on many different planes with reflection properties that differ in both kind and degree, i.e., they can be both diffusely and specularly reflecting to different extents. Because these surfaces are in different orientations to the observer, the illumination falling on each surface will be different. All this means that many targets found on roads will have a wide range of luminances and hence will have multiple luminance contrasts. Further, targets of significance to drivers are not always seen against the road surface. A pedestrian in the middle of the road will be seen against the road surface but one at the edge of the road may not. This variability in target and background luminances implies that most three-dimensional targets have multiple visibility levels (see Figure 4.2). Introducing road lighting will change the balance of visibilities between different parts of the target. If this happens it is unlikely that the target will disappear completely. It is much more likely that the appearance of the target will change as the balance between road vehicle forward lighting and road lighting changes. Provided drivers are familiar with such changes and hence know what to expect, such changes are unlikely to cause any problems in perception and hence produce few complaints.

Attempts have been made to extend the visibility approach to deal with multiple luminance contrasts within the same target (Hentschel 1971; Frederiksen and Rotne 1978), but with little impact. This is probably because the resulting complexity is too much to digest. Even if it were to be developed, there are two difficulties lying in wait. The first is that while a more sophisticated visibility approach would be useful for specific targets in specific locations under specific conditions, for determining the benefits of introducing road lighting in general it raises the additional complexity of having to choose the nature of the targets and the conditions under which they should be measured from a multitude of possibilities. The second concerns the concentration on visibility level and hence on detection. Hall and Fisher (1977) argue that by focusing the evaluation of road lighting on detection alone, there is a tendency to ignore other important requirements for driving that are necessary after detection has occurred.

Despite these concerns, there can be no doubt that introducing road lighting does have advantages for visibility and for other aspects of driving. By lighting the road far beyond the range of low- or high-beam headlamps alone, negative luminance contrast targets at greater distances will certainly be made more visible and positive luminance contrast targets may be. Greater visibility promotes a higher probability of detection. Once a target has been detected, it can be fixated and recognised. Yerrell (1976) quotes measured distances for recognising a target, with and without road lighting, for vehicles with different headlamp beam patterns. In the absence of road lighting, recognition distances on straight, curved, and hilly roads range from 16 to 51 m, while on a lit road, the recognition distances ranged from 145 m to 225 m depending on the position of the target. Further, detecting and recognising targets at greater distances allows more time in which to examine other aspects of the road ahead and more time in which to select an appropriate response. In addition, lighting the road far beyond the range of low- and high-beam headlamps provides a strong cue for visual guidance, easier estimates of changes in speed and distance, and a larger optical flow field, all these being important for driving performance (Hills 1980).

9.3 EFFECTS ON DISCOMFORT

Anyone who has driven for some distance on an unlit motorway will be aware of the feeling of relief experienced when a section of lit road is reached. This feeling is primarily due to the greater distances over which targets can be detected and recognised and the consequent increase in time available in which to make decisions. But there are other factors that affect discomfort. The first is the increase in discomfort produced by the presence of road lighting luminaires themselves. The fact is, road lighting itself may cause discomfort glare. Unfortunately, research on discomfort glare from road lighting has a long history but little outcome (see Section 4.3). Today, road lighting standards do not include criteria for discomfort, the belief being that limiting the disability glare caused by road lighting to less than a 10 to 15 percent increase in threshold increment results in road lighting that is acceptable as far as discomfort glare is concerned.

This should not be taken to mean that road lighting can never cause discomfort glare. The variables that affect discomfort glare are the luminaire luminance, the solid angle subtended at the eye by the luminaire, the angular deviation of the luminaire from the line of sight, and the adaptation luminance. Discomfort glare from road lighting increases with increasing luminaire luminance and larger solid angles and decreases with increasing road surface luminance and increasing angle from the line of sight. When driving along a lit road, luminaire luminance, solid angle, and deviation from the line of sight vary rhythmically, the rhythm depending on the spacing of the road lighting luminaires and the speed of the vehicle. The result is a steady oscillation of the discomfort glare that may itself enhance the sense of discomfort (Raynham 2004). Fortunately, most road lighting on traffic routes is mounted so high that the top of the windscreen cuts off the driver's view of the luminaires before the luminaire luminance becomes high enough to cause discomfort, although the recent vehicle design trend to extend the windscreen further along the roof may cause problems in the future. The most common situation where road lighting causes discomfort is the use of "historic" luminaires in urban centres with the luminaire mounted at low heights (Figure 9.7).

While road lighting itself has the potential to cause discomfort glare, it will certainly reduce discomfort glare from headlamps. It does this because the introduction of road lighting increases the background luminance against which the headlamps

FIGURE 9.7 A "historic" urban area lighting installation in Albany, New York. Note the low mounting height relative to conventional road lighting. Such an installation is likely to cause discomfort glare to drivers.

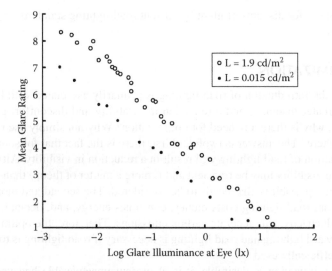

FIGURE 9.8 Mean discomfort glare (de Boer) ratings plotted against the illuminance (lx) at the eye for two different levels of average road surface luminance, one corresponding to an unlit road (0.015 cd/m²) and one to a lit road (1.9 cd/m²) (after Schmidt-Clausen and Bindels 1974).

are seen. As discussed in Section 6.6.2, Schmidt-Clausen and Bindels (1974) have developed an equation for predicting discomfort glare from headlamps containing three terms, the illuminance at the eye from the headlamps, the adaptation luminance, and the angle from the line of sight. The presence of the adaptation luminance and angle from the line of sight in the denominator of the equation means that increases in both variables reduce the magnitude of discomfort glare. Figure 9.8 shows the magnitude of the effect of introducing road lighting. Mean glare ratings are plotted against the illuminance at the eye for two different levels of adaptation luminance corresponding to two levels of road surface luminance. The deviation of the glare source from the line of sight was fixed at 1 degree. From Figure 9.8 it can be seen that changing the average road surface luminance from 0.015 to 1.9 cd/m², i.e., from unlit to the upper limits of what is recommended for road lighting, increases the glare rating by about 1.4 units on the de Boer glare scale. This nine-point scale is unusual in that higher values indicate less discomfort glare. For example, if the headlamps on an unlit road were causing a discomfort glare rating of three, corresponding to disturbing glare, introducing road lighting giving a road surface luminance of 1.9 cd/m², would move the rating much closer to a rating of just admissible glare.

There can be no doubt that introducing road lighting will reduce the level of discomfort caused by glare from headlamps. Whether or not this is enough to offset the increase in discomfort glare produced by the road lighting will depend on how the road lighting is done. Experience of current road lighting practice and the abandonment of criteria for discomfort glare by current road lighting standards suggest that it usually is.

9.4 OPTIMIZATION

Given that the introduction of road lighting is primarily associated with better visibility at greater distances and a reduction in disability and discomfort glare from headlamps, why is there any need for optimization? Why not simply use road lighting everywhere? The answer is twofold. First, there is the fact that for some targets, the introduction of road lighting will result in a reduction in visibility. Although the reduction in visibility may be transient and is more a matter of theory than practicality, it is still a possibility that needs to be considered. The second and more serious answer is that road lighting costs money, consumes energy, and, depending on the source of electricity, can generate carbon emissions. Therefore, the optimization of vehicle forward lighting and road lighting is necessary if road lighting is to be effectively and efficiently used.

While optimization is desirable, it is at present unachievable because there is insufficient information on the combined effects of vehicle forward lighting and road lighting. There are three possible approaches to collecting the necessary information. The first, and most convincing, would be to collect accident data, but while there are quite a lot of data available on the value of light for accidents of different types (see Section 1.5), these data are based on existing lighting practice and so cannot tell us much about possible innovative optimizations of vehicle forward lighting and road lighting (Rea 2001).

The second approach would be to examine the impacts of various combinations of vehicle forward lighting and road lighting have on different visual judgements believed to be important for driving. This approach faces two difficulties. The first is the variable nature of the vast range of targets likely to be found on roads. The second is the limited range of judgements that have been studied. As should be apparent from what has been said above, the judgement that has been most widely studied is visibility of simple, small, flat, diffusely reflecting targets. What should be done to improve this situation? The first thing to do would be to shift the emphasis from visibility to detectability. This is necessary because most targets have multiple visibilities and it is detectability that is important. The second would be to focus attention on the targets most at risk of death or injury on the roads at night and for whom changes in the amount of light make the most difference. As shown in Section 1.5, the most relevant target is the adult pedestrian. This suggests that the target used should be a three-dimensional pedestrian dressed in low-reflectance clothing. By examining the detectability of such a target under different vehicle forward lighting beam distributions and different road lighting systems, separately and in combination, it might be possible to develop optimum combinations of these two forms of lighting.

But measuring visibility of distant targets, even significant ones, is not enough. When traffic is heavy, the driver is not concerned with detecting targets at a distance but rather what the vehicle immediately ahead is doing. Fisher and Hall (1976) examined the effect of background luminance on the ability to detect whether a large target ahead was accelerating or decelerating, the mechanism for detection being the

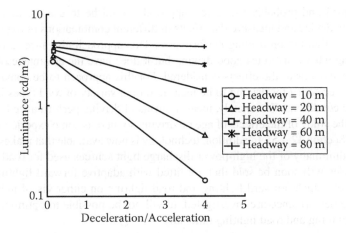

FIGURE 9.9 A nomogram showing the background luminance (cd/m^2) needed for a response to a change in speed with a reaction time within 10 percent of the minimum mean reaction time, for different decelerations/accelerations (m/s^2), for vehicles separated by different headway distances (m) (after Fisher and Hall 1976).

change in visual angle. Figure 9.9 shows a nomogram predicting the background luminance needed to respond to a given acceleration or deceleration for vehicles at different distances ahead, with a mean reaction time within 10 percent of the minimum. It is clear from Figure 9.9 that for vehicles only 10 m ahead, the background luminance makes a big difference to the ability to detect small changes in acceleration or deceleration quickly. For a vehicle 10 m ahead, the background lighting will be dominated by the headlamps. For a vehicle 80 m ahead, the situation is less clear, but a combined road surface luminance of about 4 cd/m^2 is required for a fast response. Such a luminance is higher than is recommended for road lighting alone, which indicates the need for some contribution from vehicle forward lighting. The problem with this conclusion is that there is no consideration of the need for a fast response time at different distances. With a separation distance of 10 m between vehicles, there is a clear need for a fast response, but is that still the case at a separation of 80 m? If it is not, then an average road surface luminance of 2 cd/m^2 may be enough and this is within the range for current road lighting practice.

This indicates what could be done to explore the optimization of vehicle forward lighting and road lighting. What is required is to identify a battery of judgements that are relevant to driving, the distances at which they typically need to be made and the roles of vehicle forward lighting and road lighting in those judgements. Some of the judgements that should be examined in detail are off-axis detection of high-risk targets such as pedestrians and unlit parked vehicles, changes in speed and direction of vehicles moving in the same direction, and changes in the speed and direction of vehicles on potential collision courses. These judgements are related to pedestrian fatalities, rear-end collisions, collisions while overtaking, and collisions while crossing traffic, all accident types that are sensitive to light (see Tables 1.3 and 1.4).

The third and probably the fastest approach would be to use the driver as an integrating device and measure the effects of different combinations of road lighting and vehicle lighting on driving performance. This approach also faces a difficulty in defining what constitutes good driving but it does have the advantage of adding drivers' behaviour to the effects considered. For this approach to be convincing, it would be necessary for driving to be measured on a variety of road types in different traffic conditions. Fortunately, measurements of driving performance have been used for the study of the effects of age on driving so there is some experience in this area (see Section 12.4.5). In addition, technology is now available that makes it much easier to dim many of the high-power discharge light sources used for road lighting and vehicles will soon be sold that are fitted with adaptive forward lighting. Until the roles of vehicle forward lighting and road lighting on either visual judgements or driving performance are investigated, it will not be possible to optimize vehicle forward lighting and road lighting rationally.

9.5 SUMMARY

As a general rule, road lighting is designed to promote visibility without reference to vehicle lighting, and vehicle lighting is designed to provide visibility in the absence of road lighting. This chapter is devoted to the interaction of vehicle lighting and road lighting in terms of its effects on visibility and comfort.

Both vehicle forward lighting and road lighting are designed to make what is ahead visible to the driver. The most suitable measure to determine whether or not an object ahead of the driver will be visible is its luminance contrast. The luminance contrast of an object will depend on the illuminances on and reflection properties of the object and the surfaces against which it is seen. Measurements of illuminance ahead of a vehicle on a lit road show there are three distinct zones: a near zone where vehicle forward lighting is dominant, a distant zone where road lighting is dominant, and an intermediate zone where the two forms of lighting are approximately equal. Examination of the impact on visibility of introducing road lighting for the three zones leads to three conclusions. The first is that introducing road lighting is likely to improve the visibility of most targets when they are in the distant zone. The second is that there can be no guarantees that visibility will improve for all targets, particularly in the intermediate zone. There are some objects for which the combination of light distributions from the vehicle forward lighting and road lighting and the reflection properties of the object and the road surface may lead to a reduced visibility. The third is that introducing road lighting alleviates the effects of disability glare from opposing vehicles on visibility. There can be no doubt that introducing road lighting has advantages for visibility and for other aspects of driving. Lighting the road far beyond the range of low- or high-beam headlamps alone increases the probability of detection of most targets, provides a strong cue for visual guidance, makes estimates of changes in speed and distance easier, and ensures a larger optical flow field, all these being important for driving.

The introduction of road lighting also changes the discomfort experienced by the driver at night. This feeling is primarily due to the greater distances over which targets can be detected and recognized and the consequent increase in time available in which to make decisions. On narrow roads, there is an additional reason why the introduction of road lighting reduces discomfort, the reduction in discomfort glare from headlamps.

Given that the introduction of road lighting is primarily associated with better visibility at greater distances and a reduction in discomfort glare from headlamps, why not use road lighting everywhere? The basic answer to this question is that road lighting costs money, consumes energy, and, depending on the source of electricity, can generate carbon emissions. Therefore, the optimization of vehicle forward lighting and road lighting is necessary if road lighting is to be effectively and efficiently used. While optimization is desirable, it is at present unachievable because there is insufficient information on the combined effects of vehicle forward lighting and road lighting available. There are two realistic approaches to collecting the information needed to optimize the combination of vehicle forward lighting and road lighting. One is based on the visual judgements the driver has to make. What needs to be done is to identify a battery of judgements that are relevant to driving, the distances at which they typically need to be made, and the roles of vehicle forward lighting and road lighting in those judgements. Some of the judgements that should be examined in detail are the off-axis detection of realistic targets such as pedestrians and unlit parked vehicles, the detection of changes in speed and direction of vehicles moving in the same direction, and the detection of changes in the speed and direction of vehicles on potential collision courses. These judgements are related to pedestrian fatalities, rear-end collisions, collisions while overtaking, collisions while changing lanes, and collisions occurring while crossing intersections. The other approach is to study driving performance. This approach requires a definition of what constitutes good driving but it does have the advantage of adding drivers' behaviour to the effects considered. For this approach to be convincing, it would be necessary for driving to be measured on a variety of road types in different traffic conditions. Until studies examining the roles of vehicle forward lighting and road lighting on either visual judgements or driving performance are undertaken, it will not be possible to optimize vehicle forward lighting and road lighting rationally.

10 Special Locations

10.1 INTRODUCTION

The lighting discussed in Chapter 4 is applicable to the vast majority of roads but there are still a number of locations that require special treatment, for a number of reasons. Some, such as tunnels, pose visual problems during the day. Some, such as pedestrian crossings, railway crossings, and car parks, are places where vulnerable individuals are exposed to hazards. Some, such as intersections, are high-risk areas because traffic streams cross and merge. Some, such as road works, may temporarily restrict traffic flow in an unexpected manner. Some, such as roads near docks and airports, may cause confusion to other forms of transport. This chapter is devoted to how conventional road lighting should be adapted to cope with these special locations.

10.2 TUNNELS

The first question that needs to be addressed when considering the lighting of tunnels is what constitutes a tunnel? One answer would be a section of road that is not exposed to the sky but that definition would also apply to an underpass. The difference between these two structures is a matter of the ability to see what is inside. A tunnel can be defined as a section of road that is not exposed to the sky and that requires lighting during the day for drivers to be able to see vehicles and obstructions within it. An underpass does not require daytime lighting for drivers to be able to see vehicles and obstructions inside it. One quantity that has been proposed as a means to separate tunnels from underpasses is the through-view quotient (Schreuder 1998). This is the ratio of the solid angle subtended at the driver's eyes by the visible exit aperture to the solid angle subtended by the entrance aperture. Through-view quotients greater than 0.5 are believed to be acceptable to drivers without special lighting but structures with lower through-view quotients are not. Such structures should be classed at tunnels.

The lighting of tunnels has to address two different problems. The first is the black-hole effect experienced by a driver approaching a tunnel. The second is the black-out effect caused by a lag in adaptation on entering the tunnel. Neither of these problems occurs at night, because then the average road surface luminance inside the tunnel is recommended to be a minimum of 2.5 cd/m^2, a value likely to be similar to, if not more than, the road surface luminances outside the tunnel (IESNA 2005b). By day, this is not the case. By day, the luminances around the tunnel portal will be much higher than those inside the tunnel so both the black-hole effect and the black-out effect may be experienced and driver safety may suffer (Ueki et al. 1992).

The black-hole effect refers to the perception that from the distance at which a driver needs to be able to see vehicles and obstructions in the entrance to the tunnel, that entrance is seen as a black hole. The original studies of tunnel lighting assumed that misadaptation was the major cause of the black-hole effect (Schreuder 1964). However, Narisada and Yoshikawa (1974) showed that drivers typically start to fixate on the tunnel entrance at about 150 to 200 m from the entrance. At a speed of 100 km/h (62 mph), this means the driver is fixated on the tunnel entrance for about 5 to 7 seconds before entering, which is sufficient for some foveal adaptation to occur. This finding, together with an awareness of the fast neural process of adaptation, led Adrian (1982) and Vos and Padmos (1983) to conclude that the major cause of the black-hole effect is the reduction in luminance contrasts of the retinal images of vehicles and obstructions in the tunnel entrance caused by light scattered in the eye. This is certainly the case in a clear atmosphere and with a clean windscreen, but when the atmosphere is misty and the windscreen is dirty, light scattered in the atmosphere and at the windscreen also make significant contributions to the reductions in luminance contrasts (Padmos and Alferdinck 1983a,b).

Given that the low retinal image luminance contrasts of vehicles and objects in the tunnel entrance are the problem drivers experience when approaching a tunnel in daytime, there are two possible solutions. The first is to reduce the luminance of the surroundings to the tunnel because this will decrease the amount of scattered light in the eyes. The luminance of the surroundings can be reduced given appropriate construction, e.g., by ensuring that the tunnel portal is of low reflectance, by shading the tunnel portal and the road close to the tunnel entrance with louvres designed to exclude sunlight, by using low-reflectance road surface materials outside the tunnel, and by landscaping to shield the view of high-luminance sources, such as the sky. This last possibility is particularly valuable where the tunnel is oriented east–west so that the sun can be seen immediately above the tunnel entrance at some times of the day and year. The second is to increase the luminance contrast of vehicles and obstacles inside the tunnel entrance. This can be done by a careful choice of the materials used in the tunnel entrance. The road surface inside the tunnel entrance should be of higher reflectance than that immediately outside and the walls of the tunnel, against which vehicles and objects in the tunnel are usually seen, should have a reflectance of at least 0.50. There are a number of ways to determine the luminance required in the tunnel entrance. One is by calculating the equivalent veiling luminance corresponding to the scattered light at the fovea, given the luminances that occur in different parts of the visual field around the tunnel entrance, and then estimating the tunnel entrance luminance required to keep targets of different sizes and luminance contrasts above visual threshold (Adrian 1982, 1987b; Blazer and Dudli 1993). There are also tabulated sets of recommendations of the road surface luminances required in tunnel entrances, in different settings and with different traffic flow patterns, based on these methods (CIE 1990b; IESNA 2005b). Figure 10.1 shows the luminance at the entrance of a tunnel necessary to ensure a target subtending 10 min arc and glimpsed for 100 ms is above threshold for a given target contrast and different levels of equivalent veiling luminance. It is apparent that the higher the equivalent veiling luminance and the lower the target luminance contrast, the higher the required luminance at the tunnel entrance.

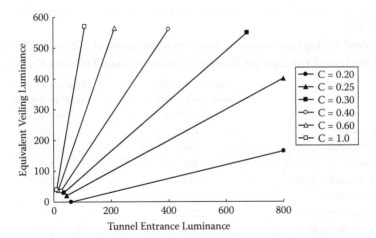

FIGURE 10.1 The luminance of the tunnel entrance (cd/m²) necessary for a target subtending 10 min arc and glimpsed for 100 ms to be visible, plotted against the equivalent veiling luminance from the tunnel surroundings (cd/m²) for different target luminance contrasts (after Adrian 1982).

As the driver approaches the tunnel, the proportion of the visual field taken up by the tunnel entrance increases and the equivalent veiling luminance caused by the scattered light is reduced so that, at the tunnel entrance, the equivalent veiling luminance is much diminished (Hartmann et al. 1986). This is evident in Table 10.1 where, for all the tunnels measured, the veiling luminance is less at 20 m from the tunnel entrance than at 200 m. The magnitude of the equivalent veiling luminance will depend on the surroundings of the tunnel. The Schipohl tunnel in Table 10.1 has an extensive view of the sky immediately above the tunnel entrance, which is likely to have a high luminance, but the Schonegg tunnel is a mountain tunnel where the surround to the tunnel entrance is rock. In the absence of snow, most rock has a low luminance.

Although the approach to the tunnel starts the process of adaptation across the retina, there is no guarantee that adaptation will be complete by the time the driver enters the tunnel (Bourdy et al. 1988). If adaptation to the luminances in the tunnel is incomplete, i.e., the driver is misadapted, the driver may not be able to see what is inside the tunnel after entering. Whether or not this black-out effect occurs depends on the driver's speed and the distance allowed for the change of luminance from the outside to the inside of the tunnel. Given a slow enough speed and a long enough distance, the visual system is well able to look after itself, so there would be no need for the tunnel lighting to be designed to allow for adaptation. However, reducing speed on entering a tunnel is not beneficial in terms of traffic flow and may be dangerous in heavy traffic, so such a behavioural solution is not often adopted. Rather, the lighting is designed to overcome the black-out problem. The approach used is to gradually decrease the road surface luminance in the tunnel, from a threshold zone starting at the entrance, through a transition zone, to the interior zone. The length of these zones is determined by what is termed the safe stopping sight distance (SSSD) (AASHTO

TABLE 10.1

Equivalent Veiling Luminances (cd/m²) from Distances of 200 m and 20 m from the Tunnel Entrance for Thirteen Tunnels in Seven Different Countries

Tunnel	Equivalent Veiling Luminance (cd/m²) from 200 m	Equivalent Veiling Luminance (cd/m²) from 20 m
Tennozen, Japan	125	50
Shin-Kobe, Japan	200	55
Hami, Japan	295	135
Plaza de Fernando, Spain	315	130
Vlake, Netherlands	200	35
Heinenoord, Netherlands	270	90
Schipohl, Netherlands	560	190
Loen, Netherlands	190	55
Castellar, France	130	30
Wiltener, Austria	200	35
Arlberg, Austria	340	150
Schonegg, Switzerland	70	40
SchloBplatz, Germany	100	45

From Adrian (1982).

2001). This is the estimated distance in which a vehicle can stop on a straight and level but wet tunnel approach when travelling at or near the speed limit. The faster the speed of the vehicle approaching the tunnel, the longer the SSSD. Figure 10.2 shows the profile to be followed for grading the road surface luminance from the threshold zone, through the transition zone to the interior zone. Different points on Figure 10.2 are determined by different conditions. The recommended road surface luminances for the threshold zone in daytime, which correspond to 100 percent luminance in Figure 10.2, cover a luminance range of 140 to 370 cd/m² (IESNA 2005b). The exact luminance recommended is determined by the tunnel orientation, the surroundings of the tunnel portal, and the speed limit, and adjusted for different traffic volumes, traffic mixes, tunnel lengths, tunnel surface reflectances, and daylight penetration. The length of the threshold zone is one SSSD less the distance from the tunnel portal at which the entrance fills a significant proportion of the visual field and adaptation is assumed to be occurring. This distance will usually correspond to the top of the tunnel entrance being 22 to 25 degrees above the horizontal at the driver's eye height. Where screening is used to filter out direct sunlight, the start of the threshold zone is taken to be the start of the screening. The length of the transition zone is determined by the assumed vehicle speed, the distance being set so as to allow 10 seconds for adaptation. The road surface luminance of the interior zone in daytime depends on the speed and density of traffic in the tunnel and covers a range of 3 to 10 cd/m², the faster the speed and the higher the traffic density, the higher the average road surface luminance recommended in the interior zone. The minimum luminance uniformity

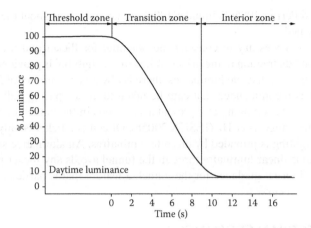

FIGURE 10.2 A road surface luminance profile from the threshold zone, through the transition zone in the interior of the tunnel, plotted against the time from entering the transition zone. The length of the transition zone will depend on the assumed speed of the vehicles entering the tunnel (after IESNA 2005b).

along each lane in the interior of the tunnel should be 0.5 (IESNA 2005b). The interior zone luminance is continued up to the exit from the tunnel, special treatment rarely being needed at the exit because of the high luminance of the tunnel exit and the faster speed of adaptation from lower luminances to higher luminances than from higher luminances to lower luminances. This approach of grading the luminance from the threshold zone through the transition zone to the interior zone is only used where the tunnel is longer than the SSSD or where the tunnel is curved. If the tunnel is shorter than the SSSD, but only a small part of the tunnel exit can be seen as the tunnel is approached, i.e., if the tunnel has a low through-view quotient, the same luminance is used throughout the tunnel, the luminance being that recommended for the threshold zone in longer tunnels.

As for the type of lighting used to provide the luminances in the tunnel, the light source most commonly used is one of the discharge sources because of their high luminous efficacy, long life, and robustness. The luminaires used in tunnels have to be of rugged construction to deal with vibration, dirt, chemical corrosion, and washing with pressure jets. Three types of light distribution are used, symmetrical, counter-beam, and pro-beam lighting. Symmetrical light distributions produce uniform luminance lighting throughout the tunnel so vehicles of different reflectances will have either positive or negative luminance contrasts with the road. Counter-beam light distributions are those where the light is directed predominately against the traffic flow. This gives a high pavement luminance so that vehicles tend to be seen in negative contrast, but there is some risk of the driver experiencing discomfort and disability glare. Pro-beam light distributions are those where the light is directed predominately in the direction of the traffic flow. This gives a low road surface luminance but high luminances for vehicles so the vehicles tend to be seen in positive contrast. Various claims have been made about the benefits of these different systems

(Novellas and Perrier 1985; Schreuder 1993) but no consensus about the best system has been reached.

Finally, it is necessary to consider the potential for flicker and the consequent discomfort and distraction to the driver. When tunnel lighting is provided by a series of regularly spaced, discrete luminaires, there is always a possibility of flicker being perceived. It is recommended that care be taken to avoid spacing individual luminaires so that drivers moving at representative speeds in the tunnel are not exposed to flicker in the range 4–11 Hz (IESNA 2005b). Of course, flicker is only a consideration if the lighting is provided by discrete luminaires. An alternative system based on a continuous linear luminaire through the tunnel avoids any flicker problem and provides good visual guidance for the tunnel, a feature that is particularly valuable where the tunnel curves.

10.3 PEDESTRIAN CROSSINGS

Pedestrians crossing the road are exposed to a high risk of death or injury by collision with high momentum vehicles. In the UK, in 2003, 22 percent of all the people killed in road accidents were pedestrians (Martin 2006). This is why special crossing points are identified where pedestrians have priority over vehicles. Pedestrians who do not use these crossing points are at much greater risk of death or injury than those who do (Martin 2006). Nonetheless, about one-quarter to one-third of pedestrian fatalities occur on or close to a pedestrian crossing (Hunter et al. 1996; OECD 1998). This is almost certainly a reflection of the level of pedestrian exposure rather than the inherent hazard of the crossing, many more pedestrians using crossings than other locations.

There are two types of pedestrian crossings, those associated with traffic signals where the signals are arranged so as to give priority to vehicles and pedestrians at different times, and those without traffic signals where a pedestrian waiting to cross or actually on the crossing has priority over vehicles at all times. The signal-controlled crossings are primarily a feature of urban areas where traffic volumes, both vehicular and foot, are high. The crossings without traffic signals are most frequently found in suburban areas where traffic of both types is lighter. To minimize the danger of using either type of pedestrian crossing, it is first necessary to make the driver aware of the existence of the crossing. The first step to meet this objective is careful site selection, the aims being to make sure that the driver's view of the crossing and people waiting to use it, as well as the pedestrian's view of traffic, are not obstructed by buildings, parked vehicles, landscaping, etc. (DfT 2005b). The second is to ensure consistency of application so that both drivers and pedestrians can identify a pedestrian crossing and know what is expected of them. The third step is to place the appropriate warning signs and signals. Where traffic signals are used, these are of the standard type for traffic control supplemented by signals directed towards the pedestrians waiting to cross. These can be in the form of text or pictograms. Where traffic signals are not used, the crossing itself is marked by some combination of side or overhead mounted signs that are externally or internally illuminated, large, high-contrast road markings, and side or overhead mounted flashing beacons (Huang et al. 2000; DfT 2005b) (Figure 10.3).

FIGURE 10.3 A pedestrian crossing made conspicuous by high-contrast road marking, additional lighting, and flashing beacons on striped poles. The white elements on the poles are retroreflective. The road is narrowed at the crossing, the promontory being marked by one of the striped poles and by marking posts fitted with retroreflectors.

In addition to these measures, special lighting is sometimes used to emphasize the presence of the crossing and to increase the visibility of a pedestrian on the crossing at night. In both the UK and the United States, pedestrian crossings are considered as a conflict area (see Sections 4.4.1 and 4.4.2). In the UK, the recommendations for such areas are given in terms of a maintained average horizontal illuminance and cover a range of 7.5 lx to 50 lx, with a minimum overall illuminance uniformity of 0.4 (BSI 2003a). In the United States, for crossings associated with intersections in high pedestrian conflict areas, the recommendations are a maintained average horizontal illuminance of 20 lx, a horizontal illuminance uniformity ratio of 0.25, and a maintained minimum vertical illuminance of 10 lx at a height of 1.5 m. (IESNA 2005a). For crossings associated with intersections in low pedestrian conflict areas, the recommendations are a maintained average horizontal illuminance of 2 lx, a horizontal illuminance uniformity ratio of 0.10, and a maintained minimum vertical illuminance of 0.6 lx. For crossings separated from intersections, a maintained average horizontal illuminance of 34 lx is recommended with a maintained minimum horizontal illuminance uniformity ratio of 0.33. Curiously, there is no recommendation for vertical illuminance for these crossings.

In both countries, where a pedestrian crossing is close to an intersection or roundabout the lighting is designed as part of the wider conflict area, but where it occurs in isolation as, for example, halfway along one side of a city block but where people

FIGURE 10.4 A pedestrian crossing with high-level supplementary lighting designed to make the crossing conspicuous. The supplementary lighting uses low-pressure sodium light sources so it differs in colour from the rest of the road lighting, which uses metal halide light sources.

wish to cross the road, there are two possibilities for lighting. One is to use the normal lighting of the traffic route, but with the road lighting luminaires arranged so that the crossing is positioned at the midpoint between luminaires. The other is to supplement the road lighting with additional lighting (Figure 10.4). The supplementary lighting approach is recommended when the average road surface luminance is less than 1 cd/m² or the crossing is located on a bend or on the brow of a hill. The supplementary lighting should illuminate the crossing to a higher illuminance than that used to produce the average road surface luminance of the road approaching the crossing. The supplementary lighting should also have a strong vertical component to ensure that pedestrians are positively illuminated, which is why it is recommended that where conventional road lighting is used, the crossing should be at the midpoint between the luminaires.

The outcome of the supplementary lighting approach is a bright stripe of light over the crossing and a higher vertical illuminance on pedestrians using the crossing. The benefits of this approach are evident in a study by Hasson et al. (2002). In this study, at two mid-block crossings in an American city, the ability of observers sitting in a car 82 m away to detect the correct number of pedestrian-sized cutouts near or on the crossing was measured, the cutouts having a diffuse reflectance of 0.18. The crossing was lit using either conventional road lighting giving a road surface luminance of less than 2 cd/m² and producing vertical illuminances in the range 8 to 11 lx, or with supplementary lighting resulting in a vertical illuminance at the crossing of 40 lx. The car in which the observers sat used low-beam headlamps. Table 10.2 shows the percentage of presentations in which the drivers were able to detect fewer than, more than, or the correct number of pedestrian cutouts in two seconds. It is evident that the supplementary lighting improves the ability to quickly detect the correct number of pedestrians, although much more at one site than the other. Whether this improvement is due to the change in light distribution implied by the emphasis given to vertical illuminance or the general increase in the amount of light in the area of the

TABLE 10.2

Percentage of Detection of Fewer Than, More Than, and the Correct Number of Pedestrian Cutouts for a 2-second Observation Period, for Two Pedestrian Crossing Sites Lit by Conventional Road Lighting with and without Supplementary Lighting

Site	Type of Lighting	Percentage (Fewer Than)	Percentage (More Than)	Percentage (Correct)
1	Conventional	50	17	33
1	Conventional + Supplementary	10	10	80
2	Conventional	20	7	73
2	Conventional + Supplementary	13	0	87

From Hasson et al. (2002).

crossing is an open question that will not be resolved until the effects of such changes on the luminance contrasts presented by pedestrians is investigated. How important luminance contrast is to visibility can be seen in a study by Edwards and Gibbons (2007). In this study, people were asked to drive a vehicle equipped with halogen headlamps over a closed test track and report when they detected a pedestrian on a crossing. The test track was lit by road lighting producing four different vertical illuminances on the crossings. The pedestrian was clothed in white, denim, or black hospital scrubs. Figure 10.5 shows the mean detection distances for the three different levels of clothing reflectance and four different vertical illuminances produced by high-pressure sodium road lighting. It is clear that the reflectance of the clothing

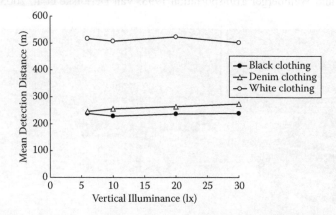

FIGURE 10.5   Mean detection distance (m) for pedestrians wearing black, denim, or white clothing on crossings plotted against vertical illuminance (lx) at the crossing (after Edwards and Gibbons 2007).

has a much greater influence on detection distance than vertical illuminance over the range examined. These findings have two implications. The first is that those concerned about pedestrian safety should concentrate their efforts on persuading pedestrians to wear light-coloured clothing. The second is that the best form of lighting for pedestrian crossings will only be found when the effects on the basic stimuli to the visual system, such as luminance contrast, are evaluated.

While the visibility of the pedestrian on the crossing is important, lighting can also be used to prepare the approaching driver for action by increasing the conspicuity of the crossing. Supplementary lighting of any type improves conspicuity by increasing the brightness of the crossing relative to the rest of the road, but using a light source of a different colour is even better. This increases the conspicuity of the crossing further because it adds another dimension on which the crossing differs from its surroundings. Janoff et al. (1977) report a study in which low-pressure sodium lighting was installed over pedestrian crossings on roads that were lit by other light sources. As would be expected, the increased illuminance on the crossing increased the distance at which a target on the crossing could be detected by an approaching driver, but observations also suggested safer behaviour by both drivers and pedestrians. This use of a different colour of light is part of the recipe for better pedestrian crossing lighting developed by Freedman et al. (1975).

Providing more light of the same or a different colour will make the crossing more conspicuous at night, but does not discriminate between an empty crossing and one in use and does nothing to increase the conspicuity of the crossing during the day. An approach that addresses both these limitations is the placing of flashing lights into the road surface at intervals across the road at the edge of the crossing (Figure 10.6). These in-road lights, which are similar in many ways to airport runway marking lights, provide a narrow beam of light towards the driver. They can be made to flash, either by a simple manual control available to the pedestrian, or automatically, by the use of a system for detecting the presence of a pedestrian entering or on the crossing (Whitlock and Weinberger Transportation 1998; Van Derlofske et al. 2003; Arnold

FIGURE 10.6 A pedestrian using a crossing with conventional road markings and in-road flashing lights.

2004). Such a system can signal a crossing in use by day or night, although the light output needs to be reduced at night if glare is to be avoided.

The question that needs to be addressed now is how effective are these different forms of lighting for pedestrian crossings in reducing accidents? Polus and Katz (1978) report a study undertaken in Israel on the impact of installing an internally illuminated sign mounted above a crossing that also provided an average horizontal illuminance on the crossing of 30 lx. As a result of installing the special lighting, accidents involving pedestrians using the crossings at night decreased by 39 percent. By comparison, daytime accidents on the crossings and accidents at any time on nearby crossings without special lighting but with similar pedestrian and vehicle traffic profiles showed no statistically significant changes. There can be little doubt that special lighting of crossings is beneficial at night because it increases both visibility and conspicuity, but that is not the complete answer. Light can also be used to provide useful information to the driver by day. To be really useful, what drivers need to know is not only that they are approaching a pedestrian crossing but also that it is in use. Van Houten et al. (1988) examined the effectiveness of a combination of signs on driver behaviour. He found that simple signs indicating the presence of a pedestrian crossing had little effect on drivers, but a combination of a sign indicating "stop when flashing" and an overhead flashing beacon activated by the pedestrian was effective in reducing pedestrian/vehicle conflicts. Another approach, that of in-road warning lights triggered either manually or automatically by the pedestrian, has similar effects. A series of before and after studies have shown a limited number of effects on pedestrian behaviour but much more significant effects on drivers (Arnold 2004). Specifically, the percentage of drivers who gave way to pedestrians increased, as did the distance from the crossing at which drivers braked, while the speeds at which vehicles approached the crossing were decreased. Consequently, pedestrian waiting time before crossing and the percentage of times a pedestrian on the crossing had to run to avoid approaching traffic were both reduced.

The message from such findings is clear—lighting has an essential but limited role to play in making pedestrian crossings safer places. Enhancing the conspicuity of the crossing and the visibility of pedestrians on and near a crossing by means of special lighting is useful at night but still relies on the driver searching the crossing and its environs before deciding if any action is necessary. To significantly improve the safety of pedestrians using crossings, by day and night, it is also necessary to signal when there are pedestrians on the crossing or waiting to use the crossing. This can be done by signals triggered manually by the pedestrians or automatically by sensors. Such a system should be more effective because it would warn approaching drivers about an actual conflict rather than a potential conflict.

10.4 RAILWAY CROSSINGS

Like pedestrian crossings, railway crossings involve the movement of one type of transport across the path of another. Unlike pedestrian crossings, on railway crossings it is the driver of the motor vehicle who is more likely to suffer death or injury unless the train is derailed in which case both the train driver and the passengers are likely to be hurt. Railway crossings are marked in various ways depending on the amount of traffic on the road and track. For remote rural crossings with low levels of traffic on both, all that is usually provided is a standard warning sign and possibly a pair of manually operated gates. These arrangements are inherently hazardous because they provide no warning of an approaching train. Even when a driver does look to see if a train is approaching, there remains the difficulty of accurately estimating the distance and approach speed of the train, particularly at night when the engine lamps will be seen as bright spots in an otherwise dark landscape showing very little angular movement. Even where gates are fitted they are subject to abuse, the temptation being to open one gate and drive onto the track and wait there while opening the other gate rather than opening both gates before driving across the track. From this description, it might be thought that such crossings are very dangerous, and they are. Fortunately, the low level of exposure due to the low traffic levels means that deaths at such crossings are rare although stupidity ensures they are also regular.

For railway crossings where there is a high level of traffic on either the road or track, much more elaborate systems are used. Specifically, the crossing is marked with road signs, two pairs of alternately lit lamps, a bell or siren, and either full or half-width barriers. The crossing is also lit at night by conventional road lighting luminaires or floodlights (see Figure 10.7). As a train approaches the crossing, the bell or siren sounds, the alternately lit lamps flash, and the barriers come down. The bell or siren is silenced once the barriers are down. The lights continue to flash until the train has passed and no other train is approaching, at which time the alternately lit lamps are extinguished and the barriers are raised. Full barriers cover the whole road and prevent drivers moving over the crossing, but anyone foolish enough to stop on the crossing may be trapped if the barriers come down before they can move off. Half-width barriers cover

FIGURE 10.7 A car stopped at a level crossing in the UK as a train approaches. The crossing is lit by two high-pressure sodium floodlights. On each side, the crossing has a full-width barrier and two pairs of alternating flashing red lamps. Both lamps of each pair appear to be lit because of the long exposure time needed for the photograph. Surrounding each pair of flashing lamps is a black shield with a red and white rectangular border made from retroreflective material. This track carries a lot of traffic.

only half the road and therefore avoid trapping the driver, but they do allow drivers with a death wish to traverse the crossing when a train is approaching.

These systems used at railway crossings are slightly different in different countries, but most follow the approach described above. In all countries, lighting, as such, has a very limited role to play, most of the information provided to drivers being through signs and signals. However, it is worth noting that the approach advocated for pedestrian crossings, namely to tell the driver when the crossing is actually in use, is already widely used for railway crossings.

10.5 CAR PARKS

Another location where pedestrians appear to be at risk because they are mixed with vehicles is in car parks. Car parks can be divided into parking lots and parking garages, the former being completely open to the sky while the latter are enclosed to some extent, sometimes totally and sometimes only with a roof. Fortunately for pedestrians, the risk of being injured by a vehicle in these locations is more theoretical than actual, probably because of the much slower speeds used by vehicles in parking lots and parking garages. Box (1981) examined the nature of vehicle accidents in parking lots in the United States and found that about two thirds involved a moving vehicle hitting a parked vehicle, less than one third involved a moving vehicle striking another moving vehicle, about 6 percent involved a moving vehicle striking a fixed object, but only 1 percent involved a vehicle hitting a pedestrian. Accidents to pedestrians are much more likely to involve tripping, slipping, and falling while walking in the parking lot (Monahan 1995).

Nonetheless, the lighting approaches used in parking lots and parking garages are markedly different from the lighting of roads and from each other. For parking lots, luminaires mounted on columns are conventionally used but the luminaires have a much wider range of luminous intensity distributions than those used for lighting roads. The specific luminous intensity distribution used depends on the shape of the parking lot and the positioning of the columns in and around it. As for the light sources used, these are typically some form of high-intensity discharge, usually high-pressure sodium or metal halide (see Section 2.5). Slightly different lighting criteria are used for parking lots in different countries. In the United States, a maintained minimum horizontal illuminance of 2 lx, a maximum horizontal illuminance uniformity ratio (maximum/minimum) of 20:1, and a maintained minimum vertical illuminance of 1 lx are recommended (IESNA 1998). These criteria are enhanced to 5 lx, 15:1, and 2.5 lx respectively for parking lots where crime and vandalism are likely. To achieve these minima, it is suggested that the installation should be designed to produce an average horizontal illuminance of 10 lx, or 25 lx for parking lots where crime and vandalism are likely. In the UK, the lighting criteria for parking lots are given in terms of mean and minimum horizontal illuminances for parking lots in different environmental zones (see Table 13.5; SLL 2006). For parking lots in zones E1 and E2 a maintained average horizontal illuminance of 15 lx is recommended with a maintained minimum illuminance of 5 lx. For parking lots in zones E3 and E4 a maintained average horizontal illuminance of 30 lx is recommended with a maintained minimum illuminance of 10 lx. In addition to these quantitative criteria, the designer is urged to give attention to controlling glare, to avoiding light pollution (see Section 13.7), and to providing enough light for any CCTV surveillance system to operate effectively. While these lighting criteria are clear they are often ignored in the direction of over-lighting. It is not uncommon to find parking lots lit to much higher illuminances, not because higher illuminances are required for driver or pedestrian visibility, but to give an impression of brightness and hence safety. This is particularly the case for parking lots attached to retail premises where the penalty in terms of lost business is particularly severe if the parking lot is perceived as dim and unevenly lit and hence unsafe.

For parking garages, the low ceiling heights and high level of obstruction caused by the structure and parked vehicles require a different approach, except on the top floor when it is open to the sky. The top floor is usually treated as a parking lot where crime and vandalism is likely. For the other floors, luminaires are mounted directly on a ceiling that is often formed from concrete. Where the ceiling is flat, a cutoff luminaire with a flat lens is used to avoid glare (see Section 4.2). Where the ceiling is constructed of coffers, a non-cutoff luminaire can be used, the glare shielding being provided by the coffer. In both cases, care has to be taken with luminaire spacing to ensure a high level of illuminance uniformity, although this is easier to achieve if the surface reflectances of the garage are high. The light sources used in parking garages are typically fluorescent or high-pressure sodium or metal halide (see Section 2.5). Regardless of the light source used, the luminaire has to be constructed to deal with a corrosive, humid, and dirty atmosphere. In the United States, a maintained minimum horizontal illuminance of 10 lx, a maximum horizontal illuminance ratio (maximum/minimum) of 10:1, and a maintained minimum vertical illuminance of 5 lx are recommended (IESNA 1998). The horizontal illuminance is calculated or measured without any shadowing from vehicles or columns. To achieve these minima, it is suggested that the installation should be designed to produce an average horizontal illuminance of 50 lx. These illuminances are recommended for use by day and night with one exception: during the day the maintained minimum horizontal illuminance in the entrance should be increased to 500 lx and the maintained minimum vertical illuminance increased to 250 lx so as to form a transition zone from daylight to the interior of the parking garage. Further, the maintained minimum horizontal illuminance on the ramps should be increased to 20 lx and the maintained minimum vertical illuminance increased to 10 lx. In the UK, the lighting criteria for the enclosed floors of parking garages are given in terms of maintained average and minimum horizontal illuminances (SLL 2006). For parking bays and access lanes an average horizontal illuminance of 75 lx is recommended, with a minimum horizontal illuminance of 50 lx. For ramp corners and intersections an average horizontal illuminance of 150 lx is recommended, with a minimum horizontal illuminance of 75 lx. For entrance and exit areas, an average horizontal illuminance of 75 lx is recommended at night and 300 lx during the day. Again, attention has to be given to providing enough light without glare for any CCTV surveillance system to operate effectively.

10.6 RURAL INTERSECTIONS

Rural roads are surprisingly dangerous. In the United States, in 2004, the number of road fatalities/100 million vehicle miles travelled was 2.3 for rural areas but only 1.0 for urban areas (NCSA 2004). The most likely explanation for this is the higher speeds achieved on rural roads. Of the 24,975 fatalities reported on rural roads in that year, 46 percent occurred after dark. Most rural roads are unlit, the driver having to rely on headlamps alone for visibility. One of the most hazardous parts of the rural road network is the intersection where traffic streams merge and cross. One approach to reducing accidents at rural intersections at night is to provide lighting. Isebrands et al. (2004) present an analysis of the effect of providing lighting at rural intersections in Minnesota. Data were collected on accidents occurring over a period of two years at 3622 rural intersections, 223 with lighting and 3399 without. The crash rate/million vehicles entering the intersection was calculated for each intersection. Table 10.3 shows the mean crash rates for the lit and unlit intersections, by day and night. This metric takes traffic flow through the intersection into account, this being much higher for the lit than the unlit intersections. This higher level of exposure at the lit junctions may explain why the average number of crashes/million vehicles is more for the lit than unlit junctions during the day. What is interesting is that the increase in average number of crashes/million vehicles at night is much greater for the unlit than the lit intersections, a fact reflected in the night/day ratio of crashes per million vehicles. This suggests that providing lighting at rural intersections leads to a reduction in the number of accidents at night.

Isebrands et al. (2004) also carried out a before and after study on the effect of installing lighting at 34 rural intersections. Accident data were examined for two to three years before and after installation of the lighting. The crash rate, i.e., the number of accidents/million vehicles entering the intersection, was calculated for day and night conditions, before and after the installation of the lighting. Table 10.4 shows the mean crash rates before and after the installation of the lighting, by day and night. There are two interesting points about this table. The first is that the mean crash rates are much higher by night than in the wider survey shown in Table 10.3, suggesting that the intersections where lighting has been installed recently have been identified as hazardous. The second is that after the installation of the lighting at these intersections the mean crash rate by day increased, but the mean crash rate

TABLE 10.3

Mean Crash Rates per Million Vehicles Entering an Intersection, for Lit and Unlit Rural Intersections, by Day and Night

Condition	Lit Intersections	Unlit Intersections
Mean day crash rate/million vehicles	0.40	0.29
Mean night crash rate/million vehicles	0.57	0.59
Night/day ratio of crash rates/million vehicles	1.43	2.03
After Isebrands et al. (2004).		

Table 10.4

Mean Crash Rates per Million Vehicles Entering a Rural Intersection, by Day and Night, Before and After the Installation of Lighting

Condition	Before Installation of Lighting	After Installation of Lighting
Mean day crash rate/million vehicles	0.30	0.39
Mean night crash rate/million vehicles	1.12	0.73
Night/day ratio of crash rates/million vehicles	3.73	1.87

After Isebrands et al. (2004).

by night decreased. The result of this is that the night/day crash rate ratio decreased from 3.73 before the installation of the lighting to 1.87 after. Again, the installation of lighting at these rural intersections has led to a reduction in accidents at night.

Given that lighting is an effective accident countermeasure for rural intersections, does the form of the lighting matter? Bruneau and Morin (2005) describe a comparison of the effects of standard and non-standard road lighting at rural intersections in the province of Quebec, Canada. Standard lighting consists of specific lighting columns and luminaires located around the intersection so as to meet the recommendations of the Ministère des Transports du Quebec that are based on the recommendations in use in the United States (IESNA 2005a). Non-standard lighting is lighting installed by municipalities that makes use of existing power utility poles to mount a luminaire. While standard lighting is designed to provide specified photometric conditions (see Section 4.4.1), the same is not true of nonstandard lighting. The lighting conditions provided by non-standard lighting can vary widely depending on where the power utility poles are in relation to the intersection and what type of luminaire and light source are used. Accident data for 376 rural intersections were examined and expressed as crash rates per million vehicles entering the intersection. Table 10.5 shows the mean crash rates for unlit intersections and intersections with standard and non-standard lighting, by night and day. Also shown are the resulting night/day ratios of crash rates per million vehicles. As would be expected, Table 10.5 shows little difference in the mean daytime crash rate per million vehicles for the three lighting types. However, there is a clear increase in mean crash rate per million vehicles at night at all types of intersection, but the increase is less at the intersections with either standard or non-standard lighting. As a result, the night/day ratios of crash rates per million vehicles are lower for intersections with lighting than intersections without lighting.

All types of accidents are considered in Table 10.5. Table 10.6 shows the night/day ratios of crash rates per million vehicles for all accidents, accidents involving personal injury, and those producing property damage only. The interesting point about these results is that the only statistically significant changes in night/day crash rate ratios occur for the property-damage-only accidents. For property-damage-only accidents, both the night/day ratios for crash rates per million vehicles for intersections with standard lighting and non-standard lighting are statistically significantly

TABLE 10.5

Mean Crash Rates per Million Vehicles Entering a Rural Intersection, by Day and Night, for Intersections without Lighting and with Standard or Non-Standard Lighting

Condition	No Lighting	Standard Lighting	Non-Standard Lighting
Mean day crash rate/million vehicles	0.9	0.7	0.8
Mean night crash rate/million vehicles	2.8	1.3	1.7
Night/Day ratio of crash rates/million vehicles	3.2	2.0	2.3

After Bruneau and Morin (2005).

TABLE 10.6

Night/Day Ratios of Crash Rates per Million Vehicles Entering Rural Intersections, for Intersections without Lighting and with Standard or Non-Standard Lighting, for Accidents Causing Personal Injury, Property Damage Only, and for All Accidents

Condition	No Lighting	Standard Lighting	Non-Standard Lighting
Night/day ratio of crash rates/million vehicles for accidents causing personal injury	1.94	1.74	1.87
Night/day ratio of crash rates/million vehicles for accidents causing property damage only	3.87	2.02	2.45
Night/day ratio of crash rates/million vehicles for all accidents	3.24	1.96	2.30

After Bruneau and Morin (2005).

less than for intersections with no lighting. However, there are no statistically significant differences between the night/day ratios for crash rates per million vehicles for intersections with the two forms of lighting. This suggests that almost any form of lighting has a beneficial effect on accidents at rural intersections where the approach roads are unlit.

Given that non-standard lighting is effective in reducing accidents at many rural intersections, the question that arises is why should this be? This question can be answered by considering what lighting can do. One thing that almost any form of road lighting at an intersection of otherwise unlit roads will do is capture the driver's attention. This means that drivers approaching a lit intersection are more likely to be prepared to take action. To determine if action is necessary drivers may carry out a visual search of the surrounding traffic and of the intersection. How important lighting is to the effectiveness of these visual searches will depend on the nature of the traffic. In heavy traffic, the main concern will be what other vehicles nearby are doing, in which case the role of the intersection lighting will be limited. In light traffic, the main concern will be anything that is moving into the intersection, particularly anything on a collision course. In this situation the form of lighting of the intersection is likely to matter. Lighting that provides good visibility over the whole intersection and all the roads that enter it will be more effective than something that only covers part of the intersection, i.e., standard lighting should be more effective than nonstandard lighting. This is because to avoid a collision it is necessary for a driver to be able to judge the velocity and direction of movement of other vehicles on intersecting roads. Wertheim (1981) has shown that velocity and direction of movement can be finely judged if the object of interest is seen against a fixed frame of reference. If there is no frame of reference, it is still possible to judge velocity and direction of movement but with much less precision. At night, good quality lighting of the intersection and the approach roads will provide a frame of reference against which the movement of other vehicles can be judged. This is likely to be most important at intersections where the time for response is limited and such intersections are likely to have high accident rates. This may be why Bruneau and Morin (2005) found that standard lighting had significantly lower nighttime crash rates than nonstandard lighting for intersections with high accident rates.

There can be little doubt that providing lighting at rural intersections is a useful accident countermeasure, albeit one that mainly affects property damage accidents. But what form the lighting should take is a more open question. Almost any lighting will be useful in attracting attention to the intersection. Beyond that, the form of the lighting needed will depend on the nature of the intersection and the traffic flow after dark. The essential points are that the lighting should provide enough light and a wide enough coverage area to allow fast and accurate judgements of velocity and direction of movement. Oya et al. (2002) carried out a statistical analysis of night accidents resulting in death or injury before and after the installation of lighting at intersections on major roads in Japan. Their results suggest that an average road surface illuminance of 30 lx or more is necessary to ensure a statistically significant reduction in nighttime accidents. Current guidance for lighting intersections does recommend such an average illuminance where the intersection involves a major road (see Tables 4.6 and 4.12), which suggests that the recommendations are reasonable, but whether or not the coverage area is sufficient is open to question.

10.7 ROAD WORKS

Road works are inherently hazardous to those working on them and to drivers moving through them. The hazard for workers comes from the proximity of traffic, while for drivers it is the change in road layout. Road works can take many different forms, varying in size, duration, stability, and timing. The methods used to inform drivers about road works can also take many forms, from variable message signs to traffic signals, manually controlled stop/go signs, cone barriers defining the road layout, flashing beacons on cones, and large flashing arrows mounted on trucks. Road workers are sometimes protected from traffic by crash barriers, but they almost always have their visibility enhanced by wearing fluorescent, retroreflective vests, jackets, or suits. Attempts to influence driver behaviour are made by setting lower speed limits, sometimes enforced through speed cameras and higher fines for speeding in road work sites.

What combination of safety measures is used depends on the form of the road works. Among the simplest are such mobile activities as collecting road kill and clearing roadside litter, where the only measure is the wearing of high visibility clothing. Small temporary obstructions such a digging a hole in the road to repair an electricity cable are usually dealt with by marking off the work area with cones and barriers and controlling traffic with manual stop/go signs or portable traffic signals. Large-scale road works that are planned to take several weeks use advance warning signs and extensive coning to define new traffic lanes with flashing lights to make the road layout clear at night and crash barriers to protect the workers. Actually setting out the cone system is one of the more perilous activities as it involves workers changing the road layout while the road is in use. The method used to reduce this risk is to have the cone truck closely followed by another truck carrying a large, self-luminous "change lane" signal on the back.

Extensive road works rarely have special lighting unless work is planned to take place at night. This means that lit road works are only likely to be found on major roads where work is undertaken at night to avoid causing major traffic congestion. This lighting has two purposes. The first is to enable the workers to see what they are doing. The second is to make drivers aware of the changed road layout and the movements of vehicles into, out of, and around the road works. A brightly lit work site attracts attention and that in itself will warn drivers approaching the site that something unusual is happening, but the lighting technology used will depend on the work being done. For relaying the road, vehicle lighting attached to the paver is often all that is used. For constructing new lanes or rebuilding bridges, a much more extensive range of work is needed so lighting covering a much wider area is necessary. This is usually provided by a cluster of luminaires mounted on a telescopic mast and powered by a diesel generator. The light sources used can vary from tungsten halogen to high-pressure sodium or metal halide discharge. Where these units are positioned and where the luminaires are aimed is mainly determined by the needs of the work, although one or more are usually positioned wherever there is a change in direction, narrowing, or merging of lanes for passing traffic. The problem with this flexibility is that often it results in severe discomfort and disability glare to approaching drivers. This is not inevitable. Rather it is a matter of poor luminaire design, in the sense that the light distribution is too wide for the limited mounting height, and poor luminaire aiming. Both of these failings can be rectified and effective lighting for work provided by following the appropriate recommendations (Ellis et al. 2003).

What needs to be considered now is how effective all these measures are. Freeman et al. (2003) monitored personal injury accidents at 29 major road work sites on motorways in the UK for 20 months. The road works covered 730 km (453 miles) of road having an exposure of 4178 million vehicle kilometres. 423 personal injury accidents occurred in the road works in this time. Interestingly, this rate of personal injury accidents is not statistically significantly different from the personal injury accident rate for the same sites when there were no road works and is very close to the national average personal injury accident rate for motorways. Further, factors such as weather conditions and ambient lighting did not affect personal injury accident rate in the road works. What this suggests is that the safety measures used at such major sites—advance warning signs, cones marking lane changes, special lighting, etc.—have been effective in overcoming the inherent hazard of road works, although given the lower speed limits used, personal injury accidents should have been fewer than the average for all motorways if the hazard had been completely nullified. It should also be noted that this conclusion refers only to personal injury accidents. Given that the most common forms of accidents at road works are multiple rear-end collisions, collisions when overtaking, and collisions with objects other than vehicles, it would have been interesting to know how the safety measures influenced property-damage-only accidents, but these data were not collected. Despite this limitation, there can be little doubt that the safety treatment of major road work sites, which includes special lighting, is on the right track. Whether this is also true for lesser road works remains to be determined. What is certain is that with some care over glare control, the lighting of road works could be improved to the benefit of driver and worker comfort.

10.8 ROADS NEAR DOCKS AND AIRPORTS

Roads near docks and airports do not need special lighting to enhance the visibility and comfort of drivers but to avoid confusion to ships and aircraft (BSI 2003b). Ships entering a harbour and aircraft attempting to land rely on signal lamps for guidance. Often, these signal lamps have to be identified from a great distance. The risk is that road lighting luminaires might be confused with the signal lamps. This risk is minimized by restricting the type of road lighting luminaires to those that do not have a high luminous intensity in the relevant direction. For docks, the relevant directions are well known so this can be done by careful mounting or shielding of the luminaires. For airports, there are many possible directions as aircraft circle around before attempting to land so the simplest approach is to use only luminaires without any direct upward light emission, i.e., full cut-off luminaires mounted horizontally. Airports impose another constraint on the form of road lighting where the road crosses the flight path, a low mounting height. This restriction leads to the use of many closely spaced columns each carrying a lower wattage light source than would usually be the case. It is important to appreciate that the presence of docks and airports place restrictions on the technology that can be used to provide road lighting, not the photometric conditions that should be created. Photometrically, the lighting of roads near docks and airports is the same as any other road of the same class.

10.9 SUMMARY

There are a number of road locations that require special lighting, such as tunnels, pedestrian and railway crossings, car parks, intersections, road works, and roads near docks and airports. Road tunnels require special lighting during the day because of two visual effects, the black-hole effect in which the tunnel entrance is seen as a black hole by approaching drivers and the black-out effect in which drivers entering the tunnel find themselves misadapted. The black-hole effect is due to light scattered in the eye from the surroundings of the tunnel, while the black-out effect is due to the driver's visual system having insufficient time to adapt to the much lower luminances in the tunnel. Both these problems can be overcome by appropriate construction of the tunnel entrance and by following the recommendations for grading the luminances from the entrance to the interior of the tunnel.

Pedestrians crossing the road are exposed to a high risk of death or injury. There are two types of pedestrian crossings, those associated with traffic signals where the signals are arranged so as to give priority to vehicles and pedestrians at different times, and those without traffic signals where a pedestrian waiting to cross or actually on the crossing has priority over vehicles at all times. Special supplementary lighting is sometimes used on crossings without traffic signals. The supplementary lighting is designed to provide a higher illuminance than that used to produce the average road surface luminance of the road approaching the crossing with a strong vertical component to ensure that pedestrians are positively lit. Such lighting will increase the visibility of pedestrians on the crossing and increase the conspicuity of the crossing. The conspicuity of the crossing can be further increased by using supplementary lighting of a different colour to that used for the adjacent road lighting. Of course, supplementary lighting does not discriminate between an empty crossing and one in use and does nothing to increase the conspicuity of the crossing during the day. An approach that addresses both these limitations is the insertion of lamps into the road surface at the edge of the crossing that flash when a pedestrian enters and is on the crossing. This has the advantage of warning the driver that there is an actual conflict rather than just a potential conflict.

Like pedestrian crossings, railway crossings involve the movement of one type of transport across the path of another. Unlike pedestrian crossings, at railway crossings it is the driver of the motor vehicle who is more likely to suffer death or injury. Railway crossings are marked in various ways depending on the amount of traffic on the road and track. For remote rural crossings with low levels of traffic on both, all that is usually provided is a standard warning sign and possibly a pair of manually operated gates. For railway crossings where there is a high level of traffic on either the road or track, much more elaborate systems are used. Specifically, the crossing is marked with road signs as well as two pairs of alternately lit lamps, a bell or siren, and either full- or half-width barriers. The crossing is also lit at night by conventional road lighting luminaires or floodlights. Lighting has a very limited role to play in railway crossings, most of the information being provided to drivers through signs and signals.

Another location where pedestrians appear to be at risk because they are mixed with vehicles is in car parks. Car parks can be divided into two types, parking lots and parking garages. The former are completely open to the weather while the latter are enclosed to some extent, sometimes totally and sometimes with only a roof. The risk of a pedestrian being injured by a vehicle in these locations is more theoretical than actual, probably because of the much slower speeds used by vehicles in parking lots and parking garages. Accidents involving pedestrians are much more likely to involve tripping, slipping, and falling. The lighting approaches used in parking lots and parking garages are markedly different from the lighting of roads and from each other. For parking lots, luminaires mounted on columns are conventionally used but the luminaires have a much wider range of luminous intensity distributions than those used for lighting roads. For parking garages, the low ceiling heights and high level of obstruction caused by the structure and parked vehicles require a different approach. In parking garages, luminaires are mounted directly on the ceiling, care being taken with the choice of luminaire to avoid glare and with luminaire spacing to ensure a high level of illuminance uniformity.

One situation where accidents are unexpectedly frequent is rural intersections. Most rural roads are unlit, the driver having to rely on headlamps alone for visibility. Studies of accident data have shown that providing lighting at rural intersections is a useful accident countermeasure, albeit one that mainly affects property damage accidents. But what form the lighting should take is a more open question. Almost any lighting will be useful in attracting attention to the intersection. Beyond that, the form of the lighting needed will depend on the nature of the intersection and the traffic flow after dark. The essential points are that the lighting should provide a high enough luminance and a wide enough coverage area to allow fast and accurate judgements of velocity and direction of movement of other vehicles approaching the intersection.

Road works are inherently hazardous to those working on them and to drivers moving through them. Road works rarely have special lighting unless work is planned at night. The lighting of the work site is usually provided by a cluster of luminaires mounted on a telescopic mast and powered by a diesel generator. Where these units are positioned and where the luminaires are aimed is mainly determined by the needs of the work although one or more are usually positioned wherever there is a change in direction, narrowing, or merging of lanes for passing traffic. The problem with this flexibility is that often it results in severe discomfort and disability glare to approaching drivers. Accident records at major road works on motorways in the UK have shown that the rate of personal injury accidents is not statistically significantly different from the personal injury accident rate for the same sites when there were no road works. This suggests that the safety measures used at such major sites—advance warning signs, cones marking lane changes, special lighting, protective barriers, reduced speed limits, etc.—can be effective in overcoming the inherent hazard of road works. Whether this is also true for lesser road works remains to be determined. What is certain is that with some care over glare control, the lighting of road works could be improved to the benefit of driver and worker comfort.

Roads near docks and airports do not need special lighting to enhance the visibility and comfort of drivers but to avoid confusion to ships and aircraft. The risk is that road lighting luminaires might be confused with the signal lights used when entering a harbour or attempting to land. This risk is minimized by restricting the type of road lighting luminaires to those that do not have a high luminous intensity in the relevant direction. Airports impose another constraint on the form of road lighting where the road crosses the flight path, a low mounting height. Docks and airports place restrictions on the technology that can be used to provide road lighting, not the photometric conditions that should be created.

From these reviews of the lighting of tunnels, pedestrian crossings, railway crossings, car parks, rural intersections, road works, and roads near docks and airports it is clear that special lighting is required in some locations. Such lighting can have four different purposes: to allow for the limitations of the visual system, to make a situation that requires the driver's attention conspicuous, to make what is happening easily understood, and to avoid confusing other modes of transport.

11 Adverse Weather

11.1 INTRODUCTION

The presence of rain, snow, fog, dust, or smoke in the atmosphere makes the driver's task more difficult, but in different ways. Rain has a direct effect on visibility by absorbing and scattering light and an indirect effect by changing the reflection properties of the road surface. Snow scatters incident light and in so doing creates a lot of visual noise in the driver's visual field. In addition, if the snowfall is heavy enough, the snow will cover the road surface, obscuring lane markings and other information designed to aid the driver. Fog and heavy spray thrown up by other vehicles have their effect by absorbing and scattering light in the atmosphere, thereby reducing the luminance contrasts of everything ahead of the driver. Dust and smoke also absorb or scatter light and thereby reduce the effectiveness of all forms of lighting. It is the impacts of such adverse weather conditions on drivers and how lighting can be used to alleviate them that are the subjects of this chapter.

11.2 ADVERSE WEATHER AND ACCIDENTS

Although the presence of rain, snow, fog, dust, or smoke undoubtedly makes the driver's task more difficult, it is not clear if that difficulty is enough to lead to an increase in the frequency of accidents. Table 11.1 shows the number of road accidents in which fatalities occurred in the United States, in 2005, classified according to the weather and the lighting conditions (NHTSA 2006b). An examination of Table 11.1 reveals a number of interesting facts. The first is that the largest number of road accidents involving death, almost 90 percent, occurs when there is no adverse weather. This is partly a matter of exposure and partly a matter of driver experience and behaviour. For most states in the United States, seriously adverse weather occurs infrequently. In the United States, where adverse weather is common at certain times of the year, e.g., snow in winter in the states bordering Canada, most drivers will be familiar with the conditions and will adjust their behaviour accordingly, a change that usually involves reducing speed and keeping greater distances from other vehicles. Both these changes make fatalities in accidents less likely. The second interesting fact is that in the absence of adverse weather 50 percent of fatal accidents occur in daylight, 30 percent in darkness, 16 percent after dark but on lit roads, and 4 percent at dawn or dusk. The third interesting fact is that fatal accidents occurring in rain or snow and sleet are affected by lighting conditions in a similar manner to those occurring in the absence of adverse weather but those occurring in fog and dust, smoke, or smog are not. For fog, 24 percent of fatal accidents occur by day but 51 percent occur after dark on unlit roads. For dust, smoke, and smog, 43 percent of fatal accidents occur during

TABLE 11.1

Total Number of Road Accidents Involving Fatalities Classified according to the Weather and the Lighting at the Time

Light Conditions	No Adverse Weather	Rain	Snow and Sleet	Fog	Dust, Smoke and Smog	Total
Daylight	17,332	1,392	390	114	81	19,309
Dark	10,224	892	263	245	82	11,706
Dark but lit	5,455	528	85	71	12	6,151
Dawn and dusk	1,381	126	41	46	14	1,608
Total	34,392	2,938	779	476	189	38,774

From NHTSA (2006b).

daytime and 43 percent occur after dark on unlit roads. Some of the explanation for this difference between when fatal accidents occur in rain and snow and sleet, and when they occur in fog is probably that fog is more likely to occur after dark as the air temperature falls, but a similar meteorological explanation is unlikely to be the case for dust, smoke, and smog. An alternative explanation for the difference in the effect of lighting conditions in different adverse weather conditions is that while rain and snow and sleet have some effect on visibility, their major effect is to change the coefficient of friction of the road surface, an effect that is independent of lighting condition and that results in longer stopping distances. Conversely, fog, dust, smoke, and smog do little to affect the grip of the vehicle on the road but can have dramatic effects on the visibility of everything ahead, particularly at night. If this explanation is correct then the fact that such a high percentage of fatal accidents in fog occur after dark on unlit roads implies that the vehicle lighting then in use is inadequate.

11.3 RAIN

Rain consists of falling water droplets with diameters in the range 20 to 200 μm (Kocmond and Perchonok 1970). A photon of light striking such a droplet may be absorbed or scattered, the scattering being caused by some combination of reflection and refraction. As a result of these processes, the proportion of light emitted by vehicle headlamps or signal lamps that reaches the eyes of drivers in rain is reduced relative to what would be the case in dry conditions. In addition, the scattered light forms a luminous veil that reduces the luminance contrasts seen through the rain. Further, water left on the windscreen will be enough to degrade the quality of the retinal image of the scene outside the area served by the windscreen wipers, a degradation that may occur even within the wiped area in heavy rain. Consequently, the direct effect of rain in the atmosphere is to reduce visibility. However, unless the rain is heavy or the windscreen wipers are ineffective, this is not the main impact of rain.

FIGURE 11.1   The headlamps of a car operated on low beam reflected from a wet road. Also shown is the reflected image of a low-pressure sodium road lighting luminaire.

Rather, it is the indirect effect of the rain on the reflection characteristics of the road surface that is the more important.

When rain covers a road surface, the visual effect is to make the road surface more of a specular reflector than a diffuse reflector (Figure 11.1). This is not usually a problem during the day because the meteorological conditions that tend to produce heavy rain also tend to produce diffuse illumination from the sky rather than direct illumination from the sun. However, at night, when the road is lit either by vehicle forward lighting alone or by a combination of road lighting and vehicle forward lighting, a wet road surface dramatically alters the effectiveness of that lighting. Where only vehicle forward lighting is available to the driver, the effect of a thin film of water on the road surface is threefold. First, the average road surface luminance will be reduced as more of the light from the headlamps is specularly reflected forward rather than diffusely reflected back to the driver.

Second, light from headlamps specularly reflected from the water film on the road fails to reach the retroreflective road markings underneath. As a consequence, the visibility of the road markings used for visual guidance and accurate lane keeping is much reduced (Rumar and Marsh 1998). Third, the disability and discomfort glare from approaching vehicles will be increased as more light from the headlamps of the approaching vehicle is specularly reflected towards the driver. The magnitude of all these effects depends on the thickness of the water film. Smooth road surfaces will be affected by thinner films than rough road surfaces and retroreflective elements and road studs that stand proud of the road surface are less likely to be degraded in visibility than those that are flush with the road surface.

As for road lighting, when the lit road is wet rather than dry the average road surface luminance is increased, but this occurs because some parts of the road surface become much higher in luminance, i.e., the road surface luminance distribution becomes much more nonuniform. This is evident from the measurements of average road surface luminance, overall luminance uniformity ratio, and longitudinal luminance uniformity ratio shown in Table 11.2 (Ekrias et al. 2007). As objects on the road are seen in silhouette against the local road surface luminance, this increase in luminance nonuniformity means that the pattern of visibilities produced by other vehicles and pedestrians will change more rapidly as they are seen against different parts of the road. The movements of objects that change their patterns of visibilities from moment to moment are likely to be more difficult to identify. This is a negative effect but there are two positive effects of introducing road lighting on wet roads. One is to reveal road marking more clearly, provided the road markings have a high diffuse reflectance. The other is to diminish the impact of disability glare from opposing vehicle headlamps by increasing the background luminance.

The most successful solution to ensuring good visibility on wet roads lies in the choice of road materials. A road constructed with a high proportion of aggregates will have a more coarse texture. Such a road will be less specular in wet conditions but will take longer to dry (Sorensen 1977). The permeability of the road is also important. A road constructed of more permeable materials will reduce the probability of specular reflections occurring because water is less likely to collect on the surface. How much a road surface will change its reflection properties between dry and wet conditions and how it compares in this respect to roads constructed of other materials can be judged from the wet road reflection classification system of Frederiksen and Sorensen (1976).

The other approach to limiting the problems caused by wet roads is to modify either the vehicle forward lighting or the road lighting. For vehicle forward lighting, the proposed adaptive forward lighting system (see Section 6.7.2) has as one of its options a wet road beam. The wet road beam involves a reduction in the illumination just in front of the vehicle and increased light to the sides of the vehicle. This beam is activated when either rain is detected on the road or when the windscreen wipers are switched on. The reason for reducing the illumination just ahead of the vehicle is simple geometry. Given conventional mounting heights for headlamps, this is the part of the light distribution that is specularly reflected towards approaching drivers and hence makes the biggest contribution to the increased disability and discomfort glare experienced (see Figure 11.1).

With road lighting, it is possible to limit the deterioration in road surface luminance uniformity when the road becomes wet by changing the luminous intensity distribution of the luminaires, the spacing between luminaires, and their position relative to the road surface. Where roads are likely to be wet for a considerable proportion of the time, as in the Scandinavian countries, the national road lighting recommendations call for a minimum overall luminance uniformity of 0.15 for the most specular of the wet road surface reflectances in their classification system (CIE 1979). How effective this approach is in maintaining visibility of objects on the road is open to question, the point being that a high luminance uniformity achieved at the cost of a low road surface luminance may not represent an improvement in visibility

TABLE 11.2

Average Road Surface Luminance (cd/m²), Overall Luminance Uniformity Ratio, and Longitudinal Luminance Uniformity Ratio for Two Road Lighting Installations on a Two-Lane Dual Carriageway in Finland, under Three Different Weather Conditions

Road Conditions	Road Lighting Installation 1			Road Lighting Installation 2		
	Average Road Surface Luminance (cd/m²)	Overall Luminance Uniformity Ratio (Minimum/Average)	Longitudinal Luminance Uniformity Ratio (Minimum/Maximum)	Average Road Surface Luminance (cd/m²)	Overall Luminance Uniformity ratio (Minimum/Average)	Longitudinal Luminance Uniformity Ratio (Minimum/Maximum)
Dry	1.11	0.740	0.782	1.76	0.577	0.830
Wet	2.37	0.220	0.398	5.01	0.209	0.310
Ploughed and salted after snow	1.66	0.639	0.756	2.35	0.539	0.648

The measurements were made of the same pieces of road using an imaging photometer positioned at the side of the road at a height of 1.5 m above the road surface (after Ekrais et al. 2007).

(Lecocq 1994). This adverse trade-off can be avoided if the better road surface luminance uniformity is achieved by closer spacing of the luminaires or by including some higher reflectance aggregate, known as brighteners, in the pavement mix to increase the diffuse reflectance of the road surface (Sorensen 1977).

11.4 SNOW

Snow consists of large, high-reflectance particles falling through the atmosphere. Light from vehicle forward lighting striking a snowflake will be scattered, some of the scattered light reaching the eyes of the driver. What the result of this is will depend on the density of snowflakes. If the density is high so there are few gaps between the snowflakes, the result is a uniform veil of high luminance that restricts visibility in all directions to very short distances. This is known as a whiteout. More usually, each snowflake is seen separately against a low luminance background. This would not be too much of a problem if the snowflakes were stationary and so fixed against the background. After all, in daylight the detailed structure of the background is always visible and causes no complaint. However, snowflakes are not fixed, and their movement against the background provides a strong signal to the peripheral visual system that is difficult to ignore, the peripheral visual system being designed specifically to detect changes in the visual world.

How distracting snowflakes are will depend on their luminance contrast: the higher the luminance contrast, the more distracting the snowflakes. Maximum luminance contrast will occur when the vehicle forward lighting is the only source of light. In this situation, the background against which the snowflakes are seen will be at a minimum. During daylight or at night when road lighting is present, the background against which the snowflakes are seen is higher in luminance and thus the snowflakes are at a lower luminance contrast.

There are four approaches that can be used to reduce the distraction caused by snowflakes. The one most readily available to the driver is to use the forward lighting on low beam rather than on high beam. The high-beam condition increases the amount of light emitted by the forward lighting, distributes more light closer to the driver's line of sight, and increases the area covered by the beam, with the result that more snowflakes closer to the driver's line of sight are seen at a higher luminance contrast.

Another approach, but one that is only available to the vehicle designer, is to place the forward lighting as far away from the driver's usual line of sight as possible, although this will reduce the effectiveness of retroreflectors. Field studies have shown that mounting the forward lighting on a snowplough as far away as possible from the driver's line of sight results in better visibility and reduced perceptions of glare in snow (Bajorski et al. 1996; Bullough and Rea 1997) Further, observations of the choices of lamps made by snowplough drivers, i.e., people who are experienced at driving in snow for long periods, reveal that the lamps furthest from the driver are used more frequently in snow (Eklund et al. 1997). Displacing the light source from the driver's line of sight is effective because the intensity of back-scattered light in snow decreases with increasing angle from the direction of incidence on the snowflake (Hutt et al. 1992).

Yet another possibility is to narrow the luminous intensity distribution of the head-lamp because this increases the distance at which significant amounts of light inter-sect the driver's line of sight. Bullough and Rea (2001a) carried out an assessment of a headlamp fitted with a mosaic of hexagonal louvres to narrow the beam compared to a conventional headlamp in the same position. The assessment was again made by snowplough drivers. The results were a reduction in the average luminance of the falling snow and small but statistically significant decrease in the perception of discomfort glare for the louvred headlamp even though the louvres reduced the light output of the headlamp.

Bullough and Rea (1997) have incorporated both the displacement of the light source and the narrowing of the luminous intensity distribution into a simple model for predicting satisfaction with forward visibility in snow and fog. Forward visibility was taken as the product of the maximum luminous intensity of the lamp and the displacement of the lamp from the driver's line of sight. The maximum luminous intensity is a measure that combines both the light output and the luminous intensity distribution of the lamp as, for the same lamp wattage, the narrower the beam, the higher will be the maximum luminous intensity. Eight snowplough drivers made ratings of visibility, glare, and satisfaction on three-point scales, each running from −1 to +1, while driving at night in snow or freezing rain. These ratings were found to be highly correlated ($r^2 = 0.91$) so the individual ratings for each condition were summed into a single metric called lighting quality. Figure 11.2 shows the relationship between the lighting quality ratings and the logarithm of forward visibility. It is clear that both displacement from the driver's line of sight and the luminous intensity dis-tribution have a role to play in providing good forward visibility in snow.

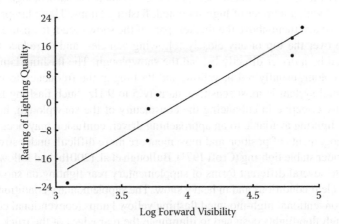

FIGURE 11.2 Ratings of lighting quality made by snowplough operators for halogen lamps with different luminous intensity distributions and displacements from the driver plotted against the logarithm of forward visibility. The rating of lighting quality is the sum of ratings of visibility, glare, and satisfaction made on three-point scales by eight drivers. The forward visibility metric is the product of the maximum luminous intensity of the lamp (cd) and the displacement of the lamp from the driver's line of sight (m) (after Bullough and Rea 1997).

A third approach is to modify the spectrum emitted by the forward lighting. As discussed in Section 4.6, there is evidence that light spectra that more effectively stimulate the rod photoreceptors of the retina ensure better off-axis detection in mesopic conditions, at the same photopic luminance. This is to the advantage of the driver in a clear atmosphere at night, but when that atmosphere contains snowflakes, ensuring better off-axis detection may be detrimental because it will make the snowflakes more distracting rather than less. Bullough and Rea (2001b) have demonstrated this effect of light spectrum for the performance of a tracking task viewed through a curtain of air bubbles moving through water, light with a spectrum that provided little stimulation to the rod photoreceptors resulting in better task performance than light that did stimulate the rod photoreceptors. This effect is less potent than displacing the light source and narrowing the luminous intensity distribution, but may be the only option if displacement and narrowing of the beam are not practical.

All this suggests an opportunity for the adaptive forward lighting system (see Section 6.7.2). When driving in snow, it should be possible to increase the luminous intensity of the headlamp on the passenger's side of the vehicle while decreasing that of the headlamp on the driver's side. If the luminous intensity distribution of the passenger side headlamp could be narrowed at the same time, the forward lighting should provide greater visibility and comfort for the driver in snow.

So far, this discussion has been solely concerned with forward lighting but rear signal lighting also deserves consideration. One type of vehicle certain to be found on the road in snow is the snowplough. Snowploughs are large, heavy trucks moving at slow speeds. Further, when ploughing they will be surrounded by a cloud of thrown-up snow that will reduce their visibility to other drivers. This may explain why about two-thirds of accidents involving snowploughs in Minnesota are rear-end collisions (Hale 1989). To make their presence conspicuous, snowploughs are commonly fitted with a number of high-mounted, flashing lamps. These lamps are high mounted to place them above the thickest part of the snow cloud, to ensure that they are visible over the top of any closely following vehicles, and to reduce the glare experienced by a driver directly behind the snowplough. The flashing lamps are of a specific colour, usually red or yellow, and flashing at the frequency to which the human visual system is most sensitive, namely 5 to 9 Hz. Such flashing lamps are undoubtedly effective in enhancing the conspicuity of the snowplough, but if they are all the lighting available to an approaching driver, confusion may occur. This is because judgements of position and movement are more difficult under strobe lighting than under stable lighting (Croft 1971). Bullough et al. (2001b) had following drivers evaluate several different forms of supplementary rear lighting on snowploughs at night in clear conditions and in heavy snow. The supplementary conditions ranged from the conventional high-mounted flashing yellow lamps (conventional configuration) through floodlights designed to illuminate the rear edges of the truck (indirect edge delineation) and two, high-mounted pairs of yellow and red flashing lamps that from a distance appear as continuously lit but alternating in colour (alternate colour changes), to continuously lit LED light strips placed vertically at the sides of the rear of the truck (continuous LED strip). Figure 11.3 shows the mean ratings of visibility and feelings of confidence about overtaking made by drivers following a snowplough fitted with the different supplementary rear lighting configurations, at night, in a

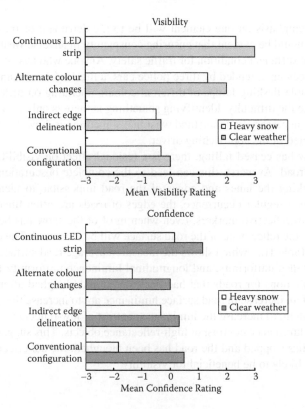

FIGURE 11.3 Mean ratings of visibility and the feelings of confidence about overtaking made by drivers following a snowplough fitted with the different supplementary rear lighting configurations at night in a clear atmosphere and in heavy snow. The mean ratings were made on a scale running from –3 (worst) to +3 (best) (after Bullough et al. 2001b).

clear atmosphere, and in heavy snow. There are statistically significant effects of rear lighting configuration and weather condition. The best rear lighting configuration in a clear atmosphere and in heavy snow, for both visibility and confidence in overtaking, was the two continuously lit LED light strips.

Bullough et al. (2001b) also report a field measurement of the ability of three drivers to detect that they were closing on the snowplough fitted with the conventional configuration and the continuously lit LED light strips. The drivers followed a snowplough at a distance of 100 m at a speed of 30 mph (48 km/h). Without warning the following driver, the driver of the snowplough was instructed to decelerate without braking, thereby ensuring that the following driver came closer to the snowplough. All three following drivers detected that they were closing on the snowplough sooner with the continuously lit LED light strips operating than with the conventional high-mounted flashing lamps operating.

What all this implies is that the rear lighting of snowploughs needs two elements, a temporally varying element to ensure conspicuity and a temporally fixed element to provide a clear and stable cue to movement. This requires some care with balance, as an undue emphasis on one element will be to the detriment of the other. More generally, it should be evident that ensuring conspicuity in adverse weather is a necessary but not sufficient condition for traffic safety. Anyone who has approached the scene of an accident attended by three police cars, a fire engine, and an ambulance, all with multiple flashing lamps of different colours operating on different phases, will recognize the difficulty. Identifying something is there is only part of the problem. It is also necessary to understand what that something is, how it is moving, and what is expected of the approaching driver.

After snow has ceased falling, there is a residual effect on visibility due to the snow on the road. At worst, this can lead to the complete obscuration of all road markings making the lanes and edges of the road impossible to identify. Where heavy snow is a regular occurrence, the edges of roads are often lined with posts fitted with retroreflective markers. Even when most of the snow has been removed by ploughing, the reflectance of the road surface will be increased. The effect of this is evident in Table 11.2, which shows the measured average road surface luminance, overall luminance uniformity, and longitudinal luminance uniformity for two road lighting installations for roads that had been salted and ploughed after a snowfall. The effects of an increase in road surface luminance are to increase the sensitivity of the visual system, to increase the luminance contrast of low-reflectance objects, but to reduce the luminance contrast of high-reflectance objects. This suggests that once the snowfall has stopped and the road has been ploughed, the net effect of snow on the ground is likely to be beneficial for visibility.

11.5 FOG

Fog consists of small water droplets with diameters in the range of 5 to 35 μm suspended in the atmosphere (Kocmond and Perchonok 1970). Photons of light incident on these droplets are absorbed and scattered. The simplest approach to quantifying these effects is to ignore the distinction between scatter and absorption and treat their combined effect on light loss as absorption alone. In mathematical terms, this approach is expressed in Lambert's law, which states that the luminous intensity of light propagating through a uniform medium is given by

$$I = I_0 e^{-\sigma d}$$

where I is the luminous intensity (cd) at distance d (m), I_0 is the unattenuated luminous intensity (cd) at the origin, and σ is the extinction coefficient (m^{-1}). The extinction coefficient of fog can range from 0.0015 m^{-1} for thin fog to 0.04 m^{-1} for thick fog. The visual outcome of this absorption and scattering is a reduction in the transmission of light through the atmosphere and the creation of a somewhat uniform veil of luminance covering the driver's visual field. The effect of the uniform luminance veil is to reduce the luminance contrast of all the things in front of the driver and hence to reduce their visibility. This general reduction in visibility has been shown to lead to drivers reducing speed in fog (White and Jeffery 1980). Hawkins (1988) found that speeds on motorways in the UK started to fall as visibility distance was reduced below 300 m so that when visibility distance had been reduced to 100 m, which corresponds to thick fog, speeds were about 30 percent less than in a clear atmosphere. Unfortunately, even these reduced speeds are not slow enough to make the stopping distance of the vehicle less than the distance the driver can see (Sumner et al. 1977). One explanation for this failure to reduce speed sufficiently when there are several vehicles travelling together is the desire to maintain contact with the vehicle immediately in front and thereby to ease the stress of having to find the way ahead. When there is no vehicle immediately ahead, drivers have to find the way ahead themselves. That this is difficult is shown by the fact that the ability to maintain lateral position on the road deteriorates in fog (Tenkink 1988).

There are a number of ways to help the driver in fog but none of them is a complete solution. One approach used in locations where fog is common is to inset retroreflective road studs into the road at regular intervals, as lane and edge markers. When illuminated by the forward lighting of a vehicle, the road studs have a high luminance and hence an increased contrast against the road. This helps the driver to keep the vehicle in the lane and provides visual guidance of the road ahead.

Turning now to the driver, the first choice faced is how to use the vehicle headlamps. It is a common experience that it is often better to use low-beam headlamps than high-beam headlamps in fog. The reason is that high-beam headlamps put more light further down the road and, as a result, project light a greater distance through the fog, thereby producing a higher luminance veil. This means that the luminance contrasts of vehicles and objects on the road are lower when high-beam headlamps are used. Whether or not this is offset by the greater contrast sensitivity induced by the higher adaptation luminance will depend on the fog density. The denser the fog, the more likely the use of high-beam headlamps will be detrimental to visibility.

One way that is claimed to help in this situation is to fit the vehicle with supplementary forward lighting intended for use in fog as specified by the ECE or the SAE. Fog lamps are rarely a legal requirement but are widely sold as an optional extra, particularly on more up-market vehicles. Where fog lamps are fitted they are usually mounted low on the vehicle, below the conventional forward lighting, and have a luminous intensity distribution that is both wide and flat, the effect being to put more light on the road immediately in front and to the sides of the vehicle and very little above the horizontal plane through the fog lamps. The low mounting position is advantageous because the fog lamps are closer to the road and further from the driver's line of sight. Also, fog is usually thinner close to the road. Minimizing the light distribution above the horizontal is desirable because light directed upwards would intersect the driver's line of sight close to the vehicle and hence increase the veiling luminance seen by the driver. There has been a long controversy associated with the best light spectrum to be used for forward lighting in fog (Schreuder 1976). For many years, France required the use of yellow forward lighting on the basis that yellow light provided better visibility in fog than white light. Whether this is true depends on the size of the water droplets forming the fog. If the water droplet size is less than the wavelength of the incident light, then scattering can occur that is wavelength dependent. However, many fogs have droplet sizes that are much larger in diameter than the wavelengths that stimulate the human visual system (Middleton 1952). In these conditions, different degrees of scattering for different visible wavelengths do not occur. It is clear that any advantage of yellow forward lighting in fog has to rely on some factor other than decreased scattering, possibly the removal of wavelengths that provide greater stimulation to the rod photoreceptors.

The visual effect of fog lamps is to provide greater visibility of road edges and nearby lane markings, thereby making lane keeping easier. Fog lamps do little to enhance the visibility of vehicles and objects further along the road. Figure 11.4 shows the calculated luminance contrasts of road markings for fog lamps alone, low-beam headlamps alone, and both fog lamps and low-beam headlamps together 10, 20, and 40 m ahead of the vehicle, in a clear atmosphere, and in light, medium, and heavy fog (Folks and Kreysar 2000). Figure 11.4 clearly demonstrates the impact of fog density on visibility by showing a marked reduction in luminance contrast with increasing fog density at all three distances. As for the best form of lighting to use, in a clear atmosphere adding fog lamps to low-beam headlamps increases the luminance contrast at all three distances, although the increase diminishes with distance. In light, medium, and heavy fogs, fog lamps alone produce the highest

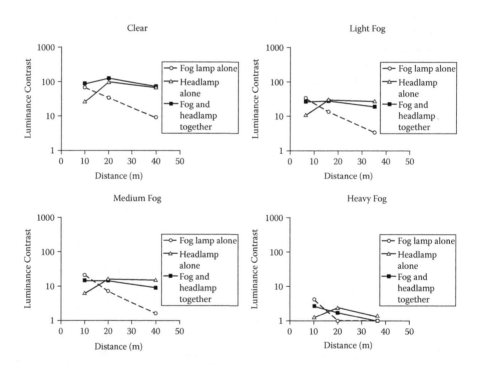

FIGURE 11.4   Calculated luminance contrasts for road markings of reflectance 0.5 in four different atmospheres for fog lamps alone, low-beam headlamps alone, and both fog lamps and low-beam headlamps together, the markings being placed at distances of 10, 20, and 40 m in front of the vehicle. The calculations are made for fog lamps mounted at 0.4 m above the road, eye height at 1.42 m above the road, and a background luminance of 0.017 cd/m^2 in a clear atmosphere. The extinction coefficients for the four atmospheres are clear = 0.00015 m^{-1}, light fog = 0.003 m^{-1}, medium fog = 0.006 m^{-1}, and heavy fog = 0.03 m^{-1} (after Folks and Kreysar 2000).

luminance contrast at 10 m but low-beam headlamps alone ensure higher luminance contrasts at 40 m.

The failure of fog lamps to light far down the road may explain why the most common form of accident in fog is a rear-end collision between vehicles, often several vehicles (Codling 1971; Koth et al. 1978). It may be that fog lamps contribute to such accidents by making lane keeping easier without contributing to visibility at a distance, the former giving drivers an unjustified confidence in their ability to see. If so, this is an example of the selective visual degradation hypothesis in action (see Section 6.4; Leibowitz and Owens 1977). What is required to increase visibility at a distance and hence diminish the risk of rear-end collisions is a high-intensity lamp with a narrow light distribution aimed down the road and mounted on the vehicle as far away from the driver's line of sight as possible (Bullough and Rea 2001a).

Given that neither fog lamps nor headlamps as presently designed are effective in avoiding rear-end collisions in fog, another approach is to fit vehicles with high-intensity rear lamps for use in fog (see Section 7.12). These are effective because they increase the distance from which the vehicle can be detected. However, if only one rear fog lamp is provided, the overestimation of distance common in fog is increased (Cavallo et al. 2001). Two well-separated rear fog lamps enable better estimations of distance to be made. Also, care is required to limit the use of rear fog lamps to foggy conditions because in a clear atmosphere their luminous intensity is high enough that they become a source of discomfort and sometimes a source of disability glare to following drivers. This is also a concern for forward lighting fog lamps. Sivak et al. (1996) carried out observations of the use of fog lamps in the state of Michigan. Table 11.3 shows the percentage of vehicles with fog lamps installed actually using them, by day and night, in different weather conditions. The pattern of use during the day is what would be expected, very few drivers using fog lamps in clear weather, but increasing percentages using fog lamps as the weather deteriorates. But at night, there is little difference in the use of fog lamps according to weather conditions, which suggests that, at night, fog lamps are primarily used to supplement conventional forward lighting. Unfortunately, the use of fog lamps with low-beam headlamps at night will certainly increase the level of glare experienced by opposing drivers, particularly when the road surface is wet, because then specular reflection of light from the road immediately in front of the vehicle makes a major contribution to glare and that is where fog lamps deliver their light.

TABLE 11.3

Percentage of Vehicles Fitted with Fog Lamps That Had Them Lit in Different Light and Weather Conditions

Light Condition	Clear Atmosphere	Moderate Rain	Light to Moderate Fog	Moderate to Heavy Fog
Day	2.8	10.4	30.8	50.0
Night	64.5	63.0	—	60.6

No data were available for light to moderate fog at night (from Sivak et al. 1996).

Vehicle lighting alone clearly has its limitations for ensuring traffic safety in fog (Flannagan 2001) so now it is necessary to consider what road lighting has to offer. Road lighting is sometimes installed in areas prone to fog on otherwise unit roads. How effective such road lighting is in enhancing visibility depends on the density of the fog. This is because adding more light will increase both the luminance of the road and objects on the road and the veiling luminance caused by scatter. The balance between these two effects depends on the density of the fog, the thicker the fog, the less the benefit of road lighting.

Of course, this assumes the road lighting is of the conventional type, mounted 8 m or higher above the road so that light has to pass through a lot of fog before it reaches the road. Girasole et al. (1998) carried out a computer simulation, based on a mathematical model of multiple scattering, of the visibility of the lower and upper parts of the back of a truck and a broken tyre on the road, under two different forms of road lighting, in fogs of different density. The two lighting systems were conventional road lighting mounted 9 m above the road and a more closely spaced, low-mounted, pro-beam system (0.9 m above the road) placed on the central reservation, the light being aimed 45 degrees across the road. Figure 11.5 shows the estimated veiling luminance from low-beam headlamps, conventional road lighting, and the low-mounted road lighting, separately, plotted against fog concentration. Figure 11.6 shows the visibility distances for the lower and upper parts of the truck and the tyre under low-beam headlamps and either the conventional road lighting or the low-mounted system, plotted against fog concentration. Visibility distance is defined as the distance at which the luminance difference between the target and its immediate background reaches threshold. The low-mounted road lighting system produces

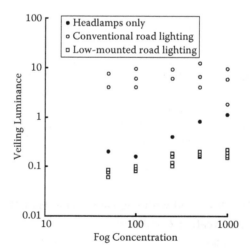

FIGURE 11.5 Veiling luminance (cd/m²) plotted against fog concentration (million droplets/m³) for low-beam headlamps, conventional road lighting, and low-mounted road lighting, separately. There are multiple data points for the two road lighting installations because the veiling luminance depends on the position of the driver relative to the road lighting luminaires (after Girasole et al. 1998).

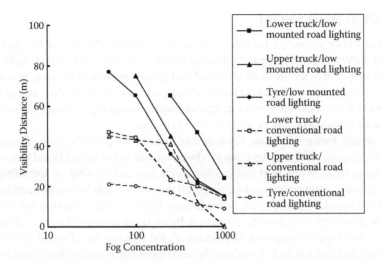

FIGURE 11.6 Visibility distance (m) for the lower and upper parts of a truck and a broken tyre seen under conventional road lighting and under low-mounted road lighting, both with low-beam headlamps, plotted against fog concentration (million droplets/m^3) (after Girasole et al. 1998).

much less veiling luminance than the conventional road lighting and hence increases visibility distances dramatically for low fog concentrations, but this benefit is much reduced at high fog concentrations because of the overwhelming effect of the low-beam headlamps on the veiling luminance. These results, and it must be remembered that they are based on a computer simulation rather than field measurement, suggest that new types of road lighting could be designed for use in areas where fog is common and would provide much more effective road lighting than is currently available. In fact, road lighting systems designed for mounting less than 1 m above road level are commercially available for use on bridges and other locations where conventional column-mounted road lighting is inappropriate. It would be interesting to see how well such a system would work in areas prone to fog and in high traffic densities.

Finally, it is important to appreciate that there are other ways to improve the flow of visual information to the driver than simply improving the lighting. Nilsson and Alm (1996) examined the impact of a vision enhancement system on a driver's ability to drive safely in fog, using a driving simulator. The vision enhancement system produced a clear image of the road ahead as a small window in the scene. With the vision enhancement system, drivers choose to drive in fog at a speed only slightly less than the speed they used in a clear atmosphere and their reaction time and the distance they moved after an unexpected hazard appeared were similar to what they were without fog. However, the lateral position of the vehicle varied more with the vision enhancement system. Technological advances in sensors and computing power are making such a vision enhancement system a real possibility for everyday use in vehicles.

11.6 DUST AND SMOKE

Dust and smoke consist of particles temporarily airborne, the size of the particles varying widely depending on the source of the dust or smoke. Photons of light incident on these particles are both scattered and absorbed. Which of these processes dominates depends on the reflectance of the particles. In black dust and smoke, absorption dominates, while in white dust or smoke, scattering dominates. Most dust and smoke is somewhere between these two extremes.

For vehicle forward lighting, black dust and smoke simply reduces the amount of light reaching the road, an effect that can be countered by using additional lamps and hence more light. White dust and smoke affects forward lighting by spreading the light distribution and creating a luminous veil that reduces the luminance contrast of everything ahead. Fog lamps will help with lane keeping in this situation, but for visibility at a distance a high-intensity narrow-beam lamp mounted far away from the driver's line of sight is required. For vehicle signal lighting, the effect of both black and white dust and smoke is to reduce the maximum luminance of the signal and to spread its apparent area, the amount of spread increasing as the amount of scatter increases. These detrimental effects can be overcome by increasing the luminous intensity of the signal, as in rear fog lamps.

If thick dust is a regular occurrence and of a specific colour, then another approach for increasing visibility is to alter the spectrum of the vehicle forward lighting. For example, a driver following a road train in the outback of Australia will often be enveloped in a cloud of red dust. Lorry drivers in the outback who regularly experience such conditions use green filters on their vehicle's headlamps, the outcome being to reduce the amount of scattered light (Wordenweber et al. 2007).

11.7 SUMMARY

The presence of rain, snow, fog, dust, or smoke in the atmosphere makes the driver's task more difficult. Rain has a direct effect on visibility by absorbing and scattering light and an indirect effect by changing the reflection properties of the road surface. Snow scatters incident light and in so doing creates a lot of visual noise. Fog, dust, and smoke have their effects by absorbing and scattering light in the atmosphere, thereby reducing the visibility of everything ahead of the driver.

Although the presence of rain, snow, fog, dust, or smoke undoubtedly makes the driver's task more difficult, it is not clear if that difficulty is enough to lead to an increase in the frequency of accidents. A study of fatal accidents in the United States shows that almost 90 percent of accidents involving death occur when there is no adverse weather. This is partly a matter of exposure and partly a matter of driver experience and behaviour. Really adverse weather occurs infrequently and, when it does, drivers tend to reduce speed and keep a greater distance from nearby vehicles, behaviours that make fatalities less likely.

Rain consists of falling water droplets. Unless the rain is heavy or the windscreen wipers are ineffective, the main impact of rain is on the reflection characteristics of the road surface. When rain covers a road surface, the visual effect is to make the road surface more specular. For vehicle forward lighting, this change in reflection properties reduces the average road surface luminance, increases the disability and discomfort glare from approaching vehicles, and reduces the visibility of the road markings. For road lighting, the effect of a wet road is to make the road surface luminance less uniform.

The problems posed by wet roads can be reduced either by changing the road surface or by modifying the vehicle's forward lighting or the road lighting. The most successful approach to the problem of wet roads lies in the choice of road materials. A road with a coarse texture or a more permeable structure will be less sensitive to rain. For vehicle forward lighting, the proposed adaptive forward lighting system has as one of its options a wet road beam. The wet road beam involves a reduction in the illumination just in front of the vehicle and increased light to the sides of the vehicle, the effect being to diminish discomfort and disability glare to other drivers. For road lighting, the increase in luminance uniformity can be minimized by changing the luminous intensity distribution of the luminaires, the spacing between luminaires, and their position relative to the road surface.

Snow consists of large, high-reflectance particles falling through the atmosphere. If the density of snowflakes is very high, the result of scattering is a uniform veil of high luminance that restricts visibility in all directions to very short distances. At lower densities, each snowflake is seen separately moving against a lower luminance background. How distracting this is depends on the luminance contrast of the snowflakes, the higher the luminance contrast, the more distracting the snowflakes. Maximum luminance contrast will occur when the vehicle forward lighting is the only source of light. During daylight or at night when road lighting is present, the background against which the snowflakes are seen is higher in luminance and thus the snowflakes are at a lower luminance contrast.

There are four approaches that can be used to reduce the distraction caused by snowflakes. The one most easily available to the driver is to use the forward lighting on low beam rather than on high beam. Another is to place the forward lighting as far away from the driver's usual line of sight as possible. Yet another is to narrow the luminous intensity distribution of the headlamp. The fourth is to modify the spectrum emitted by the forward lighting so as to reduce the stimulus to the rod photoreceptors. Of these, the most potent are to displace the light source as far away from the driver as possible and to narrow the luminous intensity distribution.

Fog consists of small water droplets suspended in the atmosphere. Photons of light incident on a droplet are absorbed and scattered. If the density of the particles is high, the outcome of these processes is a uniform veil of luminance covering the driver's visual field. The effect of this uniform luminance veil is to reduce the luminance contrasts of all the things in front of the driver and hence to reduce their visibility.

There are a number of ways to help the driver in fog but none of them is a complete solution. One approach is to inset retroreflective road studs into the road at regular intervals, as lane and edge markers. Another is for the driver to use low-beam rather than high-beam headlamps. Yet another is to use dedicated fog lamps. Where fog lamps are fitted they are usually mounted low on the vehicle and have a luminous intensity distribution that is both wide and flat. The visual effect of fog lamps is to provide greater visibility of road edges and nearby lane markings but they do little to enhance the visibility of vehicles and objects further down the road. This may explain why the most common form of accident in fog is a rear-end collision. One method to reduce such accidents is to fit vehicles with high-intensity rear lamps for use in fog. These are effective because they increase the distance from which the vehicle can be detected. However, care is required to limit their use to foggy conditions because in a clear atmosphere their luminous intensity is high enough that they become a source of discomfort and sometimes a source of disability glare to following drivers. As for road lighting, how effective this is in enhancing visibility in fog depends on the density of the fog. This is because adding more light will increase both the luminance of the road and objects on the road and the luminance of the luminous veil caused by scatter. The balance between these two effects depends on the density of the fog.

Dust and smoke consist of small particles temporarily airborne. Photons of light incident on these particles are both scattered and absorbed. Which of these processes dominates depends on the reflectance of the particles. In black dust and smoke, absorption dominates, while in white dust or smoke, scattering dominates. For vehicle forward lighting, black dust and smoke simply reduce the amount of light reaching the road, an effect that can be countered by using additional lamps. White dust and smoke affect forward lighting by spreading the light distribution and creating a luminous veil that reduces the luminance contrast of everything ahead. For vehicle signal lighting, the effect of both black and white dust and smoke is to reduce the maximum luminance of the signal and to spread its apparent area, the amount of spread increasing as the amount of scatter increases. These detrimental effects can be overcome by increasing the luminous intensity of the signal, as in rear fog lamps.

12 Human Factors

12.1 INTRODUCTION

Driving is a human activity with all the variability that such a statement implies. While driving, humans have to extract information about the world around them, decide on an appropriate course of action, and then implement that course of action. At night or in adverse weather, it may be difficult to extract the relevant information. By day or night, decisions have to be taken quickly. Sometimes, multiple events requiring attention occur simultaneously so there is competition for cognitive resources. Some people are better equipped to deal with such demands than others. Among the factors that determine how well drivers can respond to such demands are the amount of practice they have had and the capabilities of their visual and cognitive systems. How such individual differences affect driving and how lighting might be used to alleviate any consequent problems are the subjects of this chapter.

12.2 AGE AND ACCIDENTS

One factor that is connected to both driving experience and the deterioration in visual and cognitive abilities is age. Driving on public roads is not allowed until the late teens, the exact age at which driving is allowed varying from country to country. As for deterioration in visual and cognitive abilities, this is continuous from the late teens but accelerates dramatically about the sixth decade of life (Kosnik et al. 1988; Werner et al. 1990). Figure 12.1 shows the number of fatal accidents per hundred million miles traveled in the United States in 1990 for drivers in different five-year age groups, by day and night. There are two features of Figure 12.1 that deserve attention. The first is that the number of fatal accidents per hundred million miles shows an increase at both extremes of age, by both day and night. The second is that for young drivers the fatal accident rate is much greater at night than during the day but for old drivers there is little difference between day and night. Figure 12.2 shows the number of personal injury accidents per hundred million miles traveled in the United States in 1990 for drivers of different ages, by day and night. Again, both extremes of the age range show an increase in accident rate and the accident rate for young drivers shows an increase from day to night, although not as much as for fatal accidents.

Some insight into the reasons for this pattern of accident rates can be gained from the type of accidents in which people of different ages are involved. McGwin and Brown (1999) examined all the road accidents reported to the police in the state of Alabama in 1996 and looked at how responsibility for different types of accident varied with age. Table 12.1 shows the percentage distribution of the primary

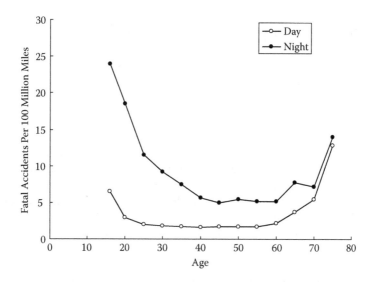

FIGURE 12.1   Fatal accidents per hundred million miles travelled for different five-year age groups by day and night. Day is taken from 0600 to 2100 hours. Night is taken from 2100 to 0600 hours. The databases used are the Fatality Analysis Reporting System for 1990 and the 1990 Nationwide Personal Transportation Survey. The age data points are plotted at the start of each five-year age group. The age group starting at 75 years includes all drivers older than 75 years (after Massie and Campbell 1993).

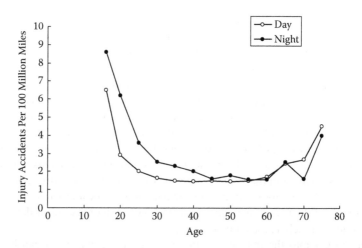

FIGURE 12.2   Personal injury accidents per hundred million miles travelled for different five-year age groups by day and night. Day is taken from 0600 to 2100 hours. Night is taken from 2100 to 0600 hours. The databases used are the 1990 General Estimate System and the 1990 Nationwide Personal Transportation Survey. The age data points are plotted at the start of each five-year age group. The age group starting at 75 years includes all drivers older than 75 years (after Massie and Campbell 1993).

Table 12.1

The Percentage Distribution of Primary Contributory Factors to Road Accidents for which Drivers in Three Age Groups Were Responsible

Primary Contributory Factor	Young	Middle-Aged	Older
Driver not in control	16.4	12.7	8.1
Misjudged stopping distance	12.0	11.2	8.4
Following too close	9.0	8.7	5.5
Avoiding object/person/vehicle	4.6	4.7	2.2
Improper driving environment	4.3	3.6	1.6
Over speed limit	3.4	1.4	0.3
Improper passing	1.5	1.4	1.1
Failed to yield right of way	16.0	15.0	28.9
Unseen object/person/vehicle	8.1	11.0	13.4
Failed to heed sign or signal	5.1	4.9	7.1
Improper lane change	2.3	2.9	4.6
Improper U-turn	1.6	1.9	3.0
Improper backing	1.3	2.0	2.6
Driving under the influence	2.2	4.9	1.8
Driver condition	1.8	1.6	1.7
Other	10.4	12.1	9.7

The database from which these data were drawn was all road accidents reported to police in the state of Alabama in 1996. The three age groups are young = 16 to 34 years, middle-aged = 35 to 54 years, and older = 55 + years (after McGwin and Brown 1999).

contributory factors to accidents for which young, middle-aged, and older drivers were responsible. All age groups are involved in accidents of all types but there is an undeniable shift in emphasis in the type of accidents for which the young and old are responsible. Specifically, young drivers tend to be responsible for more accidents where lack of control, misjudging stopping distance, following too close, avoiding an object/person/vehicle, speeding, or improper passing are the primary contributory factors. Older drivers tended to be responsible for more accidents where failure to yield right of way, failure to see an object/person/vehicle, failure to heed a sign or signal, an improper lane change, an improper U-turn, or improper backing are the primary contributory factors. What this suggests is that young drivers have no sense and old drivers have no senses. As will become evident later, this is too crude a picture, but it does contain a kernel of truth. The type of accidents in which young drivers are involved tend to be caused by a failure of judgement or an excess of ambition, while those involving older people tend to be linked to failures of perception and cognitive processing under time pressure. Why this should be and how lighting might contribute to a reduction in such accidents will now be considered.

12.3 YOUNG DRIVERS

Figures 12.1 and 12.2 suggest that it is drivers less than 25 years of age who show a dramatic increase in accident rate, particularly at night. Therefore, less than 25 years of age will be used to define what constitutes a young driver. Before considering how lighting might be used to reduce the higher accident rates of young drivers, it is necessary to consider why young drivers are involved in so many accidents. The first thing to say is that it is not likely to be due to visual or cognitive limitations. Young drivers who are in their late teens or twenties will usually have a visual system at the peak of its powers and a cognitive system capable of rapid responses to simple and complex stimuli. One possible reason why young drivers have a higher accident rate is that they may not know how to use these capabilities to allocate attention because they are inexperienced. In most countries, passing a driving test requires the applicant to show knowledge of the rules of the road, the meaning of signs, and the expected behaviour in specified situations, as well as a practical demonstration of vehicle handling in a limited number of conditions. What is only sketchily examined in the driving test is the ability to rapidly analyze the situation on the road ahead and to allocate attention appropriately, although the video hazard perception tests now used in the UK are a step in the right direction. More usually, the ability to allocate attention effectively is gained by experience of driving over the following years. This may be why there is a peak in accidents within one year of drivers passing their driving test (Maycock et al. 1991).

The effect of experience on the allocation of attention should be evident in the patterns of eye movements made by experienced and inexperienced drivers. Falkmer and Gregersen (2005) report an analysis of the fixations made by experienced and inexperienced drivers while travelling along an urban road in which there were vehicles parked on both sides, a pedestrian crossing, and a speed bump, as well as on a rural dual carriageway crossing an intersection. They showed that inexperienced drivers tended to fixate more frequently on in-car objects, keep their fixations within a narrower horizontal band, and fixate slightly more frequently on relevant traffic cues and potential hazards than experienced drivers. Such results hardly suggest that inexperienced drivers are at a disadvantage as regards the use of their visual system. However, Falkmer and Gregersen (2005) also found that experienced drivers altered their pattern of fixations between the urban road and the rural road while the inexperienced drivers did not. Such flexibility may be of advantage given the diversity of conditions that a driver is likely to experience.

Crundall and Underwood (1998) also measured the eye movements of drivers who had recently passed the driving test and young but experienced drivers when driving on a straight rural road, a busy suburban road, and a busy dual carriageway. Again, there were clear differences between the two groups with the older, more experienced drivers showing a wider spread of fixations. This was particularly evident on the dual carriageway where the experienced drivers fixated traffic in adjacent lanes while the inexperienced tended to concentrate on the lane ahead. There was also a difference in fixation duration. On the rural road where there were few hazards, the experienced drivers made longer fixations than the inexperienced drivers, but on the dual carriageway, where there were many hazards, the experienced drivers made a larger number of fixations of shorter duration than the inexperienced drivers. A proposed explanation for such differences is that experienced drivers develop a mental model of the hazards that are likely in different situations (Liu 1998). This mental model is the endogenous source that generally drives fixation patterns.

So what is the role of lighting, if any, in reducing the accident rate of inexperienced drivers? One answer is to use lighting to speed up the development of a comprehensive mental model. Road lighting could do this by emphasizing locations where hazards are common, e.g., by special lighting for pedestrian crossings and lighting intersections on otherwise unlit roads (see Sections 10.3 and 10.6). Another would be to reduce the mental load on the driver by giving more emphasis to visual guidance in road lighting design. As for vehicle lighting, there is little that can be done that is not also applicable to experienced drivers. Changing both forward lighting and signal lighting to enhance visibility and conspicuity and to more clearly inform other drivers about intentions would be of benefit to all drivers.

While a lack of experience is a factor in accident rates for some young drivers, not all young drivers are inexperienced. Anyone who gains a driving license in their late teens will have several years driving experience by their mid-twenties but would still be classed as young in accident statistics. The reasons why experienced young drivers have high accident rates are evident from Table 12.1. Basically, the types of accidents where young drivers predominate suggest that young drivers often drive with reduced margins of error, relying on their fast reactions to survive. Further, young drivers are more likely to be out late at night, are more likely to be fatigued, and more likely to be suffering from the effects of alcohol or drugs, all of which will tend to slow responses and confuse analysis. There is little that lighting can do to overcome such behavioural excesses other than to provide more time for decision-making by lighting the road further ahead.

12.4 OLD DRIVERS

Figures 12.1 and 12.2 suggest that an old driver is one greater than 70 years of age. This is the definition used here. Old drivers are of concern because ageing affects all the functions required for driving, visual, cognitive, and motor. Because the subject of this book is lighting, the changes in visual capabilities will be emphasized here but the changes in the other functions should not be forgotten. How important the limitations in the visual, cognitive, and motor functions are will depend on the circumstances. When driving at night, deficiencies of the visual system are likely to be the most important. When driving during the day, the cognitive limitations are likely to be dominant.

12.4.1 OPTICAL CHANGES IN THE VISUAL SYSTEM WITH AGE

The human visual system can be considered as an image-processing system. Like all such systems, the visual system is most effective when it is operating at an appropriate sensitivity with a clear retinal image to process. The optical factors determining the amount of light reaching the retina and hence the sensitivity of the visual system are the pupil size and the spectral absorption of the components of the eye. The area of the pupil varies as the amount of light available changes, the pupil opening to admit more light when there is little and closing when there is plenty. The ratio of maximum to minimum pupil area decreases with age, the maximum decreasing much more than the minimum. This means the old are much less able to compensate for low light levels by opening their pupils than are the young. Another change relevant to the operating state of the visual system is the amount of light absorbed in passage through the eye. Such absorption increases with age, the majority taking place on passage through the lens (Murata 1987), but this absorption is not equal at all wavelengths. Rather, the absorption at short wavelengths increases dramatically with age (Weale 1992). This goes some way to explain the diminished colour discrimination capabilities of old people. Together, the reduction in pupil size and the increased absorption of light reduces the retinal illumination of old people, so much so that it has been estimated that a 70-year-old receives only one-third of the light at the retina that a 30-year-old will receive when both receive the same illuminance at the cornea (Adrian 1995).

The optical factors determining the clarity of the retinal image are the ability to focus the image on the retina and the amounts of scattered light and straylight in the eye. In simple optical terms, the eye has a fixed image distance and a variable object distance. To bring objects at different distances to focus on the retina, the optical power of the eye has to change. The range of object distances that can be brought to focus on the retina decreases with age because of the increasing rigidity of the lens. After about 60 years of age, the eye is virtually a fixed focus optical system. Spectacles or contact lenses are used to modify the optical power of the eye, the prescription of the spectacles or contact lens changing over the years as the lens becomes increasingly rigid. As for light scatter, the amount of light scatter increases with age, due mainly to changes in the lens (Wolf and Gardiner 1965). Scattered light degrades the retinal image by reducing the difference in luminance either side of an edge, thereby reducing the magnitude of its higher spatial frequencies. Scattered light also degrades the retinal image in terms of colour by adding wavelengths from one area onto another, thereby reducing the colour difference at the edge.

Scatter can be quantified by a point spread function, which typically shows that the amount of scattered light decreases with increasing deviation from the beam of light being scattered (Vos and Boogaard 1963). Straylight within the eye is caused by light back-reflected from the retina and pigment epithelium, by transmission of light through the iris and the eye-wall, and by lens fluorescence (Boynton and Clarke 1964; van den Berg et al. 1991; van den Berg 1993). Virtually all these characteristics worsen with age (Werner et al. 1990; Weale 1992). Straylight matters because it falls uniformly across the retinal image, thereby reducing the luminance contrast of all edges and the saturation of all colours in the image.

These changes with age are the best that can be expected, but with increasing age comes a greater likelihood of pathological changes leading to visual impairment (Silverstone et al. 2000). The four most common causes of visual impairment in developed countries are cataract, macular degeneration, glaucoma, and diabetic retinopathy. Cataract is an opacity developing in the lens. The effect of cataract is to absorb and scatter more light as the light passes through the lens. This increased absorption results in reduced visual acuity and reduced contrast sensitivity over the entire visual field, as well as greater sensitivity to glare. Macular degeneration occurs when the macula, which covers the fovea, becomes opaque. An opacity immediately in front of the fovea implies a serious reduction in visual acuity and in contrast sensitivity at high spatial frequencies. Typically, these changes make seeing detail difficult if not impossible. However, peripheral vision is unaffected. Glaucoma is shown by a progressive narrowing of the visual field. Glaucoma is due to an increase in intraocular pressure that damages the blood vessels supplying the retina. Glaucoma will continue until complete blindness occurs unless the intraocular pressure is reduced. Diabetic retinopathy is a consequence of chronic diabetes mellitus and effectively destroys parts of the retina through the changes it produces in the vascular system that supplies the retina. The effect these changes have on visual capabilities depends on where on the retina the damage occurs and the rate at which it progresses.

12.4.2 Neural Changes of the Visual System with Age

The optical changes that occur with age, both normal and pathological, affect the retinal image, but for the visual system to be effective, the retinal image has to be processed by the retina and the visual cortex. There is no reason to suppose that ageing is limited to the optical elements of the visual system alone. Indeed, morphological changes have been reported in rod and cone photoreceptors in older people (Marshall et al. 1979): the density of rods and cone photoreceptors have been shown to decrease in extreme old age (Feeny-Burns et al. 1990) and the number of ganglion cells in the retina and the number of neurons in the visual cortex both decrease with increasing age (Devaney and Johnson 1980; Balazsi et al. 1984). Weale (1992) provides a useful review of these neural changes and their possible causes. The fact that the neural elements of the visual system also show changes with age is important because it implies that the compensation for visual system ageing that can be provided by lighting is inevitably limited.

12.4.3 Other Changes with Age

In addition to the changes in the visual system, ageing introduces more general limitations with respect to decision-making and motor functions. For example, reaction times to all stimuli increase with age and the greater the number of alternatives, the greater the increase in reaction time with age (Sivak et al. 1995). This implies that old drivers need more time to make a safe decision. In addition, old people are more easily distracted and have difficulties in directing attention between several different stimuli (Fildes et al. 1994; Sivak et al. 1995). As for motor functions, muscular strength and joint flexibility decrease rapidly in old age and manual dexterity is diminished (Sivak et al. 1995).

12.4.4 Changes in Visual Capabilities

As might be expected, the changes in the optical and neural characteristics of the visual system that occur with increasing age have an impact on what the visual system is capable of doing. A consideration of the effects of age on threshold performance reveals that the effect of increasing age is almost always negative, in the sense that the visual system becomes less discriminating and more sensitive to adverse conditions. Specifically, old people tend to show reduced visual field size, reduced visual acuity, reduced contrast sensitivity, increased sensitivity to glare, slower recovery from exposure to glare, and degraded colour discrimination. Figure 12.3 shows the functional field of view of one eye for a 24-year-old and a 75-year-old (Williams 1983). The boundaries of the fields are formed by contours of equal detection performance. The decrease in functional field of view with age is clear, a decrease that is greater when there is limited time for detection (Ball et al. 1993). Figure 12.4 shows the change in visual acuity plotted against age (Adrian 1995). Increasing age leads to worse visual acuity.

Figure 12.5 shows contrast sensitivity functions for four people of four different ages (McGrath and Morrison 1981). Examination shows that the effect of age is to

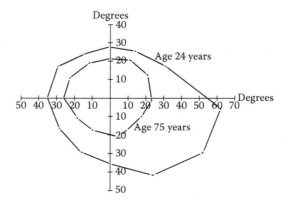

FIGURE 12.3 The functional visual fields for one eye of a 24-year-old and a 75-year-old. The fields are defined by contours of equal detection performance (after Williams 1983).

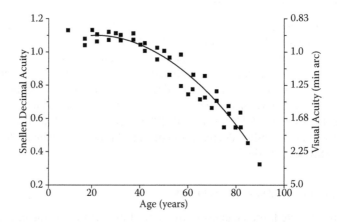

FIGURE 12.4 Visual acuity, expressed as Snellen decimal acuity (min arc^{-1}) and as angle subtended (min arc), as a function of age (years) (after Adrian 1995).

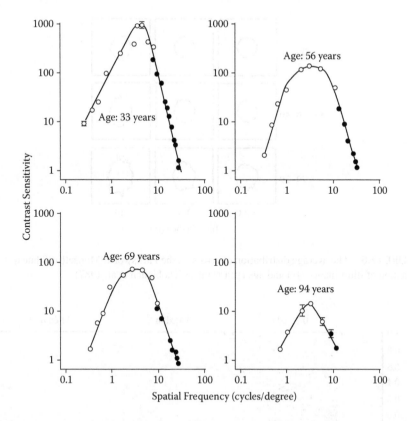

FIGURE 12.5   The contrast sensitivity functions of four observers of different ages. Open circles are for near viewing. Filled circles are for far viewing (after McGrath and Morrison 1981).

decrease the maximum contrast sensitivity and decrease the range of spatial frequencies over which resolution can occur. More extensive studies show the same trends (Owsley et al. 1983). The increased scattering of light has an effect on the magnitude of disability glare. The formula for disability glare from headlamps (see Section 6.6.2) has a term for age that results in an increase in equivalent veiling luminance with increasing age for the same stimulus. This increased equivalent veiling luminance will lead to lower luminance contrasts for old drivers. In addition, the time taken to return contrast sensitivity to what it was before exposure to glare increases with age (Schieber 1994).

Figure 12.6 shows the average distribution of errors made on the Farnsworth-Munsell 100 hue test as a function of illuminance and age (Knoblauch et al. 1987). Zero errors is indicated by a smooth circle. As the number of errors increases, the circle becomes larger and more ragged. It is clear from Figure 12.6 that old people tend to make more errors in hue discrimination, particularly at low illuminances and for hues determined by the short wavelength content of the spectrum.

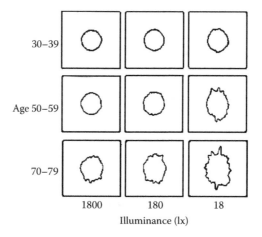

FIGURE 12.6   The average distribution of errors on the Farnsworth-Munsell 100 hue test as a function of illuminance (lx) and age (years) (after Knoblauch et al. 1987).

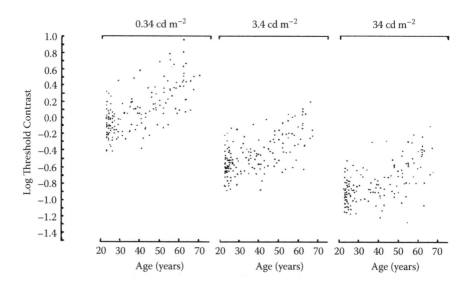

FIGURE 12.7   Log threshold luminance contrasts for individuals of different ages (years) at three different background luminances (cd/m²) (after Blackwell and Blackwell 1971).

While the trend of deterioration in visual function with age is evident in all these aspects of threshold performance, it is important to appreciate that there are wide individual differences in visual function, sometimes wide enough to overcome the average effects of age. This is evident in Figure 12.7, which shows the threshold luminance contrast measured for individuals of different ages at three different adaptation luminances (Blackwell and Blackwell 1971). While the trend of increasing threshold contrast with increasing age is obvious, it is also clear that the individual differences are large enough for some people in their sixties to have lower threshold luminance contrasts than some people in their twenties. It should not be assumed that all old drivers are visually challenged.

12.4.5 CONSEQUENCES FOR DRIVING

The changes that occur in visual capabilities with age have consequences for driving performance. Wood (2002) measured the driving performance of groups of drivers of different ages and with different degrees of visual impairment. Specifically, 139 drivers in good general health and holding a current Queensland, Australia, driving license were divided into five groups, labeled young, middle-aged, and old, this last group being subdivided into those with normal vision, mild visual impairment, and moderate or severe visual impairment. The mean ages of the three age groups were 27, 52, and 70 years for the young, middle-aged, and old groups. For the old group with normal, mildly impaired, and moderately impaired vision, the mean ages were, 69, 71, and 71 years respectively. Normal vision was defined as having a static visual acuity of 6/7.5 or better (see Section 3.4.1). Mildly impaired vision was defined as having slight clouding of the lens, early glaucoma, or early macular degeneration in one or both eyes. Moderate to severe visual impairment was defined as having cataract in both eyes or advanced glaucoma or macular degeneration in one or both eyes. All these drivers drove round a 5.1-km (3.2-mile), closed-road circuit, i.e., one closed to the public and hence free of other traffic. While driving round the circuit, the drivers were asked to report any road signs they saw; to report any large low-contrast hazards they saw in the road and to avoid them by steering around them; to judge whether or not the gap between a pair of cones was wide enough to get through and, if it was, to drive through it and, if not, to drive around it; and to respond to the onset of one of five LEDs mounted in the car in front of the driver. Having driven round the circuit, the drivers' ability to handle the vehicle was tested by having them maneuvre in and out of a row of low-contrast cones and reverse into a parking space. Table 12.2 gives the measures of driving performance that showed statistically significant differences between the groups. As would be expected, both age and visual impairment tend to produce worse driving performance but the balance between these two factors changes with the nature of the task. For tasks that involve switching attention, such as detecting the onset of the LED stimulus, age is the dominant factor. For tasks where visibility is limited, such as the detecting and avoiding low-contrast road hazards, visual impairment is more important. For other tasks, such as seeing road signs and reversing, both age and visual impairment are influential.

Wood (2002) also constructed a composite driving performance score for each individual relative to the performance of all the drivers. The performances included in the composite score were road sign recognition, cone gap perception, cone gap maneuvring, number of LEDs seen, circuit time, and road hazard detection and avoidance. For each task, a z score was calculated for each individual. The z score for the individual is the deviation from the mean of the distribution of the task performance measure for all drivers divided by the associated standard deviation. Then, the mean of the z scores for each individual over all the components of the composite

TABLE 12.2

Mean Performance Measures for Young, Middle-Aged, and Old Drivers

Driving Performance Measure	Maximum Possible	Young Drivers	Middle-Aged Drivers	Old Drivers with Normal Vision	Old drivers with Mild Low Vision	Old Drivers with Moderate or Severe Low Vision
Road signs seen	65	51.3	50.0	46.4	46.9	40.7
Road hazards seen	9	8.7	8.7	8.7	8.4	8.0
Road hazards hit	9	0.3	0.3	0.5	0.6	1.8
Number of LEDs seen	15	11.5	10.2	7.3	8.2	7.8
Correct gap maneuvres	9	8.0	7.7	7.3	7.2	6.5
Cones hit while maneuvering	9	0.5	0.2	0.3	0.7	0.4
Circuit time (s)	—	428	434	468	482	478
Maneuvre time (s)	—	38.1	38.7	41.5	49.1	48.8
Reversing time (s)	—	30.9	39.0	48.5	62.6	62.8

(After Wood 2002).

score was calculated, meaning that equal weight was given to all the various aspects of performance. Finally, the grand mean composite z score for each group was calculated from the mean z scores of the individuals in that group. Figure 12.8 shows the grand mean driving performance z score for each group. Again, it is evident that both age and visual impairment lead to deterioration in driving performance.

Given the changes in visual and cognitive capabilities that occur with age and the evident impact on driving performance, it would seem remarkable that the increase in accident rates for old drivers shown in Figures 12.1 and 12.2 are not greater, particularly at night, particularly for fatalities, and particularly so when one remembers the greater fragility of old people. One reason for the underwhelming impact of ageing on accident rates is behavioural compensation. What this rather grand term means is that old drivers drive more slowly and have greater opportunities to avoid conditions that they find difficult, and take those opportunities, i.e., old drivers try to limit the amount of driving they do at night, in bad weather, in dense traffic, and on unfamiliar roads (Hakamies-Blomqvist 1994). Such compensatory behaviour certainly has its advantages but it does have a negative side. Driving as a skill can be considered to have two components, a series of automatic routines that are used unconsciously, and conscious control undertaken in unusual circumstances (Sivak et al. 1995). The automatic routines are the more resistant to the changes that occur

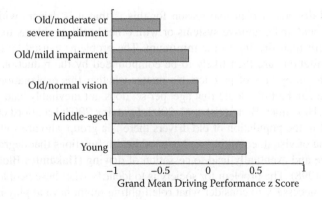

FIGURE 12.8 Grand mean composite driving performance z scores for young, middle-aged, and old drivers, the old drivers being divided into those with normal vision, mild visual impairment, or moderate or severe visual impairment (after Wood 2002).

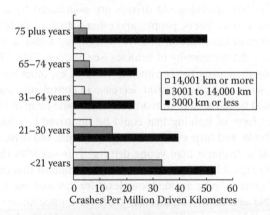

FIGURE 12.9 Crashes per million driven kilometres plotted against age group for three classes of different annual distance travelled (km) (after Langford et al. 2006).

with age. The automatic routines are developed and reinforced by continuous experience, so if compensatory behaviour reduces experience of specific conditions, such as driving at night, when the driver is required to drive at night, performance is likely to be degraded more than would have been the case — an example of use it or lose it in practice.

The consequences of low mileage are evident in Figure 12.9. This shows the number of crashes per million kilometres driven for five age groups, each of which has been divided into three mileage classes, based on data from The Netherlands (Langford et al. 2006). The interesting point is that for the intermediate and high mileage classes, there is no increase in the crash rate with increasing age. Indeed, the oldest age group has the lowest crash rate for both intermediate and high mileage groups. It is the lowest mileage class that shows the expected increase in crash rate for old drivers. A plausible reason for this is that it is drivers with the most degraded visual and cognitive systems or with other health problems that are most likely to limit their driving to the minimum. The problems caused by their physiological limitations are then likely to be compounded by the reduction in driving skill brought on by lack of practice. No matter whether this explanation is correct or not, there can be little doubt that age, per se, does not inevitably lead to a higher crash rates (Hakamies-Blomqvist et al. 2002; Fontaine 2003; Langford et al. 2006). Rather, within the population of old drivers there is a group who are well down the steep decline of visual, cognitive, and general health functions that degrade driving performance and eventually lead to cessation of driving (Hakamies-Blomqvist and Wahlstrom 1998). The problem for society is to identify who these people are.

It is now necessary to consider what role lighting might have to play in reducing the crash rates for old drivers who drive low mileages. This needs to be considered for day and night. Old drivers with limited mileage are most likely to be on the road during the day. Given this situation it might be thought that lighting has no role to play. For road lighting this is true, but there are opportunities for vehicle lighting (Rumar 1998). Crashes involving old drivers are dominated by a failure to yield the right of way or to see objects, people, and other vehicles (Table 12.1). This suggests that the universal introduction of daytime running lights would be useful as they would increase the conspicuity of vehicles (see Section 7.14). The risk with this approach is that by increasing the conspicuity of vehicles, other road users without daytime lighting, such as cyclists, would decrease in conspicuity and so be more at risk. This risk could be reduced if all road users had some form of marking lighting attached. Another form of lighting that could be improved is vehicle signal lighting, particularly brake and turn signals. As a general rule, old drivers need stronger signals to generate a response than young drivers. This suggests the use of higher luminous intensity signal lighting during the day, a proposal that implies two-level signal lighting if glare is to be avoided, one level for day and one for night. This is technically feasible and allowed under the ECE recommendations but is not currently implemented, Yet another factor that might be used is the height of the marking and signal lights above road level. Old drivers have slower responses than young drivers, so the longer notice they receive of a change ahead, the better. In traffic, signals about the intentions of vehicles some way ahead may be masked by intervening vehicles. By providing repeater signal lights at a high level, the likelihood of such obstruction occurring is reduced.

All these suggestions have the advantage of reinforcing the present system of vehicle marking and signal lights, a system that is familiar to most old drivers. This is an advantage because old people in general find it difficult to cope with very novel stimuli. There are many ideas for changing vehicle signal lighting so as to provide more information (see Section 7.15), but these should be carefully reviewed before being adopted because they may confuse old drivers rather than help them. This is also true for the idea of delivering more information to the driver through nonvisual means by using sensors and displays. There is a place for such driver assistance systems, e.g., an audible warning of a potential collision while reversing, but care needs to be taken that they make life easier for the old driver rather than more difficult. The same principle applies to signs and signals on roads. Indeed, it could be argued

FIGURE 12.10   A complex road layout by day consisting of two mini-roundabouts in close proximity but with one offset relative to the other and a pedestrian crossing between.

that one of the most valuable things that could be done to make driving easier for the old driver is to reduce the number and variety of road signs and markings, The introduction of bus lanes that stop and start at short intervals, the presence of unexpected speed bumps and chicanes for traffic calming, the widespread use of mini-roundabouts, and the increasing variety of road markings with different meanings have all served to increase the complexity of the driver's task (Figure 12.10). The introduction of a set of road signs and markings designed with the objective of creating simple and consistent guidance to the old driver would be a marked improvement over the present trend of ever-increasing complexity designed to satisfy the demands of various special interest groups.

At night, the greatest improvement would be produced by a more widespread use of road lighting. Good road lighting enables the road ahead to be seen over a greater distance, diminishes glare from headlamps, and provides visual guidance. Unfortunately, road lighting is expensive in both financial and environmental terms so its universal adoption is unlikely. What might be financially possible is to light all the locations where old drivers have problems. One such class of locations is road intersections (see Section 10.6). In rural areas, the very fact that an intersection is lit while the nearby road network is not would draw attention to the decisions that would need to be made there and, provided the approaches to the intersection are also lit, make the decisions easier to reach. In suburban and urban areas, where all roads are lit, the lighting of the intersections could be different to the lighting of the adjacent roads, in either the amount or colour of light.

Another aspect of the road environment that could be improved for old drivers is the visibility of road signs from a distance (Sivak et al. 1981). This is basically a matter of sign size, luminance contrast, and luminance, all of which could be enhanced (Kline 1995). At night, providing individual sign lighting rather than relying on the illumination from vehicle headlamps and retroreflection would be a step forward. Other possibilities are the wider use of antiglare screens where there is a median between the carriageways and the use of self-luminous road studs for visual guidance. For vehicle forward lighting, the proposed adaptive forward lighting system would appear to be of advantage as it changes the headlamp light distribution to match the environment automatically (see Section 6.7.2). If this is too exotic, Rumar (1998) suggests that old drivers would benefit from having a low-beam headlamp luminous intensity distribution that provided more light to the sides of the road, more light on the road beyond 50 m, and a softer cut-off. Other helpful changes would be more widespread use of automatic systems for changing between low- and high-beam headlamps, for reducing the reflectance of interior and side mirrors when a vehicle comes close behind, and for cleaning and leveling headlamps.

All these suggestions are technically feasible and would not have any deleterious effects on younger drivers, but the question that needs to be asked is if the expenditure can be justified. From Figure 12.9 it appears that the old drivers who would benefit most are those who drive the fewest miles. Whether or not attempting to encourage drivers whose visual and cognitive systems are in decline onto the roads at night is a responsible use of resources is a matter of judgment. Even if the answer to this question were to be positive, it would still be necessary to carry out a number of small-scale demonstration projects to determine the effectiveness of the proposed measures. A much stronger case can be made for trying to make driving by day easier for old drivers. The proposed changes could be easily implemented and the expenditure limited. If a review of the current system of road signs and markings were to lead to a simplified system, there might even be some savings. The one certain thing about this discussion is that it will not go away. The demographics of many developed countries predict a greatly increased number of old people and hence a greatly increased number of old drivers. It would be a good idea to consider their needs now before fatalities and injuries start to rise.

12.5 RESTRICTIONS ON DRIVING

Some people are not allowed to drive, some have driven but are then forced to stop, and some stop driving voluntarily. In modern society, not being allowed to drive imposes severe limitations on opportunities for employment and for socializing. Being forced to stop driving or stopping driving voluntarily is a blow to independence, not quite as extreme as going into a care home, but a definite step in that direction. Different countries have different rules for determining who is allowed to drive and who should stop driving. These rules are based on age, level of training, and medical status. In this last category are visual capabilities.

12.5.1 STARTING DRIVING

In the UK, obtaining a license to drive requires the applicant to demonstrate some basic visual capabilities, the requirements being more stringent for drivers of commercial vehicles than those of private vehicles. The requirements for people with normal vision in both eyes involve visual acuity and visual field size. For private vehicle use, the applicant has to be able to read a new-style UK vehicle number plate (see Figure 3.18) at a distance of 20 m in daylight, a demonstration that is done in the field at the time of the practical driving test and that corresponds to a visual acuity of about 6/10 as measured using a conventional Snellen chart (Drasdo and Haggerty 1984). For commercial vehicle use, the applicant has to have a visual acuity of at least 6/9, with correction, in the better eye, and at least 6/12 in the worse eye (see Section 3.4.1). Without correction, the applicant has to have a visual acuity better than 3/60 in both eyes. For both private and commercial vehicle use, the applicant's visual field has to be at least 120 degrees about the fixation point on the horizontal meridian and at least 20 degrees vertically, above and below the fixation point. There are two aspects of vision that might be expected to be required but are not. The complete loss of vision in one eye is acceptable for driving private vehicles provided the remaining eye can meet the acuity and visual field requirements, but is not acceptable for driving commercial vehicles. Similarly, normal colour vision is not required for driving private vehicles, although the ability to discriminate red, green, and yellow is needed for driving some types of commercial vehicle. In addition to the requirements for visual acuity and visual field size there are a number of visual problems for which a specialist's opinion is required with regard to the extent of the problem that the condition causes for driving, whether the problem is permanent, temporary, or progressive, and, ultimately, whether the individual should be allowed to drive. Other countries have similar but not identical requirements (Casson and Racette 2000).

Few would disagree that some level of visual capability is necessary for driving, but the basis of the current requirements is unclear (Gilkes 1988). Attempts to relate specific visual functions to the frequency of accidents have met with mixed success (Burg 1967, 1971; Johnson and Keltner 1983; Ball et al. 1993; Higgins and Bailey 2000) probably because accidents occur infrequently and have many different causes, and drivers who are aware of their limitations modify when and where they drive.

An alternative approach to establishing how good vision has to be to permit an individual to drive has been to examine the effect of specific limitations of vision on driving performance.

A study by Higgins et al. (1996) examined the effect of reduced visual acuity on driving performance. As might be expected, drivers with reduced visual acuity drove more slowly and had more difficulty in identifying traffic signs and hazards in the road simulating speed bumps.

Racette and Casson (2005) report a study of the assessment of drivers who had slight to moderate visual field loss. The extent of their loss was measured prior to their driving over a variety of public roads accompanied by a driving instructor who made an assessment of their ability to drive safely based on their performance. As might be expected, there was a lot of individual variability but there was also a trend for drivers with greater visual field loss to be graded as unsafe. Bowers et al. (2005) have used a similar approach to examine the ability of drivers with mild to moderate reduction in visual field to make different maneuvres on public roads. They found that drivers with more restricted visual fields were worse at maintaining position in a lane while going round a curve and matching speeds when changing lanes, activities that rely on peripheral vision to some extent. Conversely, they found no relationship between visual field size and lane keeping on a straight road, maintaining distance from the vehicle ahead, and crossing intersections, activities that primarily rely on foveal vision.

McKnight et al. (1991) examined the visual capabilities and driving performance of truck drivers with binocular and monocular vision. There were clear differences between the two groups in visual capabilities. Specifically, the monocular drivers were inferior to the binocular drivers in contrast sensitivity, visual acuity at low light level and in the presence of glare, and depth perception. However, these differences in capability made little difference to the measured driving performance. There were no differences between monocular and binocular drivers for lane keeping, clearance judgement, gap judgement, and hazard detection. The only aspect of driving for which a statistically significant difference was detected was the distance at which a sign requiring action could be read. For this the binocular drivers read the sign about 12 percent further away. Both groups of drivers showed wide individual differences in driving performance, much bigger differences occurring within each group than between the groups.

Defective colour vision in some form affects about 8 percent of males and 0.4 percent of females (Boyce 2003). Cole and Brown (1966) examined the difference in reaction time to a red traffic signal for people with normal colour vision and for protanopes, a group who do not have a long-wavelength-sensitive cone photoreceptor (see Section 3.2.5). They found that the protanopes had longer reaction times than colour normals to the same red traffic signal and were much more likely to miss the signal at lower signal luminances. Nathan et al. (1964) also found longer reaction times for people with colour-defective vision as well as more mistakes in identifying the signal colour. That such observations can have behavioural consequences is shown by the finding that people with defective colour vision had more accidents involving failure to halt at a red traffic signal than people with normal colour vision (Neubauer et al. 1978).

In a sense, such results are a statement of the obvious. Drivers who have limited visual capabilities tend to have problems when driving, the nature of the problem depending on the nature of the visual limitation. While this is interesting, it does not get us much further forward in answering the basic question, what limits should be set for a license to drive? What can be said is that the current system of vision testing used in most countries is very crude. How crude can be seen from the results of Wood (2002). As part of this study of the effects of age and visual impairment on driving performance, measurements were made of the individual driver's static and dynamic visual acuity, useful field of view, static and kinetic fields of view, contrast sensitivity, sensitivity to glare, and motion sensitivity. A step-wise multiple linear regression analysis revealed that 50 percent of the variance in the composite z scores (see Section 12.4.5) could be explained by a combination of motion sensitivity, useful field of view, contrast sensitivity, and dynamic visual acuity. Interestingly, the visual capability measure most widely used to determine suitability to drive, static visual acuity, did not increase the variance explained by a statistically significant amount. Such findings suggest that a more sophisticated visual screening system for testing applicants for driving licenses could be developed (Clay et al. 2005). Alternatively, given the developments in driving simulators, it should not now be too difficult to produce a programme to test driving potential, a programme that examines the various visual and cognitive abilities required of the driver (Ball et al. 1993). As long as the current crude test remains, there is a risk that people who are capable of driving safely will be refused permission to drive and people who are a menace on the road will be given permission to drive.

The question that now needs to be addressed is what is the role of lighting in obtaining a license to drive. The answer is very little. Tests of visual function are made during daytime or in conditions simulating daytime. Tests of driving performance are also made during daytime so driving at night is ignored. It is interesting to consider that anyone driving at night on an unlit road has a narrower horizontal field of view than that required for the issuing of a license. This neglect of vision at low light levels means that the main role of lighting in determining whether or not someone gets a license to drive comes through signs and signals. Signs are designed to have a high luminance contrast and sized so as to be visible from the required distance (see Section 5.3) by a driver with the normal visual acuity. Larger signs would make driving easier for someone with poor visual acuity. Traffic signals are designed to have colours within restricted colour boundaries set out in the CIE 1931 chromaticity diagram (see Figure 2.3). These boundaries are intended to make signal colours easier to discriminate by people with the more common forms of colour-defective vision. It would be better still if colour were not the only difference between traffic signals with different meanings. With present traffic signal designs, people with colour-defective vision can often use the relative positions of the red, yellow, and green signals to supplement the colour information they find hard to extract. It would be interesting to incorporate other means of conveying the necessary information, such as the shape of the signal (Whillans 1983). Then, as long as the driver could tell which signal was lit, the meaning would be clear.

What should be clear from the above is that the current methods of determining whether or not people have adequate visual capabilities to drive are crude and inadequate. That there have to be some limits is obvious but where those limits should be and whether or not they should be based on accident rates or on driving performance are open questions. The demographics of many countries are leading to more older drivers so there is great deal of pressure to permit drivers with a wide range of visual disabilities to continue to drive, even extending as far as published advice on driving with visual impairment (Peli and Peli 2002). There is a definite need to give more thought to the visual aspects of issuing a license to drive and the conditions under which those visual aspects are tested. At the very least, some attention should be given to driving at night.

12.5.2 STOPPING DRIVING

In modern society, most people seek to gain a driving license as soon as they are legally permitted and to retain it as long as possible. There are those who lose their license through bad or foolish behaviour on the road, but for most the factor that determines when driving is stopped is a decline in health associated with increasing age. It is for this reason that many countries have introduced regular reviews of an individual's ability to drive after a given age, typically 70 years of age. The rigour of this review varies from country to country. Some countries, such as Finland, make older drivers undergo a medical screening process, while others, such as Sweden and the UK, do not, but rather rely of confessions of inadequacy to identify drivers who should not be driving. The significance of general ill health for the decision to stop driving is evident from a study by Hakamies-Blomqvist and Wahlstrom (1998), who conducted a survey of all the drivers in Finland who had not renewed their license when they reached 70 years of age and a sample of those who had. Questions about the reasons for not renewing their license revealed that most were suffering from poor health. The most common form of ill health that differentiated between those who renewed their driving licenses and those who did not were neurological conditions, depression, and glaucoma. Glaucoma is a form of partial sight that reduces the visual field and, if untreated, ultimately leads to blindness (Silverstone et al. 2000). Interestingly, only 7 percent of people who had stopped driving had been advised to do so by the physician treating their underlying health problem. Most had stopped voluntarily. Such voluntary cessation of driving may also explain the results of Hakamies-Blomqvist et al. (1996), who found that the trends in population-adjusted accident rates for old drivers were the same in Finland and Sweden, the former with medical screening of old drivers and the latter without.

The picture that is emerging is of old drivers gradually retreating from what they perceive to be stressful and dangerous situations until they eventually surrender another part of their independence. This picture is certainly evident in the results of Freeman et al. (2005), who examined the relationship between visual function and time of stopping driving for a large sample of older drivers in the United States. They found that drivers with poorer visual acuity, reduced contrast sensitivity, and smaller visual fields at the time of measurement were more likely to have stopped driving 8 years later than those with better vision. Freeman et al. (2006) tried to determine if particular types of visual failings could be linked to particular changes in driving behaviour within the following 2 years. What they found was that poorer visual acuity, degraded contrast sensitivity, and reduced visual field all led to a lower mileage being driven. Poorer contrast sensitivity and more limited visual fields were associated with a cessation of night driving, and poor visual acuity alone was linked to an unwillingness to drive in unfamiliar areas.

Given that deterioration of vision is linked to at first modifying and then stopping driving, it is appropriate to ask if better lighting might have a role to play in keeping older drivers driving safely for longer. The answer is the same as that given earlier for old drivers in general (see Section 12.4.5). The main point to take from this discussion of the reasons why people stop driving is that it is not age, per se, that is the cause of driving cessation but rather the consequences of ageing for health. This implies that attempts to reduce accident rates are more likely to be successful if attention is focused on the health and not simply on the age of the driver. As part of that focus, it would be interesting to consider the use of a graduated license to drive to which various restrictions could be added according to the individual's health. For example, conditions limiting driving to daytime only or within a set distance from home could be added. At the moment, withdrawing a license to drive is often an all-or-nothing matter. A graduated license would have the advantage of formalizing what is actually happening informally.

12.6 SUMMARY

Some people are better equipped to deal with the demands of driving than others. Among the important factors that determine a driver's ability are the amount of practice and any limitations in visual or cognitive systems. The influence of these factors can be seen in the number of fatal and personal injury accidents per 100 million miles travelled by day and night, for different age groups. The highest accident rate of both types occurs for drivers at the extremes of the age range, the youngest and the oldest. However, the types of accidents associated with the extremes of age are different. Young drivers tend to be responsible for more accidents where lack of control, following too close, misjudging stopping distance, avoiding an object/person/vehicle, speeding, or improper passing are the primary contributory factors. Older drivers tended to be responsible for more crashes where failure to yield right of way, failure to see an object/person/vehicle, failure to heed a sign or signal, an improper lane change, an improper U-turn, or improper backing are the primary contributory factors.

Young drivers are not likely to be involved in accidents because of visual or cognitive limitations. Young drivers who are in their late teens or early twenties will usually have a visual system at the peak of its powers and a cognitive system capable of rapid responses to simple and complex stimuli, but they may be inexperienced. Inexperienced young drivers have different patterns of eye movements to experienced drivers. Specifically, they tend to fixate where they have been taught to and to maintain that pattern for all types of road. Experienced drivers adjust their fixation pattern to the road type.

While a lack of experience is a factor in accident rates for some young drivers, not all young drivers are inexperienced. Even when they are experienced the nature of the type of accidents in which they are involved suggests that young drivers drive with reduced margins of error, relying on their fast reactions to survive. Further, young drivers are more likely to be out late at night, are more likely to be fatigued, and more likely to be suffering from the effects of alcohol or drugs, all of which will tend to slow responses and confuse analysis. There is little that lighting can do to overcome such behavioural excesses other than to provide more time for decision-making by lighting the road further ahead.

Old drivers show increased accident rates because ageing affects the visual, cognitive, and motor functions required for driving. As the visual system ages, a number of changes in its structure and capabilities occur. With increasing years the ability to focus over a wide range of distances is diminished, the amount of light reaching the retina is reduced, more of the light reaching the retina is scattered, the spectrum of the light reaching the retina is changed, and more straylight is generated inside the eye. Such changes with age are the best that can be expected. With increasing age comes a greater likelihood of pathological changes in the eye leading to visual impairment in which the normal deterioration in visual capabilities with age is accelerated. These are all optical changes in the eye but there are also neural changes. This means that the compensation for visual system ageing that can be provided by lighting is inevitably limited.

The consequences of such optical and neural changes with age are reduced visual field size, increased threshold luminance, reduced visual acuity, reduced contrast sensitivity, reduced colour discrimination, and greater sensitivity to glare. The consequence of these changes in visual capabilities is a deterioration in driving performance, but not necessarily an increase in accident rate. This is because old drivers tend to compensate for their reduced capabilities by avoiding situations that they find difficult and stressful, such as driving in bad weather, in dense traffic, at night, and on unfamiliar roads. Such compensatory behaviour certainly has its advantages, but it does have a negative side. Driving as a skill can be considered to have two components, a series of automatic routines that are used unconsciously, and conscious control undertaken in unusual circumstances. The automatic routines are the more resistant to the changes that occur with age. The automatic routines are developed and reinforced by continuous experience. This may be why it is the old drivers who cover the lowest mileage who are responsible for the increase in accident rate ascribed to old drivers as a whole.

Lighting could be used to help old drivers. By day, there is no role for road lighting but vehicle lighting could be modified to advantage. The universal introduction of daytime running lights would be helpful, as would the provision of higher-luminous-intensity vehicle signal lamps for use by day. By night, a wider use of road lighting would be beneficial. Unfortunately, road lighting is expensive so its universal adoption is unlikely. What might be financially possible is to light all the locations where old drivers have problems, such as road intersections, and to light all road signs. As for vehicle lighting, the proposed adaptive forward lighting system would appear to be of advantage as it changes the headlamp luminous intensity distribution to match the environment automatically. All these suggestions are technically feasible and would not have any deleterious effects on younger drivers, but the question that needs to be asked is whether the expenditure can be justified. One thing is certain and that is that the problem of what to do about old drivers will not go away. The demographics of many developed countries predict a greatly increased number of old people and hence a greatly increased number of old drivers.

This raises the question as to how permission to drive is given and withdrawn. Different countries have different rules for determining who is allowed to drive and who should stop driving. These rules are based on age, level of training, and medical status. Getting a license to drive always involves some test of visual capability, usually some simple measure of static visual acuity. Unfortunately, attempts to relate individual visual functions to the accident rates have met with mixed success, probably because accidents occur infrequently and have many different causes, and drivers who are aware of their limitations modify when and where they drive. What is clear is that people with poorer motion sensitivity, useful field of view, contrast sensitivity, and dynamic visual acuity show poorer driving performance.

Lighting has little involvement in obtaining a license to drive. Tests of visual function are made during daytime or in conditions simulating daytime. Tests of driving performance are also made during daytime so driving at night is ignored. It is clear that the current methods of determining whether or not people have adequate visual capabilities to drive are crude and inadequate. That there have to be some limits is obvious but where those limits should be and whether or not they should be based on accident rates or on driving performance are open questions.

As for stopping driving, most people seek to retain a driving license as long as possible. The picture that emerges from studies of cessation of driving is of old drivers gradually retreating from what they perceive to be stressful and dangerous situations until they stop driving altogether and surrender another part of their independence. In many situations the decision to stop driving is taken voluntarily. It is not age, per se, that causes people to stop driving but rather the consequences of ageing for health. This implies that attempts to reduce accident rates for old drivers are more likely to be successful if attention is focused on the health and not simply on the age of the driver. As part of that focus, it would be interesting to consider the more widespread use of a graduated license to drive to which various restrictions could be added according to the individual's health. At the moment, withdrawing a license to drive is often an all-or-nothing matter. A graduated license would have the advantage of formalizing what is actually happening informally.

13 Constraints

13.1 INTRODUCTION

All forms of lighting are subject to constraints and lighting for driving is no exception. These constraints range from the legal through the financial to the environmental, some of them being weak and some strong. Vehicle lighting meeting specific photometric conditions is a legal requirement in almost all countries. Road lighting is not a legal requirement but rather a matter of policy, the nature of which will vary from country to country and from time to time. Both vehicle lighting and road lighting imply costs. For vehicles, the first costs of lighting are included in the cost of the vehicle. Costs-in-use are borne by the driver and depend on the timing and hours of use of the vehicle. For road lighting, both first costs and life cycle costs can be involved in the decision whether or not to install road lighting and certainly influence the amount and form of road lighting provided. Recently, environmental factors have also entered into consideration, particularly for road lighting. These environmental factors can take four forms: the amount of energy consumed, the carbon dioxide emissions produced when generating the electricity used, the disposal of the materials used at the end of product life, and the problem of light pollution. This chapter is devoted to a consideration of these constraints as they apply to road lighting. The legal requirements applicable to vehicle lighting have been discussed in Sections 6.3 and 7.3. The financial and environmental aspects of vehicle lighting are not considered separately from those of the vehicle itself. Even the light pollution produced by vehicle lighting is ignored, probably because it is considered to be transient rather than permanent.

13.2 ASSESSING THE NEED FOR ROAD LIGHTING

The lighting of roads costs money. When the roads are privately owned, whether or not to install lighting is a matter for the owner, but the vast majority of roads are publicly owned and their lighting is a matter for public authorities. To justify decisions, public authorities have adopted specific guidelines for determining when expenditure on road lighting is warranted. In the United States, many states use the guidelines set out by the American Association of State Highways and Transportation Officials (AASHTO 1984). These guidelines consist of a series of quantitative and qualitative criteria. For example, for determining if a linear section of freeway should be lit, the quantitative guidelines are that the average daily traffic flow should be more than 30,000 vehicles, the average distance between three or more adjacent intersections should be 1.5 miles (2.4 km) or less, and the night/day accident ratio should be more than 2.0. The qualitative guidelines concern the brightness of the surroundings

295

through which the freeway passes. Meeting the quantitative and qualitative criteria is a necessary but not sufficient condition to ensure the installation of road lighting. Even if the criteria are met, most public authorities allow themselves some discretion by stating that they may need to conduct a full safety evaluation of the specific site as well as considering what funds are available and where they are best spent.

An alternative approach used in the UK is a calculation method based on the costs and benefits of accident prevention, expressed in monetary terms (DfT 2007a). The first costs of road lighting are easy to obtain, although assumptions are needed when life cycle costs are used. The assumptions required relate to hours of use, electricity costs, maintenance costs, and disposal costs. Hours of use can be identified fairly exactly, given that the rising and setting of the sun is quite predictable, so the only unknown is the variation in switch-on or switch-off times caused by cloud cover. The same is not true for electricity costs, maintenance costs, and disposal costs. The economics of road lighting are usually assessed over a period of 30 or 40 years, so different assumptions about electricity costs, monetary inflation, and interest rates can make a large difference to the life cycle costs.

Even more contentious is the cost of benefits. This is because it is necessary to make assumptions about the reduction in accidents likely to occur if the road lighting is installed and the savings to be made by preventing accidents. In the UK, the currently assumed percentage reductions in personal injury accidents at night produced by lighting motorways and all-purpose dual carriageways is 10 percent, while for all-purpose single carriageway roads it is 12.5 percent (DfT 2007a). These percentage reduction assumptions are dramatically lower than the previous assumption of 30 percent reduction, based on CIE (1992b), and are only relevant to new roads and existing unlit roads. Smaller percentage reductions are assumed when the decision is about renewing or upgrading existing lighting. The number of personal injury accidents prevented by installing road lighting is then estimated by multiplying the number of personal injury accidents occurring during darkness by the appropriate percentage reduction due to lighting. For new roads, where there are no accident records, the national average personal injury accident rate for roads of the same type is used. For existing roads, the average annual number of personal injury accidents occurring during darkness on the road for the past 5 years is used. If data on personal injury accidents is not classified by night and day, it is assumed that 28 percent of personal injury accidents on strategic roads occur by night (DfT 2005c).

Having obtained the annual number of personal injury accidents that will be prevented by installing lighting, it is then necessary to estimate the value of preventing these accidents. This includes both measurable elements, such as the cost of hospital treatment, police costs, and material damage, and imponderable elements such the costs of lost productivity and the value of human pain and suffering. In the UK, the Department for Transport (DfT) publishes values accruing from the prevention of accidents of different degrees of severity that occur during the hours of darkness (Table 13.1). Once the costs of lighting and the financial benefits associated with a reduction in accidents are estimated, it is possible to calculate if an investment in road lighting gives a reasonable rate of return.

A hybrid approach has been suggested for the United States (Deans et al. 2003). The first step is a monetary balance between the costs and benefits of installing road

TABLE 13.1

Average Value of Preventing an Accident during Hours of Darkness in the UK in 2007, by Severity and Road Type, Expressed in Pounds (£)

Severity	Built-Up Roads (£)	Non-Built-Up Roads (£)	Motorways (£)	All Roads (£)
Fatal	1,613,970	1,754,950	1,789,030	1,689,270
Serious	184,850	212,940	219,460	194,490
Slight	18,560	21,790	25,680	19,460
All injury	64,220	137,000	119,950	84,770
Property damage only	1,550	2,310	2,220	1,660
Average	91,610	156,020	136,820	109,360

Built-up roads are those roads other than motorways with speed limits of 40 mph or less. Non-built-up roads are those roads other than motorways with speed limits greater than 40 mph (from DfT 2007b).

lighting based on the reduction in accidents occurring during the hours of darkness, the costs of these accidents, and the installation, maintenance, and energy costs of the proposed road lighting. The second step is a prescription method in which the specific site is assessed for risk. Among the factors considered are the percentage of trucks in the traffic, the brightness and use of the surroundings, the geometry of the road layout, and the speed limits.

Each of the three methods summarized has different virtues. The prescriptive approach using the AASHTO guidelines has the virtue of simplicity, although how the prescription is derived is not at all clear. The monetary approach used in the UK presents a high-precision façade but is largely smoke and mirrors. By manipulating the assumptions about interest rates, the values of accidents prevented, and accident reductions, the policy makers can reach almost any decision they desire. The suggested screening process of Deans et al. (2003) is the most attractive option, being a mixture of the general and the specific. The general monetary balance can be used to weed out the hopelessly uneconomic proposals, while the risk assessment of the specific site can be used to determine where an investment is likely to be most effective.

13.3 COSTS OF ROAD LIGHTING

The amount of money available for a project is always a constraint. When examining any proposed road lighting installation, there are several different costs to be considered: first cost, energy cost, and life cycle costs. What these three costs are has been calculated for six different road lighting installations made up of three road layouts and two different light source types, all designed to meet the current British road lighting luminance recommendations (BSI 2003a) for the representative British road surface (see Section 4.5). For the single carriageway installations,

TABLE 13.2

Assumed First Cost in Pounds (£) and Power Demand (W) of Different Components of a Road Lighting Installation in the UK in 2004

Component	First Cost (£)	Circuit Power Demand (W)
8 m steel galvanized and painted lighting column, installed	205	—
10 m steel galvanized and painted lighting column, installed	297	—
12 m steel galvanized and painted lighting column, installed	382	—
150 W high-pressure sodium light source on standard control gear and typical luminaire	Light source = 7.35 Luminaire = 168	172
150 W metal halide light source on low loss control gear and typical luminaire	Light source = 31.36 Luminaire = 168	167
Connecting a column to the electricity supply in urban area (column close to general service cable)	299	—

the minimum average road surface luminance is 1 cd/m^2 and the minimum overall luminance uniformity ratio and minimum longitudinal uniformity ratios are 0.4 and 0.5 respectively. For the dual carriageway installations, the minimum average road surface luminance is 1.5 cd/m^2 and the minimum overall luminance uniformity ratio and minimum longitudinal uniformity ratio are 0.4 and 0.7 respectively.

The first cost per kilometre of the installation is calculated as the sum of the cost of lamp, luminaire, column, and connection to the electricity supply, multiplied by the number of columns along a kilometre of road. Table 13.2 sets out the assumed costs of columns of different heights, of luminaires and light sources, and of the connection of a column to the electricity supply for an urban area where proximity to a general service cable ensures a stable cost.

To calculate the annual energy cost per kilometre it is necessary to know the power demand of the lamps, the number of hours of use, and the price of electricity. Table 13.2 shows the actual power demands of 150 W high-pressure sodium and metal halide light sources according to the Balancing and Settlement Code Procedure used in determining charges for unmetered road lighting in England and Wales (Elexon Limited 2005). The hours of use are taken to be 4100 hours for dusk-to-dawn operation. This is based on the predictions made using a method developed by Tregenza (1987) for latitudes and climate similar to Nottingham, UK. The price of electricity is assumed to be 4.5 p/kWh although market trading will cause this price to vary.

To calculate the life cycle cost per kilometre of installation, it is necessary to know the cost of light sources and luminaires (see Table 13.2), the life of the installation, and the discount rate. The life of the columns, and hence the installation, is assumed to be 40 years, with luminaires being replaced after 20 years, high-pressure

TABLE 13.3

First Cost/Kilometre in Pounds (£), Annual Energy Cost/Kilometre in Pounds (£), and 40-Year Life Cycle Cost/Kilometre in Pounds (£) for the Lighting of Three Different Road Layouts Using 150 W High-Pressure Sodium (HPS) or 150 W Metal Halide (MH) Light Sources and Producing the Minimum Luminance Conditions Specified in BSI (2003a) from a Representative British Road Surface

Road Layout	Lighting Layout	Light Source	First Cost/ km (£)	Annual Energy Cost/ km (£)	40-Year Life Cost/ km (£)
Single carriageway	Staggered	HPS	20,577	762	41,356
Single carriageway	Staggered	MH	24,692	955	56,843
Single carriageway	Single sided	HPS	19,312	793	40,939
Single carriageway	Single sided	MH	21,215	1,017	60,519
Dual carriageway	Opposite	HPS	47,585	2,221	108,155
Dual carriageway	Opposite	MH	54,896	2,403	135,794

sodium light sources after 4 years, and metal halide light sources after 3 years. The assumed discount rate, which is the difference between the interest rate and the rate of inflation, is taken to be 3 percent as recommended by HM Treasury. The life cycle cost represents the sum of the capital cost of the installation, the cost of replacing lamps and luminaires, and the cost of electrical energy over 40 years, all future costs being converted to their present values.

Table 13.3 shows the calculated first costs, annual energy costs, and 40-year life cycle costs per kilometre of lit road for three different road lighting layouts using 150 W high-pressure sodium and metal halide light sources to provide the maintained minimum luminance conditions specified in BSI (2003a). Examination of Table 13.3 reveals a number of interesting facts. First, the costs of the dual carriageway installations are much greater than those of the single carriageway installations. This is partly due to the higher average road surface luminance required for the dual carriageway and partly because the dual carriageway has columns on both sides of the road opposite each other. Second, for all three road and lighting layouts, a metal halide installation costs more than a comparable high-pressure sodium installation, in first costs, annual energy costs, and 40-year life cycle costs. This is due to the fact that metal halide lamps have a lower luminous efficacy and a shorter life than high-pressure sodium lamps (see Table 2.4). This explains the reluctance to use metal halide for road lighting except in town and city centres where the appearance of people and buildings is given greater priority. Third, an important determinant of first cost is the spacing between columns. The closer the spacing, the greater the number of columns and electrical connections required, and as is apparent in Table 13.2, columns and connections are more expensive than light sources and luminaires. This, in turn, raises the question of the relative importance of different road lighting criteria.

The average road surface luminance is the criterion that gets the most attention because it obviously influences first cost, but the two uniformity criteria, minimum overall luminance uniformity and minimum longitudinal luminance uniformity, are also important for first cost. This is because luminance uniformity is critical for determining the spacing between columns. Fourth, the first cost of an installation represents between one-third and one-half of the 40-year life cycle cost. Fifth, a comparison between the 40-year life cycle costs of the lighting installations and the costs of personal injury accidents given in Table 13.1 suggests that it is only necessary to prevent one or two personal injury accidents per kilometre of lit road over a period of 40 years for the lighting installation to be financially justified.

The costs contributing to Table 13.3 are typical UK prices in 2004 and assume the road lighting is to be mounted on columns. In urban areas, some countries adopt a system of suspending luminaires on wires strung between buildings, the power coming from one of the buildings on a separately metered circuit. This eliminates the cost of columns and allows the luminaires to be positioned over the road. All countries will have different prices for different elements and all prices may vary over time. Finally, the joker in the pack is the cost of connection to the electricity grid. In rural areas, the cost of connection can increase dramatically. Despite these uncertainties, the above analysis serves to indicate the broad costs of road lighting and to demonstrate that important variables can be identified by a simple analysis of costs.

13.4 ENERGY CONSUMPTION

There are three reasons why the amount of energy consumed might become a constraint on the design of road lighting. The first is the cost. The annual energy costs per kilometre of typical road lighting installations in the UK are shown in Table 13.3. These costs will vary from country to country depending on the cost of electricity. The higher the electricity cost, the more likely it is that energy costs will become a constraint on road lighting. The second is the need to reduce the maximum demand on the electricity network. This is very unlikely to be relevant except in countries in high latitudes, as maximum demand usually occurs at times of maximum activity by the population, which is usually during the day, but road lighting is operating at night. For countries in high latitudes, for some part of the year, darkness may coincide with high levels of activity, in which case, using road lighting to reduce maximum demand might be considered. The third is the desire to reduce the use of such ultimately limited resources as oil, gas, and coal. Given this desire, any use of electricity is to be avoided unless the electricity is generated from renewable resources such as the sun, wind, or waves.

Table 13.4 gives the annual energy consumption of the six typical UK road lighting installations discussed in Section 13.3. As before, the metal halide installations consume more energy than the equivalent high-pressure sodium installations and the lighting of dual carriageways takes more energy than the lighting of single carriageways. Whether or not such energy consumptions are considered significant will depend on the commitment to minimizing resource depletion in the particular country, the proportion of all the electricity generated that is used for road lighting, the

TABLE 13.4

Annual Energy Consumption/Kilometre (kWh/km) for the Lighting of Three Different Road/Lighting Layouts Using 150 W High-Pressure Sodium (HPS) and Metal Halide (MH) Light Sources and Producing the Minimum Luminance Conditions Specified in BSI (2003a) from a Representative British Road Surface

Road Layout	Lighting Layout	Light Source	Annual Energy Consumption/km (kWh/km)
Single carriageway	Staggered	HPS	16,993
Single carriageway	Staggered	MH	21,222
Single carriageway	Single sided	HPS	17,622
Single carriageway	Single sided	MH	22,600
Dual carriageway	Opposite	HPS	49,356
Dual carriageway	Opposite	MH	53,400

fuel mix used to generate the electricity, and the ease with which political pressure can be applied to the decisions about road lighting as compared to the decisions of individual householders. The greater the commitment to minimizing resource depletion, the higher the proportion of electricity generated that is used for road lighting, the greater the amount of fossil fuels used in electricity generation, and the easier it is to apply political pressure, the more likely road lighting is to be the focus of attempts to reduce electricity use.

13.5 CARBON DIOXIDE EMISSIONS

Carbon dioxide is one of several gases that contribute to global warming. It is not the most potent but it does have the biggest impact because it is produced in large amounts by the burning of fossil fuels for heat and for the generation of electricity. Given a desire to limit global warming by really reducing carbon dioxide emissions rather than just covering embarrassment with the fig leaf of carbon offset payments, road lighting is an obvious target. It consumes electricity, it is conspicuous, and it is in use at times when there are few people about. However, before cutting road lighting it is important to remember its value for traffic safety and that the amount of carbon dioxide emitted will depend on how the electricity used to power the road lighting is generated. Different countries have very different fuel mixes when it comes to electricity generation. New Zealand generates 65 percent of its electricity from hydropower, most of the rest coming from gas. Its neighbour, Australia, generates almost 80 percent of its electricity by burning coal, making it the highest per capita emitter of carbon dioxide in the world. France generates 80 percent of its electricity through nuclear reactors, which produce no carbon dioxide emissions. The UK fuel mix for electricity generation is 33 percent coal, 39 percent gas, and 20 percent nuclear, with the rest coming from oil and renewables. The carbon emissions of different fuels, expressed in kilograms of CO_2 per kilowatt hour of electricity produced, range from zero for nuclear, wind, and hydro, to 0.363 for natural gas and 0.873 for coal (DTI 2006). If reducing carbon dioxide emissions is the aim, there is little point in limiting road lighting in countries where the fuel mix used to generate electricity is dominated by nuclear or renewables, but where gas and coal are extensively used for electricity generation, the need for road lighting should be carefully considered. One answer to this problem lies in using photovoltaic panels linked to batteries to store electricity during the day and to use the stored electricity to power the road lighting at night. Such a system is technically possible and commercially available for parking lots, but its use is presently limited to locations where connecting to the electricity grid is prohibitively expensive. Of course, this suggestion ignores the energy implications of the manufacture and distribution of the necessary photovoltaic panels and batteries. It may be that energy consumed in these activities exceeds the savings that might be made.

13.6 WASTE DISPOSAL

Another factor that is beginning to be considered as a constraint on road lighting is the ease of disposal at the end of life. Questions about waste disposal can take two forms, the hazard to human health posed by the materials in the product and the ease of recycling. Some elements of a road lighting installation present little problem. For example, aluminium and steel are widely used for columns, but present no hazard and can be recycled. Luminaires are usually constructed of some combination of aluminium, glass, and plastic, all of which present little hazard and can be recycled. Where there are problems is in the light source and control gear. The light sources most widely used for road lighting are low-pressure sodium, high-pressure sodium, and metal halide discharge (see Section 2.5). As for control gear, there are still capacitors containing PCBs present in the oldest part of the road lighting stock and while electromagnetic control gear is widely used, it is slowly being replaced by electronic control gear. Without getting into the specifics of individual products, it is not possible to identify problems with waste disposal, so the best that can be done here is to point out some general trends. For example, light sources that contain mercury are likely to be subject to stricter regulation than those that contain sodium. This is evident through the existence of EU directive 2002/95/EC, which prohibits the use of mercury, but not sodium, in electrical and electronic equipment to less than a low threshold level. Similarly, electronic control gear is likely to be more expensive to dispose of than electromagnetic control gear because it contains more expensive materials that are more difficult to isolate and recycle. This will be the case because of the existence of EU directive 2002/96/EC, widely known as the waste electrical and electronic equipment (WEEE) directive. This is currently being implemented in many EU countries in an attempt to reduce the size of the electrical and electronic waste stream. What this directive requires is that anyone who manufactures, brands, or imports electrical and electronic equipment be responsible for the collection, treatment, recycling, and disposal of the product at the end of life. In the UK, various trade groups have set up schemes to ensure the collection and appropriate treatment of both lamps and control gear. How effective these will be remains to be seen, but there can be little doubt that questions about waste disposal are now on the agenda for the designers of road lighting.

13.7 LIGHT POLLUTION

Light can be considered a form of pollution. Complaints about light at night can be divided into three categories: light trespass, sky glow, and glare (LRC 2007). Light trespass is local in that it is associated with complaints from individuals in a specific location. The classic case of light trespass is a complaint about light from a road lighting luminaire entering a bedroom window and keeping the occupant awake.

Sky glow is more remote than light trespass in that it can affect people over great distances. Complaints about sky glow originate from many people, ranging from those who have a professional interest in a dark sky, i.e., optical astronomers (McNally 1994; Mizon 2002), through those who are concerned with the effect of too much light at night on flora and fauna (Rich and Longcore 2006) to members of the general public who simply like to be able to see the night sky. For night sky viewers, both professional and amateur, the problem sky glow causes is that it reduces the luminance contrasts of all the features of the night sky. A reduction in luminance contrast means that features that are naturally close to visual threshold will be taken below threshold by the addition of the sky glow. As a result, as sky glow increases, the number of stars and other astronomical phenomena that can be seen is much reduced. As for the flora and fauna, the effects of sky glow are widespread, ranging from blooming at the wrong time through a change to the balance between predator and prey to misguided behaviour that leads to population decline.

Glare has been a feature of outdoor lighting for many years and is associated with the presence of a luminaire projecting a high luminous intensity towards an observer. If a luminaire produces a desire to shield the eyes, glare may be said to be occurring. As far as light pollution is concerned, glare is local in that it is caused by a particular luminaire seen from a particular direction by someone outside the lit area. Glare can be produced by nearby luminaires or those far from the observer and may or may not be linked to light trespass. Glare can disable vision by reducing the luminance contrasts in the retinal image of the scene or it may simply cause discomfort (see Sections 4.3 and 6.6).

13.7.1 LIGHT TRESPASS

Light trespass is not well defined. For some people, light trespass occurs when light enters the rooms of their dwelling. For others, light trespass occurs when light extends over their property boundary into their garden. In principle, light trespass can be avoided by the careful positioning and aiming of luminaires with appropriate luminous intensity distributions. For road lighting, this means choosing luminaires that direct the vast majority of the light emitted onto the road surface and minimize the amount of light emitted beyond that surface, on both sides of the road. If complaints of light trespass still occur, then the classic response is to fit what is called a house-side shield to the luminaire. This is a baffle that prevents light traveling direct from the luminaire to windows immediately behind the luminaire. Figure 13.1 shows

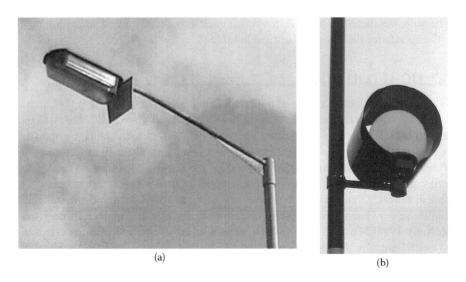

(a)

(b)

FIGURE 13.1   Two attempts to reduce light trespass: (a) an external house-side shield fitted to a deep bowl road lighting luminaire and (b) a cylindrical baffle fitted to a flashing beacon marking a pedestrian crossing.

two forms of baffle. One is a simple, external house-side shield fixed to a luminaire to prevent light trespass into the bedroom of an adjacent house. The other is a cylindrical baffle fitted to a flashing beacon installed to mark a pedestrian crossing. The regular flashing of the beacon makes any light trespass particularly disturbing.

Fitting baffles adds to the cost of the lighting installation so it is as well to have some guidance as to whether or not a complaint of light trespass is justified. Advice in the form of maximum illuminances that should be allowed to fall on windows is available (ILE 2000; CIE 2003). This advice is presented at four different levels, these levels corresponding to the four qualitative environmental zones suggested by the CIE (1997). Different zones are necessary because the ambient light level is very different in different locations. It makes no sense to specify one maximum illuminance on a window for both a city centre and a rural village. A maximum illuminance limit based on the city will be too high for a village, while a criterion based on a rural village will be unachievable in a city. Table 13.5 identifies the four environmental zones suggested by the CIE. Table 13.6 sets out the maximum vertical illuminance that should be allowed to fall on a window in each of the four environmental zones, before and during a curfew (see Section 13.7.2).

While specifying the maximum vertical illuminance on a window is an adequate criterion to determine if complaints of light trespass based on light entering a window are justified, the outcome is specific to the site and does not do anything for complaints based on the effects of light entering the garden. A more comprehensive approach to light trespass based on property rights is being developed (Brons et al. 2007, 2008). This approach, called the outdoor site-lighting performance (OSP) method, uses a virtual, transparent "shoebox" surrounding the property. The virtual

TABLE 13.5

The Zoning System of the Commission Internationale de l'Eclairage

Zone	Zone Description and Examples of Such Zones
E1	Areas with intrinsically dark landscapes: National parks, areas of outstanding natural beauty (where roads are usually unlit)
E2	Areas of "low district brightness": Outer urban and rural residential areas (where roads are lit to residential road standard)
E3	Areas of "middle district brightness": Generally urban residential areas (where roads are lit to traffic route standard)
E4	Areas of "high district brightness": Generally, urban areas having mixed recreational and commercial land use with high night-time activity

After CIE (1997).

TABLE 13.6

Maximum Vertical Illuminance (lx) on a Window in the Four Environmental Zones, before and during a Curfew

Environmental Zone	Maximum Vertical Illuminance on a Window before Curfew (lx)	Maximum Vertical Illuminance on a Window during Curfew (lx)
E1	2	1
E2	5	1
E3	10	2
E4	25	5

After ILE (2000).

"shoebox" has vertical sides at the property boundary and a flat "ceiling" 10 m above the highest mounted luminaire in the installation or the highest point of the property illuminated. For road lighting, the "shoebox" is extended along a representative section of road, the vertical planes of interest for light trespass being the planes located along the edges of the right-of-way. Conventional lighting design software can be used to calculate the illuminances falling on these planes. By identifying the location and magnitude of the maximum illuminances on these planes, the potential for light trespass can be established. Of course, this simply indicates where a light trespass problem is likely to occur, not what the solution might be. Possible solutions include choosing a luminaire with a more appropriate luminous intensity distribution, moving the luminaire further into the site and away from the property boundary, planting screening vegetation, and fitting some form of baffle to the luminaire.

The OSP "shoebox" approach is consistent with the way people think about property rights. Owning property confers considerable freedom of action on the owner within the property boundaries, provided those actions do not impinge negatively on others nearby or on the public good. The OSP "shoebox" approach is both flexible and realistic. It is flexible in that different maximum illuminance limits can be set by different communities, using the CIE environmental zones if desired. It is realistic in that it uses widely available software to make the necessary calculations; decisions on actions necessary to avoid light trespass can be made at the design stage; it does not require detailed knowledge of what surrounds the site being lit, knowledge that is often not available to the designer; and it includes the contributions of both direct and reflected light. It will be interesting to see if the OSP "shoebox" approach gains widespread acceptance.

13.7.2 Sky Glow

Sky glow is evident over most cities and towns in the form of glowing, flattened dome of light (Figure 13.2). Sky glow has two components, one natural and one due to human activity. Natural sky glow is light from the sun, moon, planets, and stars that is scattered by interplanetary dust, and by molecules and aerosols in the Earth's atmosphere, and light produced by a chemical reaction of the upper atmosphere with ultraviolet radiation from the sun. The luminance of the natural

FIGURE 13.2 Sky glow over the city of Canterbury, UK, a city with a population of 135,000.

sky glow at zenith is of the order of 0.0002 cd/m². The contribution of human activity is produced by light traversing the atmosphere and being scattered by the air molecules and aerosols therein. Aerosols are suspended water droplets and dust particles. Air molecules scatter light forward and back with a little to the side. This Rayleigh scattering is much greater for short visible wavelengths, which is why the sky appears blue. Aerosols scatter light predominantly forward. This Mie scattering is independent of wavelength in the visible region, which is why clouds appear white. Where there are few aerosols and few air molecules, there is very little sky glow, which is why major new optical telescopes are built in such areas as the Atacama Desert of the Chilean Andes, where the population is small, the air pollution is negligible, and the air is very thin and dry. Established optical telescopes at low altitude and near large cities are of diminishing value, despite rearguard actions fought to minimize sky glow.

The magnitude of the contribution of electric lighting from a city to sky glow at a remote location can be crudely estimated by Walker's Law (Walker 1977). This can be stated as

$$I = 0.01 \times P \times d^{-2.5}$$

where I is the proportional increase in sky luminance relative to the natural sky luminance for viewing 45 degrees above the horizon in the direction of the city, P is the population of the city, and d is the distance from the remote location to the city (km).

This empirical formula assumes a certain use of light per head of population. Experience suggests the predictions are reasonable for cities where the number of lumens per person is between 500 and 1000. More sophisticated models based on the physics of light scatter have been used to generate light pollution maps, these models making allowances for the curvature of the earth and allowing predictions to be made for different altitudes and azimuths of viewing and different atmospheric conditions (Garstang 1986; Baddiley and Webster 2007; Kocifaj 2007).

The problem in dealing with sky glow is not in measuring or predicting its effects on the visibility of the stars, but rather in agreeing what to do about it. The problem is caused by two facts. The first is that sky glow is caused by multiple light sources, some private, some public. The second is that what constitutes the astronomer's pollution is often the business owner's commercial necessity and sometimes the citizen's preference. Residents of cities like their streets to be lit at night for the feeling of safety the lighting provides. Similarly, many roads are lit at night to enhance the traffic safety. Businesses use light to identify themselves at night and to attract customers. Further, the floodlighting of buildings and the lighting of landscapes are methods used to create an attractive environment at night. The fundamental problem of sky glow is how to strike the right balance between these conflicting desires.

One of the earliest attempts to achieve a balance involved the use of low-pressure sodium light sources for road lighting in cities adjacent to observatories. This approach was effective because astronomers could easily filter out the very limited range of wavelengths produced by the low-pressure sodium light source (see Figure 2.5). Unfortunately, this approach has fallen by the wayside, for two reasons. The first is that the nonexistent colour rendering properties of the low-pressure sodium light source make it an unattractive prospect for towns and cities. The second is that the growth in the use of light outdoors by commercial and residential property owners, using a wide range of light sources, has simultaneously increased the amount of light being emitted and undermined the effectiveness of the use of low-pressure sodium light sources by local authorities. Today, the most common advice given on how to reduce light pollution is to use what is called a full cutoff luminaire (see Section 4.2). Following such advice is straightforward because luminaires of this type are widely available. But this advice is flawed, for two reasons. The first is because restricting the upward light output from luminaires only limits the proportion of light emitted directly upward but ignores the light scattered after reflection from the illuminated surfaces. Light scattered after reflection is a major contributor to sky glow when viewing the sky from a position close to the sources of light.

The second reason why simply using full cutoff luminaires may not be the best solution to the problem of light pollution is that the advice considers the luminaire in isolation and not as part of a lighting system. Keith (2000) calculated the upward luminous flux produced by a roadway lighting installation, per unit area of road illuminated, including both light directly emitted upward and light reflected from the road and its surroundings. Figure 13.3 shows the upward luminous flux density for many different road lighting installations. The calculations were done for a collector road of 10 m width and of a specific reflectance, lit by lighting installations using a 250 W high-pressure sodium light source in full cutoff, cutoff, and semi-cutoff luminaires (see section 4.2), arranged in a staggered pattern and spaced so as to meet average road surface luminance and luminance uniformity criteria. What is interesting in Figure 13.3 is the fact that the upward luminous flux density increases as the lighting power density increases. What this implies is that if the use of full cutoff luminaires demands a closer spacing of the luminaires to meet the road lighting criteria, the lighting power density will be increased and, consequently, the upward luminous flux density will increase. Figure 13.3 also implies that a full cutoff luminaire is not necessarily the best option. In Figure 13.3, it appears that a semi-cutoff luminaire will give the lowest installed power density and the lowest upward luminous flux density, despite the fact that it may send a significant proportion of the lamp luminous flux directly up into the sky (Bullough 2002a). This conclusion should not be too surprising given the small differences in the amount of uplight produced by different road lighting luminaires (see Table 4.1).

However, this conclusion assumes that all directions of light emitted above the horizontal produce the same level of scattered light. For a site remote from major sources of light, in clear sky conditions, which describes an observatory in operation, this assumption is not true (Baddiley and Webster 2007). For these conditions, light emitted close to the horizontal causes more sky glow for most viewing directions than reflected light, because reflected light is usually obstructed by the buildings around it and because light emitted close to the horizontal has a greater path length through an atmosphere that contains a greater density of molecules and aerosols. This is particularly significant for road lighting. Road lighting luminaires are designed to project light along the road at shallow angles. When the road surface is dry and diffusely reflecting, the reflected light will be evenly distributed over the upward hemisphere, but when the road is polished by traffic or wet from rain, strong specular reflections will occur at angles close to the horizontal. Thus, road lighting

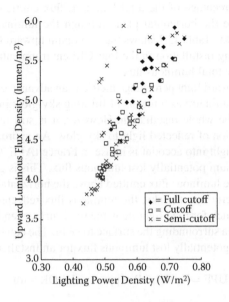

FIGURE 13.3 Upward luminous flux density (lm/m²), plotted against lighting power density (W/m²), for full cutoff, cutoff, and semi-cutoff luminaires. The values plotted are for a collector road of 10 m width, reflection class R3, lit by luminaires arranged in a staggered layout, and using a 250 W high-pressure sodium light source (after Keith 2000).

TABLE 13.7
Maximum Upward Light Recommended for a Complete Lighting Installation, Expressed as a Percentage of the Total Luminaire Luminous Flux (lm), in Different Environmental Zones

Environmental Zone	Maximum Permitted Upward Light for a Lighting Installation (%)
E1	0
E2	2.5
E3	5.0
E4	15.0

After ILE (2000).

might make a greater contributor to sky glow than would be expected from the total light output, a contribution that is enhanced by the forward lighting of vehicles using the road. This is particularly important when a major road points directly towards a sensitive site such as an observatory (Waldram 1972).

An alternative approach to limiting sky glow is to specify a maximum upward light ratio for the complete installation (ILE 2000). The upward light ratio of an installation is the percentage of the total luminous flux emitted by the luminaires that is emitted above the horizontal plane through the luminaires, i.e., that goes directly up into the sky. Table 13.7 shows the maximum upward light recommended for a complete lighting installation in the four CIE environmental zones, expressed as a percentage of the total luminous flux.

The maximum upward light percentage metric is an advance on the simple advice to use a full cutoff luminaire as a means of limiting sky glow in that it is quantitative and applies to the whole installation. However, it is still inadequate in that it ignores the contribution of reflected light to sky glow. An approach that takes both direct and reflected light into account is in use in France (AFE 2006). This method calculates the maximum potentially lost luminous flux. This is given as the sum of three components: the luminous flux emitted above the horizontal plane through the luminaires, i.e., directly into the sky, the luminous flux reflected from the surface to be illuminated, which for road lighting is the road surface, and the luminous flux reflected from the area surrounding the surface to be lit. The equation used for calculating the maximum potentially lost luminous flux for an installed luminaire is

$$UPF = F_{la} \, (ULOR + p_1u + p_2(DLOR - u))$$

where UPF is the maximum potentially lost luminous flux (lm), F_{la} is the luminous flux emitted by the light source in the luminaire (lm), ULOR is the upward light output ratio, i.e., the proportion of luminous flux emitted by the light source that leaves the luminaire above the horizontal plane through the luminaire, p_1 is the reflectance of the surface to be illuminated, u is the utilization factor for the surface to be illuminated, i.e., the proportion of the luminous flux emitted by the light source that reaches the surface to be illuminated directly, p_2 is the reflectance of the surface surrounding the surface to be illuminated, and DLOR is the downward light output ratio, i.e., the proportion of luminous flux emitted by the light source that leaves the luminaire below the horizontal plane through the luminaire.

The UPF is a maximum because it assumes that the area to be lit and its surroundings form a completely flat plane. It therefore ignores any reduction in luminous flux reaching the sky because light is absorbed or blocked by vegetation, buildings, or even the topography of the site and its surroundings. The UPF describes potentially lost flux because in urban situations, it may be that light reaching the surrounding area is itself useful in enhancing the safety of pedestrians and the quality of the visual environment.

The UPF is a minimum when the ULOR is zero and the utilization factor, u, is the same as the DLOR, i.e., all the luminous flux emitted by the lamp goes directly to the surface to be illuminated and nowhere else. The ratio of the actual UPF to the minimum UPF is called the upward flux ratio (UFR) and can be used as a measure of the efficiency of the installation.

Table 13.8 shows the results of calculations of UPF for a road lighting installation, using three different luminaire types representative of those used for road lighting in France. The road is 7 m wide, 1 km long, and has a reflectance of 0.08. The surroundings have a reflectance of 0.10. The lighting is designed to produce an average initial illuminance of 20 lx, with an overall luminance uniformity ratio of 0.4,

TABLE 13.8

Installation Details, Achieved Photometric Conditions, Maximum Potentially Lost Luminous Flux (UPF), and Annual Electricity Consumption for Three Road Lighting Installations of the Same Kilometre of Road, Using Flat Lens, Shallow Bowl, or Deep Bowl Luminaires

Quantity	Flat Lens Luminaire	Shallow Bowl Luminaire	Deep Bowl Luminaire
Luminaire mounting height (m)	8	8	10
Luminaire spacing (m)	26	32	35
Tilt (degrees)	0	5	20
Overhang (m)	0	0	1
Utilization factor	0.38	0.43	0.30
ULOR	0	0.01	0.03
DLOR	0.79	0.81	0.79
Number of luminaires/km	38	31	29
Average initial illuminance (lx)	22	20	20
Overall luminance uniformity	0.4	0.4	0.4
Longitudinal luminance uniformity	0.70	0.72	0.72
Threshold increment (%)	8	10	12
Maximum potentially lost luminous flux/ km (lm/km)	28,500	26,815	49,300
Direct upward luminous flux/km (lm/km)	0	3,255	14,355
Luminous flux reflected by road and surroundings/km (lm/km)	28,500	23,560	34,945
Annual electricity consumption/km (kWh/ km)	17,267	14,086	19,604
After AFE (2006).			

a longitudinal luminance uniformity ratio of 0.7, a maximum threshold increment of 15 percent, and is to be used for 4000 hours per year. Three different luminaire types are used, one with a flat lens, one with a shallow bowl, and one with a deep bowl, enclosures that approximate full cutoff, cutoff, and semi-cutoff luminaires, respectively. All three luminaires are used in a single-sided arrangement. The flat lens and shallow bowl luminaires are equipped with 100 W high-pressure sodium light sources, which have an initial luminous flux of 10,500 lm. The deep bowl luminaire is fitted with a 150 W high-pressure sodium light source, which has an initial luminous flux of 16,500 lm.

Close examination of Table 13.8 is worthwhile. Despite achieving very similar photometric conditions on the road, there are significant differences in maximum potentially lost luminous flux/kilometre between the three luminaire types. Not surprisingly, the deep bowl luminaire with its 3 percent upward light output ratio (ULOR) produces the greatest maximum potentially lost upward luminous flux/km.

What is interesting is that the shallow bowl produces the lowest maximum potentially lost upward luminous flux/km, lower than that of the flat lens luminaire. This is due to the fact that to achieve the same photometric conditions, a closer spacing is required with a flat lens luminaire than with a shallow bowl luminaire. As a result, the flat lens luminaire has more lamps per kilometre than the shallow bowl luminaire. This also explains why the annual electricity consumption/kilometre is greater for the flat lens luminaire than for the shallow bowl luminaire. Also of interest are the relative amounts of upward luminous flux/kilometre emitted directly and after reflection. The three luminaire types show different amounts of direct upward luminous flux, as would be expected from their different ULOR values, but, for all three, the amount of luminous flux reflected upward is greater than that emitted directly upward. The folly of ignoring reflected luminous flux when attempting to limit sky glow for observers close to the source of light should be apparent.

The OSP "shoebox" system described in the discussion of light trespass (see Section 13.7.1) can also be used to examine the amount of luminous flux lost from a property, directly and by reflection. For this purpose, the average illuminance is calculated on all five planes of the shoebox. The average illuminance for each plan is multiplied by the area of the plane and the five resulting luminous flux values are summed. This sum gives an estimate of the total luminous flux leaving the property. This is a worst-case estimate of the contribution to sky glow because it is unlikely that all the luminous flux leaving the property will be directed upwards. Nonetheless, the ability to handle light trespass and sky glow with the same calculation system is a definite plus. Further, the sides of the shoebox could be subdivided to emphasize light emitted around the horizontal should this be desired.

All these metrics, upward light ratio, maximum potentially lost luminous flux, and the total flux from the "shoebox," are attempts to quantify the amount of upward light generated by individual lighting installations. But if the number of exterior lighting installations continues to increase and illuminances used in those installations continue to increase, then sky glow will inevitably increase. This situation is analogous to what is happening with carbon emissions from motor vehicles. The engines of individual motor vehicles have become more and more efficient but total carbon emissions continue to increase because the number of vehicles continues to increase. To effectively limit the contribution of human activity to sky glow there are two complementary options.

The first option is to limit the amount of light used at night, accepting that if taken too far such limits will increase the risk to life and limb through decreases in visibility and consequent increases in accidents and/or crime. The International Dark-Sky Association (IDA) has attempted to influence the amount of light used at night by providing a pattern code of practice that contains, along with much other useful advice, a maximum luminous flux density for outdoor lighting in large developments, expressed in lumens per acre (Table 13.9). These maximum luminous flux densities are applicable to any new commercial or residential development, in each of the four CIE zones, plus one additional zone called Zone E1A and defined as a Dark-Sky Preserve (IDA 2002).

These maximum luminous flux densities encourage the use of fully shielded luminaires, i.e., luminaires with no light output above the horizontal plane through

TABLE 13.9

Maximum Total Luminous Flux Density, in Lumens/Acre, for Commercial, Industrial, and Residential Exterior Lighting in Different Environmental Zones

Application	Zone E4	Zone E3	Zone E2	Zone E1	Zone E1A
Commercial and Industrial — all luminaires	200,000	100,000	50,000	25,000	12,500
Commercial and Industrial — unshielded luminaires only	10,000	10,000	4,000	2,000	1,000
Residential — all luminaires	20,000	10,000	10,000	10,000	5,000
Residential — unshielded luminaires only	5,000	5,000	1,000	1,000	0

After IDA (2002).

the luminaire, the limits being much more stringent for light provided by unshielded luminaires. While such maximum luminous flux densities have the advantage of simplicity, they may be over-simple because they take no account of the fact that different activities in a zone may require different amounts of light to be performed safely. A better approach would be to adopt maximum lighting power densities for different activities in each zone. This is the approach adopted by the California Energy Commission in 2005 (CEC 2005). Maximum lighting power densities are specified for many outdoor lighting applications around buildings, such as parking lot lighting, retail display areas, façade and canopy lighting, and advertising signs. Unfortunately, the lighting of public roads is presently excluded from these limits, although why this should be so is not clear.

Limiting power density will, unless there is a sudden major improvement in light source luminous efficacy, limit the installed luminous flux but it will not guarantee minimum upward light unless the design of the lighting is also controlled in some way. The Californian regulations attempt to do this in an ad hoc way by prescribing the equipment that should be used, e.g., all exterior luminaires using light sources of more than 175 W shall have a luminous intensity distribution conforming to the cutoff category (see Section 4.2), even though the cutoff category allows anything from zero to 16 percent of the light source luminous flux to be emitted directly up into the sky (Bullough 2002a; LRC 2007). A better approach, in that it would allow the lighting designer more freedom while still limiting sky glow, would be to use the OSP "shoebox" system to specify the maximum amount of luminous flux/unit lit area that should be allowed to leave a site, a different maximum being used for each of the different environmental zones.

The second option for reducing the sky glow is to pay careful attention to the timing of the use of light. Light pollution is unlike most other forms of pollution in that when the light source is extinguished, the pollution disappears very rapidly. This suggests that a curfew defining the times when lighting should not be used, or should

TABLE 13.10

Maximum Luminous Intensity (cd) to Control Glare for the Four Environmental Zones, before and during a Curfew

Environmental Zone	Maximum Luminous Intensity before Curfew (cd)	Maximum Luminous Intensity during Curfew (cd)
E1	0	0
E2	20,000	500
E3	30,000	1,000
E4	30,000	2,500

After ILE (2000).

only be used at a reduced level, could have a dramatic effect on sky glow. Curfews are commonly attached to the planning consents for sports lighting installations. The possibility of dimming road lighting in the middle of the night when traffic densities are low is a variation on this theme (see Section 14.4.2).

13.7.3 Glare

Both road lighting and vehicle lighting can cause both disability and discomfort glare (see Sections 4.3 and 6.6). When glare is the cause of a complaint of light pollution, it is usually associated with road lighting rather than vehicle lighting and will be made by someone outside the lit area. Drivers, who are in the lit area, may complain about glare from the road lighting or from vehicles, but that glare is not light pollution. Bullough et al. (2008) have shown that discomfort glare from outdoor lighting is influenced by the illuminance received at the eye directly from the glare source, the ratio of this illuminance to the illuminance received at the eye from the immediate surround to the glare source, and the ambient illuminance at the eye. The greater the illuminance received directly from the glare source, the higher the ratio of source/surround illuminances and the lower the ambient illuminance, the greater will be the discomfort. This understanding has recently been incorporated into the OSP "shoebox" with the result that one approach can deal with light trespass, sky glow, and glare (Brons et al. 2008).

The solution to a complaint of light pollution due to glare is to reduce the luminous intensity of the luminaire that is the subject of the complaint in the direction of the complainant. This can be done by changing the luminaire type, by re-aiming the luminaire or by fitting baffles to the luminaire to confine the luminous flux emitted by the luminaire more closely to the area that has to be lit. To indicate when such actions might be justified, maximum luminous intensities for each of the four environmental zones, before and during curfew, have been suggested by the ILE (2000). Table 13.10 shows these maximum luminous intensities. These maxima should be applied in whatever direction is relevant to the complaint, i.e., from the luminaire to the complainant.

13.8 SUMMARY

The constraints that limit lighting for driving range from the legal through the financial to the environmental. Vehicle lighting meeting specific photometric conditions is a legal requirement in almost all countries. Road lighting is not. Both vehicle lighting and road lighting imply costs. For vehicles, the first costs of lighting are included in the cost of the vehicle. For road lighting, both first costs and life cycle costs can be involved in the decision whether or not to install road lighting and certainly influence the amount and form of road lighting provided. As for environmental factors, these can take four forms: the amount of energy consumed, the carbon dioxide emissions produced when generating the electricity used, the disposal of the produce at the end of life, and light pollution. This chapter is devoted to a consideration of these constraints as they apply to road lighting.

The first decision that has to be made about road lighting is whether to install it or not. There are two approaches to making this decision. One is prescriptive and involves meeting a series of quantitative and qualitative criteria related to traffic flow, night/day accident ratio, and road layout The other is a comparison of the costs and benefits of road lighting, over a number of decades. Over such a time scale, both costs and benefits are subject to large uncertainties so the apparent precision of this method is illusory.

Having decided to install road lighting, the next constraint is its cost. This comes in three forms: first cost, energy cost, and life cycle cost. Calculations of these three costs for a number of typical UK column-mounted installations reveal that the first cost is about one-third to one-half of the 40-year life cycle cost, which includes energy costs. An important determinant of first cost is the spacing between columns. The closer the spacing, the greater the number of columns and electrical connections required, and columns and connections are more expensive than lamps and luminaires. This raises the question of the relative importance of different road lighting criteria. The average road surface luminance is the criterion that gets the most attention because it obviously influences first cost, but the two uniformity criteria, minimum overall luminance uniformity ratio and minimum longitudinal luminance uniformity ratio, are also important for first cost because they are critical for determining the column spacing.

Road lighting can also be constrained by four aspects of the environment: the use of natural resources, carbon dioxide emissions, waste disposal, and light pollution. Minimizing the use of natural resources is directly linked to the amount of electricity used for road lighting and the fuel mix used to generate that electricity. A desire to reduce carbon dioxide emissions makes road lighting an obvious target for cuts. It consumes electricity, it is conspicuous, and it is in use at times when there are few people about. However, before cutting road lighting it is important to remember its value for traffic safety and that the amount of carbon dioxide emitted will depend on how the electricity used to power the road lighting is generated. Different countries use very different fuel mixes for electricity generation. There is little point in limiting road lighting in countries where the fuel mix used to generate electricity produces little carbon dioxide, but where gas, oil, or coal are extensively used for electricity generation, then the need for road lighting should be carefully considered. Another growing constraint on road lighting is the ease of disposal of equipment at the end of life. Questions about waste disposal can take two forms, the hazard to human health posed by the materials in the product and the ease of recycling. The elements of a road lighting installation that cause the most concern are the light source and the control gear. Light sources that contain mercury are likely to be subject to stricter regulation than those that contain sodium. Electronic control gear is likely to be more expensive to dispose of than electromagnetic control gear.

There is also growing concern about light pollution. Light pollution is evident as localized light trespass and glare and a more general increased sky luminance called sky glow. Light trespass and glare can be avoided by the careful positioning and aiming of luminaires with appropriate luminous intensity distributions. For road lighting, this means choosing luminaires that direct the vast majority of the light emitted onto the road surface and minimize the amount of light emitted beyond that surface, on both sides of the road. If complaints of light trespass or glare still occur, then the classic response is to fit what is called a house-side shield to the luminaire.

Sky glow is evident over most cities and towns in the form of a glowing, flattened dome of light. The effect of sky glow is to make many stars invisible because their luminance contrast is reduced to below threshold. There is a small natural contribution and a variable human contribution to sky glow. The magnitude of the human contribution is dependent on the density of air molecules and aerosols in the atmosphere and the amount of light traversing the atmosphere. The greater the amount of scattering material and light in the atmosphere, the greater the sky glow.

There are a number of approaches being introduced to road lighting practice to minimize light pollution. The most common is the simplest, using only full cutoff luminaires that do not emit light directly up into the sky. This approach is good for limiting glare but less useful for reducing sky glow because it ignores light reflected from the road surface. It may also increase costs because the use of full cutoff luminaires often requires closer spacing to meet the luminance uniformity criteria. Another approach takes both direct and reflected light into account and measures the efficiency of a road lighting installation by calculating the proportion of light emitted by the luminaires that directly illuminates the road surface. By maximizing this proportion, the amount of wasted light that contributes to sky glow is minimized. Yet another approach is based on the idea of a virtual, inverted "shoebox" being placed around and over a property. For road lighting, the "shoebox" is extended along the axis of the road as far as desired. For roads, the idea is to specify the maximum luminous flux that should be allowed to cross the vertical planes parallel to the road and the "ceiling" plane over the road. This approach can be used to control sky glow, light trespass, and glare.

All three of these approaches are attempts to quantify the amount of upward light generated by individual lighting installations. But if the number of exterior lighting installations continues to increase and the illuminances used in those installations continue to increase, then sky glow will inevitably increase. Attempts have been made to limit the amount of light used outdoors through the planning system by classifying locations into different environmental zones and then setting either maximum luminous flux density limits or power density limits for each zone. An alternative approach that is sometimes used for sports facilities and could be used for road lighting is a curfew that restricts the use of the lighting to certain times. Long term, limiting the amount of light that can be used and the times when the lighting can be used are the only approaches to reducing sky glow that are likely to be successful.

All forms of lighting are subject to constraints. Road lighting is no exception. Cost constraints will always be present but the importance attached to the environmental constraints will vary over time as their political impact changes. Given the increasing evidence for the contribution of carbon emissions to climate change and the emotional appeal of such slogans as "Our children will not see the stars" it is reasonable to predict that, in the future, road lighting will be under greater pressure to justify its existence than in the past.

14 Envisioning the Future

14.1 INTRODUCTION

This chapter is different from all the others in that it is an attempt to envision the future so it contains more opinions than facts. These opinions are based on a synthesis of the research and practice that has been the foundation of the previous chapters but they are still opinions. The future can be envisioned at two levels. One attempts to predict what is most likely to happen. The other attempts to set out what should happen. The former is relatively straightforward because it is limited in time and is likely to be driven primarily by costs and emerging technology. What these two factors suggest for lighting for driving is that over the next decade vehicle forward lighting will become more sophisticated with elements of the adaptive forward lighting system most attractive to vehicle purchasers trickling down from up-market vehicles to more modestly priced vehicles. Vehicle signal lighting will change little apart from the wider use of LEDs and the adoption of daytime running lamps as a requirement by more countries. Road lighting is likely to see much greater use of metal halide light sources, much less use of low-pressure sodium light sources, and the eventual introduction of the LED as a practical general lighting light source (McSweeney 2007). In addition, electronic control gear and monitoring networks that enable road lighting to be dimmed on demand are likely to be more widely adopted. As for road markings and signs, these are likely to continue to grow in number and complexity as combinations that reinforce perceptual cues have been found to be effective in influencing drivers' behaviour (Charlton 2007).

Attempting to set out what should happen to lighting for driving is simultaneously more difficult and less likely to be successful. It is more difficult because it involves a longer time scale so it is more open to unforeseen influences. It is less likely to be successful because what should happen is a fundamental review of current practice by vehicle manufacturers, road engineers, and illuminating engineers working together, but these are groups that do not normally interact. Nonetheless, without ideas about how lighting for driving might be made more effective in the future, any action is unlikely. The approach taken here starts by identifying the fundamental problems facing the lighting used while driving and then seeks to identify solutions to those problems, solutions that are effective and without unintended consequences. Attention will be given first to vehicle lighting because it is used whenever the vehicle is in motion after dark, and sometimes during the day, regardless of whether the road is lit or not. This will be followed by a discussion of the future of road lighting and possible developments in road markings, signs, and signals. Finally, the factors that may determine whether or not these opinions become facts are considered.

319

14.2 VEHICLE FORWARD LIGHTING

14.2.1 PROBLEMS

Vehicle forward lighting has changed dramatically over the last decade but it still faces two fundamental problems. The first is the limited range of visibility available when using low-beam headlamps on unlit roads, which are the majority of roads. This is a problem because stopping distances at the speeds usually driven are much longer than the visibility distance (see Section 6.4). Whether or not stopping distances matter will depend on the opportunity for maneuvre. For a pedestrian at the edge of a wide road, the driver may have the opportunity to swerve rather than stop but on a narrow road or a road with heavy traffic swerving may not be possible and when encountering a large vehicle stationary in the lane of a busy road, stopping may be the only option. Given that an emergency stop is necessary but the stopping distance is longer than the visibility distance, a collision is inevitable, although the consequences may be reduced from death to injury if the vehicle's speed can be lessened before impact.

It is important to appreciate that visibility distances vary greatly with the size and luminance contrast of the target. Targets of small size and low luminance contrast have much shorter visibility distances than targets of large size and high contrast. This partly explains why the type of accidents that show the biggest effect of the removal of daylight are those involving small, low-contrast targets such as pedestrians who tend to wear dark clothing (see Section 1.5). The number of miles driven on unlit roads where limited visibility distances occur is increased by the underuse of high-beam headlamps (see Section 6.4). This underuse arises from the second problem that vehicle forward lighting faces, glare between opposing vehicles. When facing an approaching vehicle drivers are taught to change from high-beam to low-beam headlamps in order to limit the extent to which the vision of the other driver is degraded. If such switching is required frequently, there is a tendency to keep the headlamps on low beam regardless of the reduced visibility distances resulting. Of course, the impact of glare from headlamps is not confined to the choice between low- or high-beam headlamps. The main effect of glare from oncoming headlamps is to reduce the luminance contrasts of all that lies ahead (see Section 6.6). This reduction is due to the veiling luminance created by light from the oncoming headlamps being scattered in the eye, at the windscreen, and in the atmosphere. This is particularly a problem for old drivers who experience much more scatter in the eye than young drivers (see Section 12.4). Further, because the additional light from the oncoming headlamps changes the state of adaptation of the retina, immediately after the oncoming vehicle has passed, the driver is likely to be misadapted and it takes a finite period of time to recover the necessary sensitivity. Recovery of sensitivity after exposure to glare is again a particular problem for old drivers who, because of increased absorption of light on passage through the eye, are effectively working at a lower light level than young drivers and hence take longer to recover from exposure to glare. This conflict between visibility and glare has been a feature of the design of vehicle forward lighting ever since its inception.

14.2.2 SOLUTIONS

The first thing to say is that tinkering with existing headlamp luminous intensity distributions and spectra is unlikely to solve the problem of limited visibility distances caused by the use of low-beam headlamps when it is necessary to reduce glare to other drivers (Sivak 2002). What is required are much more dramatic shifts in approach. There are four possible solutions to the problem of the short visibility distances available with low-beam headlamps. The first is to have road lighting on all roads. Good road lighting extends visibility distances, but this is a counsel of perfection. The cost of lighting every remote country lane would be colossal and the complaints of those concerned with light pollution would be deafening.

A more realistic option is to do nothing. This might be called the Darwinian solution because it effectively considers pedestrians foolish enough to walk on unlit roads at night in dark clothes as in the shallow end of the gene pool and as expendable. Such an approach is implicit in the system for deciding whether or not to invest in road lighting (see Section 13.2). The number of accidents involving death or injury occurring at night within a set time is a factor to be considered when deciding whether or not to install road lighting. In addition to the adverse consequences for pedestrians, doing nothing really does do nothing to reduce the risk of accidents involving other unexpected low-contrast obstacles such as animals, unmarked vehicles, or unusual road configurations.

Doing nothing is unlikely to be acceptable to nations devoted to health and safety. Much more likely is a legislative solution. This could take two forms. One possibility would be to introduce lower speed limits on unlit roads after dark. Given a slow enough speed, the visibility distance of even low-contrast objects illuminated by low-beam headlamps can be made greater than the stopping distance. Even if the speed limits required to achieve this objective were to prove to be too low to be politically acceptable, any reduction in speed at night would reduce the damage to people and property should a collision occur. The disadvantage of setting lower speed limits on unlit roads after dark or in conditions of poor visibility would be the increase in journey times and the risk of increased traffic congestion, although this might be offset for some authorities by the possibility of increased revenue derived from the speed cameras used to enforce the lower speed limits.

Another legislative possibility would be to demand that unlighted users of the roads, such as pedestrians and some cyclists, should wear clothing fitted with a minimum area of retroreflective material at night. Luoma et al. (1995b) measured the distances at which pedestrians in dark clothing with and without retroreflective material could be recognized while walking beside or crossing the road. The observers were passengers in a car being driven with low-beam headlamps on unlit rural roads who had been instructed to respond only to pedestrians. Figure 14.1 shows the mean recognition distances for different forms of retroreflective material for pedestrians walking towards the car at the edge of the road and crossing the road. Clearly, almost any form of retroreflective material increases the recognition distances dramatically, but those mounted on parts of the body that are in frequent motion when walking are more effective than those that are not, especially for crossing the road. Such findings explain why those employed to work on and near the roads, by day or night, use high

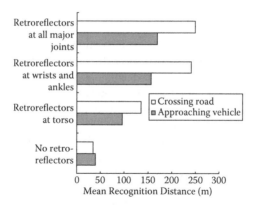

FIGURE 14.1 Mean recognition distances (m) for pedestrians approaching a vehicle while walking along the edge of the road or crossing the road. The recognition was made by passengers in a car being driven using low-beam headlamps, on a rural road at night. The pedestrians wore dark clothing, some being fitted with retroreflective material in various places (after Luoma et al. 1995b).

visibility suits, jackets, or vests, clothing that contains fluorescent and retroreflective materials. Such clothing has not yet been adopted by the general public for use at night, but given the example of the dramatically increased use of cycling helmets over recent years it is not beyond the realm of possibility that promotion of the idea would make legislation on this matter unnecessary.

The third solution would involve technology applied to the vehicle. Technology to automatically switch between high-beam and low-beam headlamps depending on the presence of approaching vehicles is already commercially available. The widespread adoption of this technology would gradually eliminate the problem of underuse of high-beam headlamps but do nothing for visibility distances when low-beam headlamps were in use because of the proximity of approaching vehicles. This problem might be solved by the provision of low-mounted side lighting for use when two opposing vehicles approach at night. A wide, flat luminous intensity distribution, similar to that of a fog lamp, emitted from the side of each vehicle so as to illuminate the adjacent lane would go some way to overcoming the effects of disability glare and the limited range of low-beam headlamps. The operation of the side beam could be linked to the automatic switching between high- and low-beam headlamps so that the side beam was only lit when low-beam forward lighting was in operation and an approaching vehicle had been detected.

Some forms of adaptive forward lighting (see Section 6.7.2) are also becoming more widely available but they do little to address the two fundamental problems. Rather, they ensure a better match of forward lighting to the prevailing conditions, conditions that can range from manoeuvring around curves, driving on motorways, in towns, on wet roads, or in fog. It is thought that a better match of forward lighting to prevailing conditions will help reduce some types of accident involving pedestrians (Sullivan and Flannagan 2007).

The solution that is most likely to be successful in addressing the two problems faced by vehicle forward lighting is the use of supplementary information to the driver, based on sensors. As discussed in Section 6.7.3, both passive and active infrared systems can be used to generate an image of the road ahead and to display that image on a screen so that it is accessible to the human visual system. The passive infrared system uses the long wavelength infrared radiation associated with the different temperatures of the various elements of the scene to create the image. The active infrared system emits short wavelength infrared radiation and uses the differences in reflection from different elements of the scene to create the image. As the passive infrared system does not emit any additional electromagnetic radiation and the active infrared system only emits electromagnetic radiation outside the visible wavelength range, neither can cause glare to the human visual system, which means both can be kept perpetually in a state approximating high beam, even when other vehicles are approaching or close ahead.

How effective such supplementary information can be is shown by the work of Sullivan et al. (2004b). They had groups of young and old drivers drive a car equipped with a passive infrared system around a closed, unlit circuit using low-beam headlamps. The passive infrared system was mounted in either a heads-up position at the base of the windscreen or lower down and to the right, close to the central console. While driving round the circuit, there were targets of three types to be detected, pedestrians and warmed simulations of deer and small animals, all located at the edge of the road and all having a reflectance of 0.10. Table 14.1 shows the percentage of targets detected and the mean distance at which they were detected, using unassisted vision and vision assisted by the passive infrared system in the two positions. Table 14.1 shows that the old drivers, whose mean age was 68 years, detected fewer of the targets than did the young drivers, whose mean age was 25 years, with and without the passive infrared system. When the targets were detected, the old drivers detected them at shorter distances than did the young drivers, with and without the passive infrared system. This is to be expected given the decline in visual and

TABLE 14.1

Percentage Detection and Mean Detection Distance (m) of Pedestrians, Simulated Deer, and Small Animals Beside the Road by Young and Old Drivers Using Unassisted Vision and Using Assisted Vision, the Assistance being Provided by a Passive Infrared System with a Head-Up Display (HUD) and a Display Mounted Down and to the Right (HDD)

Driving Performance	Unassisted Vision		Assisted Vision with HUD Display		Assisted Vision with HDD Display	
	Young	Old	Young	Old	Young	Old
Percent detection (%)	100	77	95	80	94	75
Mean detection distance (m)	57	32	85	51	115	48

From Sullivan et al. (2004b).

cognitive capabilities of old people. What is interesting is that, for both age groups, having the passive infrared system did not increase the percentage of detection but did increase the distance at which detection occurred, when it occurred. It is clear that providing supplementary information to the driver has the potential to alleviate the limitations of low-beam headlamps but not to overcome the effects of age.

While increasing the distance at which dark objects are detected on unlit roads is useful, there are a number of potential problems associated with the use of such imaging systems. The first is that of creating an awareness of the limitations of such systems, limitations that can cause discrepancies between what can be seen by the driver's visual system and through the displayed image. Discrepancies where the displayed image gives information above and beyond what is visible to the eyes of the driver are expected. After all, what is the purpose of an imaging system that does not reveal more than can be seen with the eyes alone? What is disconcerting is when the eyes detect something that the displayed image does not show. This is possible with the passive infrared system when the temperature of the object of interest is the same or similar to its immediate background, because then there is no or little luminance contrast for the object in the displayed image. This means that the passive infrared system is good for revealing animate objects, such as pedestrians or animals, but not good at showing up inanimate obstacles in the road, such as debris. Even animate objects can disappear from the image if the background to the object has a similar temperature to the object itself, e.g., a human standing against a wall that has been warmed by the sun. Similar disconcerting discrepancies can occur with the active infrared system, but now their occurrence depends on the relative radiances of the object and its immediate background, values that depend on the radiant intensity distribution of the emitter on the vehicle and the reflectances of the objects in the relevant wavelength range. Awareness of the limitations of such systems is important if undue reliance on the resulting images is to be avoided. They are best considered as a means to provide advance warning of something ahead of the visual range that needs the attention of the driver's visual system.

Another potential problem is that of cognitive overload. Presenting information as a displayed image requires the driver to fixate that image, during which time the driver will not be looking at the road. How long it takes to extract information from the display will depend on where it is positioned, how similar the display is to what can be seen through the windscreen, and how much information the display contains, as well as how much practice the driver has had with the system. How important such considerations are is open to question. Sullivan et al. (2004b) found no statistically significant differences in percentage detection nor mean detection distances for the two forms of passive infrared system that differed in where the display was located (see Table 14.1). Further, indirect measurements of workload, obtained from either the NASA Task Load Index questionnaire or the high-frequency component of steering, failed to show any difference in workload when the passive infrared imaging system was used to supplement the driver's vision, although there was an increase in workload when the driver was told to search for targets. It would appear that, in this case, the workload was determined by the task defined rather than the means used to do it. Of course, these results were collected in undemanding conditions. There was no other traffic and the road layout was simple, both situations that would make performance insensitive to time spent looking at the display. Whether the position of the display becomes more important in more demanding conditions remains to be determined. As for the form of the display, the increasing use of satellite navigation displays is fortuitous as it has provided a lot of experience of the best ways to deliver information to drivers through displayed images. The images used for satellite navigation tend to be schematic rather than actual and the visual information is often augmented by auditory information. This suggests that there is a lot to be said for delivering the information about what lies ahead in a predigested form, in two sensory modes, rather than as a simple image that the driver has to analyze. Predigested information should be limited to what is essential even if it is restricted to a warning that there is something beyond the range of the headlamps that deserves attention (Tsimhoni et al. 2005).

Finally, there remains a question about the effect of such systems on driver behaviour. For example, when workload is increased, drivers tend to reduce speed (Lansdown et al. 2004), but when risk is reduced there is a tendency to increase speed (Stanton and Pinto 2000). It seems likely that supplementary information provided through displayed images will increase workload in some situations and, if effective, decrease risk, so what speed drivers will choose when driving on low beams is unclear. What can be said is that Sullivan et al. (2004b) found that the mean speed was reduced slightly when the drivers were asked to search for the targets, but there was no difference in mean speeds when searching using vision alone and when aided by the night vision system. However, Nilsson and Alm (1996), using a driving simulator, found that a display giving a small clear window through fog resulted in an increase in speed. By now it should be apparent that while providing supplementary information to the driver through infrared systems has the potential to overcome the problems of vehicle forward lighting, it is not a panacea. The success of such systems will depend very much on how the information is presented, how drivers are trained in their use, and how they choose to use them.

Yet another possibility for improving forward lighting, but one that is presently at the experimental stage, is active lighting (Wordenweber et al. 2007). Active forward lighting consists of three elements, a surround sensing system that searches what lies ahead of the vehicle, a computer system that looks at the results of the search so as to detect other vehicles and hazards and then to decide how to alter the light distribution of the headlamps, and a headlamp system that is capable of a wide range of luminous intensity distributions. The surround sensing systems can be based on microwave radar, infrared laser radar, or artificial vision using video cameras, the different systems varying in their range, resolution, field of view, sensitivity to weather conditions, and the amount of processing power required. The headlamp can consist of multiple light sources, such as LEDs, so that by modulating the current supplied to each LED, the luminous intensity distribution can be varied over a wide range. The great advantage of such a system is that it has no moving parts. Alternative systems based on digital micro-mirrors or on a scanning mirror do have moving parts. In the former, light from a single source is reflected from an array of micro-mirrors. Information from the surround sensing system is used to adjust each of the micro-mirrors so as to shape the headlamp luminous intensity distribution to the required conditions. In the latter, light from a single source is reflected from a single mirror that is turned and tilted to scan the road ahead. By cutting off the light from the light source at appropriate times, the desired light distribution can be created. Two examples of how active forward lighting might be used are glare-free high-beam lighting and marker lighting. For glare-free high-beam lighting, the surround sensing system detects the approach of opposing vehicles and their location and eliminates the part of the headlamp high beam that directly reaches the opposing driver's eyes. Figure 14.2 shows the illuminance reaching the opposing driver's eye at different distances from a vehicle equipped with an active lighting system using a digital micro-mirror headlamp. For the marker light, the surround sensing system is used to detect unlit objects of interest beyond the range of the low-beam headlamps. Once detected, a very narrow beam is used to spotlight the object, thereby drawing the attention of the driver to the object and giving the driver more time to respond.

Active forward lighting represents a triumph of technology, but whether it will ever be widely adopted will depend on its reliability in detecting, classifying, and identifying the appropriate response for the diverse objects of interest to the driver. One advantage active lighting does have over the supplementary information approach using visual displays is that of immediacy. The supplementary information approach requires the driver to consciously scan the display and reach the appropriate decisions. Active lighting does this for the driver so the response time to an unexpected stimulus should be shorter. However, it is possible that the supplementary information display approach will be developed to such an extent that the information collected will be predigested sufficiently to eliminate the difference in response time. Time will tell, along with cost and reliability.

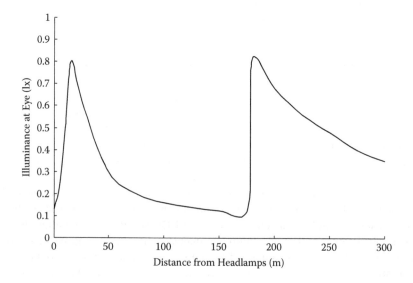

FIGURE 14.2 The illuminance (lx) reaching the opposing driver's eye at different distances from a vehicle equipped with an active lighting system using a digital micro-mirror headlamp. The detection range of the sensor system is from 25 m to 180 m (after Wordenweber et al. 2007).

The ultimate solution to the problems of vehicle forward lighting is to do away with the driver altogether. This is one aim of competitive research programs into driverless vehicles being undertaken in the EU and the United States. Different programs have different objectives. Some are for civilian use and apply to vehicles on paved roads. Some are for military use and need to be capable of operating on a wide range of surfaces. Some apply only to specific sections of road where the infrastructure has been modified to match the capabilities of the driverless vehicles. Others assume vehicles moving in groups, called platoons, under some form of central control. The most adventurous programs are seeking to develop a truly autonomous driverless car that can finds its way through traffic, anywhere, by day or night, using only its sensors, software, and actuators. Given the increase in the sophistication of vehicle systems that has occurred over the last 50 years, the driverless vehicle is not science fiction, but whether or not it will ever be acceptable to the market is open to question.

14.3 VEHICLE SIGNAL LIGHTING

14.3.1 PROBLEMS

Vehicle signal lighting as currently practiced has only one problem, most of its limitations being more a matter of missed opportunities than problems. Vehicle signal lighting is designed to inform other drivers of the presence of the vehicle or changes in the movement of the vehicle. The problem faced by vehicle signal lighting is adaptation to prevailing conditions. This is best exemplified by the use of rear fog lamps. Rear fog lamps have a much higher luminous intensity than rear position lamps so as to ensure they are visible through dense fog (see Section 7.12). The problem occurs when drivers use their rear fog lamps in faint mist, the result being glare for drivers immediately behind. As for the missed opportunities, these are essentially forms of signal lighting that might be used to convey more information to other drivers, either the same information as presently given but in a more stable form, or information above and beyond that currently supplied.

14.3.2 SOLUTIONS

The problem of matching the signal lighting used to indicate the presence of the vehicle to prevailing conditions can be readily solved with existing technology. Many vehicles today have a photocell sensor that detects changes in ambient light levels so that forward lighting can be turned on or off as needed. It does not call for too great a leap of imagination to see the same system being used to provide a continuous adjustment of luminous intensity rather than to make a simple on/off switch. Such a system could be applied to front and rear position lamps, side marker lamps, rear fog lamps, and daytime running lamps so as to ensure conspicuity without glare. Of course, a similar adjustment would need to be applied to intermittently lit signal lighting designed to convey an intended change in movement, such as turn lamps, hazard flashers, and stop lamps, to ensure they could still be easily discriminated from the continuously lit signal lighting.

As for the opportunities, the easiest to implement would be lighting to ensure stable perception of movement in varying conditions. As shown in Figure 4.2, a vehicle moving on a road at night has multiple luminance contrasts. In fact, there are so many that it is very unlikely that the vehicle will ever become invisible. What is much more likely is that the pattern of luminance contrasts associated with the vehicle will change as it moves along the road, particularly if the road is lit and wet, because then the background luminance against which the vehicle is seen will vary much more than when the road is unlit and dry. The opportunity is to use signal lighting to produce a stable pattern of luminance contrasts. There are two possible approaches. The first is to use contour lighting (Wordenweber et al. 2007). Contour lighting involves the outlining of the vehicle by self-luminous means, such as regularly spaced marker lamps or side emitting fibre optics, or by retroreflective material. Contour lighting is sometimes used on heavy trucks but is rarely used on other vehicles. The other approach is diffusing lighting placed under the vehicle. The idea behind this approach is to ensure that the luminance of the road against which the vehicle is seen is relatively stable so that the pattern of luminance contrasts associated with the vehicle is itself relatively stable. This is something that is already available in the after-market, although it has acquired an unfortunate social connotation and has been banned in some jurisdictions. For under vehicle lighting to be effective, regulation would be required to fix the amount, distribution, colour, and stability of the lighting.

Both these approaches seek to provide a stable outline for the vehicle. The value of such a stimulus is that drivers can detect outlines faster through the magnocellular channel of the human visual system than information that passes through the parvocellular channel, such as colour (see Section 3.2). Further, an outline enables the type of vehicle to be identified and that, in turn, enables the experienced driver to have some idea of the type of movement of which the vehicle is capable. Of the two approaches, contour lighting is probably the more useful as it will be visible from many different directions over greater distances and, when provided by retroreflective material, can signal the presence of the vehicle to other vehicles even when there is no power available for lighting. Such information may be useful in reducing the number of rear-end collisions at night, particularly collisions with vehicles broken down on the road (Morgan 2001; Sullivan and Flannagan 2004).

Another area where vehicle lighting could be used to provide more information important to other drivers is related to the perceptions of speed and distance. These perceptions require changes in the stimuli presented to the visual system relative to each other. By day this is not a problem because the vehicle can be seen against a structured background. By night, on an unlit road, the perceptions of speed and distance are much more difficult, particularly when the vehicle is coming directly towards or away from the driver. In this situation, perceptions of speed and distance for vehicles approaching and vehicles ahead but moving in the same direction are governed by changes in the relative position of pairs of headlamps and rear position lamps, respectively. The contour lighting discussed above would make speed and distance easier to judge, as do well-separated pairs of headlamps and rear position lamps. What is not helpful is having a single headlamp or rear position lamp, as is effectively the case for motorcycles, or a single rear fog lamp or a single rear reversing lamp as is allowed by some vehicle lighting regulations. Judging speed and distance from the expansion or contraction in the angular size of a single small lamp is much more difficult than using the change in separation between pairs of lamps. This suggests that single rear fog lamps and reversing lamps on cars and vans should be forbidden. For such vehicles, all signal lamps should be in pairs mounted on opposite sides of the vehicle.

Another aspect of signal lighting that deserves attention is the ability of drivers behind the immediately following vehicle to receive the signal. Vehicle signal lighting that is mounted close to the road may be masked from other drivers by the vehicle immediately following. One of the claimed advantages of the centre high-mounted stop lamp (CHMSL) is that it allows the operation of the brakes to be detected by drivers when there is another vehicle between because it can be seen through the windows of the immediately following vehicle. This is correct when the immediately following vehicle is a car but may not be true if it is a van or truck. This limitation could be alleviated somewhat if the conventional rear signal assemblies consisting of pairs of rear position, stop, turn, fog, reversing, and hazard warning lamps were to be duplicated high up on the vehicle, at least on vehicles with a height of more than about 2 m.

The main opportunity that is missed with current vehicle signal lighting is to provide other drivers with additional information. For example, the application of the brakes is shown by the lighting of the stop lamps, but that does not discriminate between a gentle touch and heavy braking. For that to be discriminated, the drivers of the vehicles behind have to perceive the magnitude of the deceleration of the vehicle ahead. It would certainly be possible to indicate the force with which the brakes had been applied by linking the area, luminous intensity, or flash rate of the stop lamps to that force. A similar but less common opportunity applies to reversing lamps. The intention to reverse is shown by the lighting of the reversing lamps. Modifying the CHMSL so that it would also operate as a high level, white reversing lamp that would attract attention by flashing would be advantageous.

All these suggestions rely on lighting to convey the message, but automatic vehicle-to-vehicle communication is surely imminent. This will allow information of greater complexity to be conveyed by means other than looking through the windscreen. An example is a radio message broadcast over a limited range whenever a vehicle's hazard warning lamps are activated. Such a message could be integrated into a satellite navigation system to indicate to approaching drivers where the hazard is. Similar messages might be generated when a crash was detected either by those involved or by passing vehicles or through the emergency services. What this suggests is that the days of vehicle signal lighting as the only means to indicate the presence of a vehicle or the intentions of the driver are numbered. The fact is, as a means of conveying detailed information, vehicle signal lighting is rather primitive, closer to Morse code than the mobile phone. It cannot be long before additional sources of information operating through wireless networks are adopted. Such systems already exist for avoiding distant traffic congestion. What is required is for information to be exchanged between vehicles in close proximity about intentions such as braking or changing lanes. This should not be taken to mean that all signal lighting would disappear from vehicles. There are other people on and near the road who need to know about the presence of vehicles and the intentions of their drivers, such as cyclists and pedestrians, and who are unlikely to be equipped with the necessary technology. Rather, it means that vehicle-to-vehicle communication is most likely to supplement vehicle signal lighting rather than replace it entirely.

14.4 ROAD LIGHTING

14.4.1 PROBLEMS

Road lighting, as currently practiced, does not have a problem in the sense that it fails to do what it sets out to do. Rather, the problems it does have are whether or not what it sets out to do is what it should be doing and if what it sets out to do represents the best use of resources. The problems that surround road lighting can be considered under three headings: technical, financial, and environmental. The technical problems arise from the way road lighting practice has developed. The fact is, road lighting has evolved in isolation. This is most evident in the fact that road lighting does not consider the contribution of vehicle forward lighting to visibility (see Chapter 9). Where such metrics as resolving power and small target visibility (see Section 4.3) have been used as criteria for road lighting, the assumption is that road lighting alone should be sufficient to make small, low-reflectance targets visible. Further, the distances at which these targets should be visible range from 80 to 100 m, only slightly more than should be possible by low-beam headlamps alone. This suggests that if vehicle forward lighting is taken into account and the role of road lighting is limited to ensuring the visibility of larger targets beyond the range of low-beam headlamps, standards for road lighting will change.

The other technical problem of road lighting arises from the growth in traffic densities. The studies that have examined the effectiveness of road lighting in making targets visible have mainly been concerned with an empty road. Given that traffic density is one of the factors considered when deciding whether or not to install road lighting, such a situation is unrealistic, yet there has been little research concerning what road lighting may have to contribute to road safety at different levels of traffic density. It may be that in very dense traffic, road lighting has little contribution to make because the vehicle lighting dominates what needs to be seen, while in very light traffic, road lighting may be only be significant beyond the range of low-beam headlamps and it is in medium density traffic that road lighting is most important.

The financial problems associated with road lighting are concerned with all three costs relevant to lighting a road: first costs, operating costs, and life cycle costs. There is no doubt that road lighting makes life easier for the driver at night. Driving on a lit road enables the driver to see further ahead and hence to have more time to respond to what lies ahead. It also reduces glare from the headlamps of opposing drivers and provides some visual guidance. In urban and suburban areas, road lighting has ancillary benefits for pedestrians and householders in terms of crime reduction, although this benefit might be more easily achieved through lighting specifically designed for crime reduction (Painter and Farrington 2001). Given these benefits, it might be thought that road lighting should be provided over the whole road network. Unfortunately, this would be prohibitively expensive with the result that there are established procedures for deciding where and when road lighting should be installed (see Section 13.2). These procedures are based on such variables as traffic flow, road geometry, accident history, and so on. Different countries strike different balances between the costs of road lighting and the benefits they provide and hence have different criteria for deciding where and when to install road lighting. Interestingly, once road lighting is installed, it usually remains unchanged apart from the inevitable decline in road surface illuminances and luminances brought about by the changes in luminous flux of the light source and dirt deposits on and in the luminaire. This means road lighting is operating in the middle of the night when there are few drivers about to receive its benefits. Is this a good use of resources?

The environmental problems of road lighting are primarily those of light pollution (see Section 13.7). Light trespass has been an occasional problem with road lighting for many years but one that can be easily dealt with by shielding the luminaires. Sky glow is a more recent and more serious problem in that it is an inevitable consequence of road lighting.

14.4.2 SOLUTIONS

Recent advances in technology have the potential to reduce the financial and environmental problems of road lighting. One such technology is the photovoltaic generation of electricity during the day that is then stored in a battery and used to operate the road lighting at night. Such systems already exist but are largely confined to locations where connection to the electricity supply network is expensive. Further development of photovoltaic technology to increase its efficiency and reduce its cost will make such systems more attractive.

Another technology is remote monitoring of road lighting luminaires, using either mains signaling or wireless communication to connect a large number of luminaires to a local transmitter that in turn is linked to a central server through a mobile phone network or by landline. The central server provides a web portal through which authorized individuals gain access to monitor and control the luminaire network. Monitoring of the status of each luminaire allows light source failures to be identified quickly. Even when operating as expected, monitoring provides a record of the number of hours of operation of the light source so preventive maintenance can be planned. In addition, monitoring supplies information on the amount of energy used and when it was used. Where real-time pricing of electricity is applied, knowing when the energy was used is as important as knowing how much has been used. As for control, the simplest control consists of adjusting the times when the luminaires are switched on and off but to make full use of the system, the light source in each luminaire should be connected to electronic control gear that allows the light source to be dimmed. Given this, there is the possibility of adjusting the amount of light used for road lighting according to the traffic flow and weather conditions. At present, the most usual approach to control of road lighting is step dimming with slow transitions between each step (Guo et al. 2007). The step dimming protocol used invariably leads to lower road surface luminances being provided at low traffic densities, which is ironic given that the current road lighting recommendations are often linked to the ability to detect small targets far away, something that will only be possible in low traffic densities. Nonetheless, the longer the time spent operating in a dimmed state, the less will be the energy consumption of the road lighting. Using such remote control systems has resulted in energy savings in the range 25 to 45 percent in the UK, China, and Finland (Guo et al. 2007; Walker 2007). Remote monitoring systems undoubtedly increase the first costs of an installation, but when used as described, should markedly reduce operating costs and life cycle costs as well as sky glow at some times of the night.

The remote monitoring systems summarized above can be considered as a form of adaptive lighting, where the amount of light supplied is adjusted according to the demand (McLean 2006). This is a useful improvement in road lighting technology, but it does not address the major questions about road lighting: Where should it be provided and what form should it take? To answer the question about where road lighting should be provided means identifying those areas where it will do most good. The daylight saving time method can be used to identify the type of accidents most sensitive to a change in light level from day to night (see Section 1.5). These are accidents involving pedestrians and animals. Accidents involving wild animals are most likely to occur in rural areas. Given the extent of the rural road network and the costs of road lighting, it is not realistic to expect road lighting to be used to reduce accidents involving such animals; use of one of the infrared imaging systems discussed in Section 14.2.2 is much more likely. What would be appropriate would be to concentrate road lighting where pedestrians are most common and to use it at its highest level when pedestrians are most active. Of course, this approach would leave the rare pedestrian on rural roads exposed to a higher level of risk, but the use of retroreflective clothing would reduce that risk.

The first step towards such a rational application of road lighting would be to revise the road classification system giving greater emphasis to the level of pedestrian activity. The existing road classification systems used in the UK and the United States do consider pedestrian activity but treat it as a subsidiary criterion after the road type has been identified according to the road layout and traffic density (see Section 4.4). The classification system should still be based on three types of road: residential roads, traffic routes, and conflict areas. In residential areas, where traffic is slow moving, on-street parking is common, and pedestrians are primarily local residents, the lighting should be designed for the resident, although care should be taken to avoid glare to drivers. On traffic routes where there are very few pedestrians or cyclists, the lighting should be designed to serve the interests of the driver. In conflict areas, vehicular traffic, pedestrians, and cyclists all come together, hopefully not too intimately, and the lighting should be designed to give good visibility for all three user groups. It is important to note that in this classification system, conflict areas extend beyond roundabouts and intersections to include all roads where there are significant numbers of cyclists and pedestrians, no matter whether the cyclists are in dedicated lanes or mixed with the traffic and the pedestrians are restricted or free range.

The basis for road lighting design would be different for these three road classes. For roads in residential areas, the road lighting is primarily there to meets the needs of pedestrians and cyclists (BSI 2003a). One of these needs is likely to be a feeling of safety at night. Research in urban and suburban locations of the United States suggests that a horizontal illuminance on the sidewalk in the range of 10 to 50 lx is necessary to ensure a perception of safety at night by pedestrians (Boyce et al. 2000). This range of illuminances is large and much of it is higher than current recommendations for residential areas (see Tables 4.3 and 4.13), but it does provide some flexibility to match the lighting to the ambient conditions. This matching is important because an illuminance that is much less than the ambient to which people are adapted will appear dark. Conversely, an illuminance that is much higher than the ambient that people are adapted to will appear bright. It may be that illuminances lower than 10 lx could provide a perception of safety provided the ambient lighting is lower than 10 lx and people have sufficient time to adapt to it, but for this to occur would require some sort of planning control over a large area, as is implicit in the outdoor power density limits used in California. Of course, there is some illumi-nance below which the visual system cannot function effectively. Studies of visual performance and perception in isolated rural areas where the ambient lighting is provided by moonlight suggest that this limit is about 1 lx (Brandston et al. 2000).

For traffic routes, road surface luminance is a reasonable criterion for road light-ing design as the relevant directions of view are limited. The problem with lighting for traffic routes is the basis on which the average road surface luminance should be determined. At the moment, average road surface luminances are a matter of consensus rather than proof, based largely on experience rather than evidence, and are different in different countries. There is no doubt that higher luminances would be better for vision, but the judgment is that the average road surface luminances recommended are adequate. Most of the scientific evidence available is concentrated on the visibility of small, low-reflectance targets. The question that arises is whether on not this is the right sort of target on which to base road lighting recommendations for all traffic densities? At high traffic densities, it can be argued that detection of the relative movements of nearby vehicles matters much more than the ability to see detail. At the other extreme of traffic density, it may be more important to be able to estimate the speed and distance of what vehicles there are and to detect pedestrians than to see small targets, particularly pedestrians beyond the range of low-beam headlamps. The problem this approach poses is that the same traffic route may have a high traffic density at one time and a low traffic density at another. The remote monitoring systems discussed above could be used to adjust the lighting to what is required for different traffic densities. The problem with this is that there is little evidence about what is required. There are a number of approaches that could be used to identify suitable values. One would be to use a remote monitoring system to systematically change average road surface luminances on a number of roads while counting all the accidents of different types that occurred at different times of the day and night, in different traffic densities. The average road surface luminance to be recommended could be that at which the number of accidents started to increase above the level found in daylight, for the same traffic density. Alternatively, the ref-erence point could be taken as the number of accidents occurring at night when the road was lit to current recommendations. Investigations that rely on accident counts to arrive at a conclusion are likely to take a long time as accidents are quite rare events and might be considered unethical in the sense that drivers would be taking part in an experiment without giving consent. Another approach that would reach an answer sooner is to use the same remote monitoring system to adjust the average road surface luminance but to measure driving performance in traffic. This approach requires the identification of the relevant performance measures and these are likely to differ with traffic density. Whatever approach is used, there needs to be more research in this area if road lighting for traffic routes is be put on a rational footing.

Finally, it is necessary to consider conflict areas. By definition, there are areas where many vehicles, cyclists, and pedestrians are trying to use the same piece of road at almost the same time, e.g., roundabouts; intersections; pedestrian crossings; and bus, taxi, and bicycle lanes. The presence of drivers, pedestrians, and cyclists in different locations on and beside the road means there are multiple viewing directions, so average road surface illuminance is a reasonable metric for recommendations. Again, the problem is that there is little evidence as to what average illuminance is required to reduce accidents in conflict areas. The same approaches described above for traffic areas could be used, but extending the performance measures to pedestrians and cyclists as well as drivers. However, given the diverse range of activities undertaken by drivers, pedestrians, and cyclists, determining the illuminance required in conflict areas by this method is likely to take a long time and ultimately rely on some form of consensus over the weight to be attached to all the different measures. An alternative would be to assume that as conflict areas are where more people are making more decisions more quickly than any other part of the road system, the lighting provided should be sufficient to allow the human visual system to operate without serious degradation. The question then changes to what background luminance is needed to avoid serious degradation of the human visual system. There are a number of ways of answering this question. One is to examine the vast literature on human vision so as to identify the changes that occur with decreasing background luminance [see Boff and Lincoln (1988) for a summary]. The problem with this approach is that most visual functions decline gradually at first but at an increasing rate as luminance is reduced (see Section 3.4), which leaves anyone wishing to make a simple recommendation with the problem of deciding where to draw the line, a problem that also exists when visual performance is used as a basis for interior lighting recommendations (Boyce 1996). An alternative would be to argue that there is an obvious decline in foveal visual functions where photopic vision changes to mesopic vision, which, by convention, is at an adaptation luminance of 3 cd/m^2. Another measure used to define this change is the illuminance provided at civil twilight, which is defined as that period from when the upper edge of the sun is at a tangent to the horizon to when the centre of the sun is 6 degrees below the horizon (Andre and Owens 1999). The illuminance range for civil twilight in the centre of the United States is from approximately 320 to 3.2 lx, a range so large that one is again left with the problem of deciding where to draw the line. Yet another approach would be to ask drivers their opinion of the adequacy of the amount of light. Measurements of observers' perceptions of brightness of a road lit by headlamps have been made in Germany (Wordenweber et al. 2007). These measurements led to the conclusion that a minimum average road surface luminance of 1 cd/m^2 is required for acceptable road illumination. Yet another approach is to look at the illuminances at which people consider they need additional light. Measurements of the illuminances on the road at which drivers in New York State turn on their headlamps have shown a wide distribution of values (Boyce and Fan 1994). The illuminances at which 50 percent of drivers have their headlamps on ranged from 445 lx to 1119 lx, values that are much higher than the illuminances recommended for road lighting in the United States, which range from 8 to 34 lx (see Table 4.3). This discrepancy may have occurred because the drivers are turning on their headlamps to be seen by rather than to see by, i.e., to increase their conspicuity. It would be interesting to determine if drivers of vehicles with daytime running lamps turn on their headlamps at the same ambient illuminance. Another possibility is to consider when existing road lighting is turned on. This is usually done by a photoelectric control unit that can be set to different illuminances, ranging from about 35 lx to 200 lx. The current most popular setting in the UK appears to be 70 lx. Experience of driving when road lighting is first lit suggests that such an illuminance is more than sufficient and using headlamps at this time makes little difference to visibility. By now it should be clear that in the absence of data on the impact of road lighting on accidents in conflict areas that there are many ways to identify a suitable amount of light, many of which lead to answers that differ significantly in their implications for the costs of road lighting. Again, there needs to be more research in this area if road lighting for conflict areas is to be put on a rational footing.

14.5 MARKINGS, SIGNS, AND TRAFFIC SIGNALS

14.5.1 PROBLEMS

Road markings or road studs can be considered as the minimum required for visual guidance (CIE 2007). Road signs can be considered as the simplest method to provide more detailed information to drivers. Road markings, road studs, and road signs that rely on retroreflection of light from headlamps to be seen at night have a common problem. This is the decline in the luminances of the markings, studs, and signs in the presence of a layer of water or dirt. These reductions in luminance tend to reduce the luminance contrast of the marking or studs or the detail of the sign so that the distances from which the markings, studs, or signs are visible are reduced.

Road signs that are illuminated, either internally or externally, are much less sensitive to water or dirt than those that are not. The problem faced by some of these signs is how to match the luminances of the sign to the ambient conditions. The signs for which this problem occurs are those where light is used to convey the message, e.g., changeable message signs using LEDs or miniature incandescents as pixels from which letters and numbers are formed. The problem is that much higher luminous intensities are needed for the message to be visible by day than at night. Unless the luminous intensities of the sign are reduced at night, the sign itself may become a source of glare to approaching drivers.

This problem does not occur for changeable message signs and for road signs where the sign contains detail with inherent luminance contrast. This means that such signs need only be illuminated after dark and in adverse weather. Unfortunately for energy consumption, many such signs have the light source operating continuously because the cost of providing photoelectric control is more than the cost of the electricity consumed. Whether or not this wasteful practice will continue in the face of rapidly increasing electricity prices and demands to reduce carbon emissions remains to be seen.

A specific problem for externally illuminated road signs is the degree of light pollution produced by light that is reflected from the sign or that misses the sign altogether. As for traffic signals, these have no fundamental problems. They have been around in the same form for many years and are well understood by drivers. The biggest change to affect traffic signals in recent years has been the use of LEDs as the preferred light source. There is an opportunity to add shape to colour as a distinguishing feature between the different signals, thereby making life easier for drivers with colour-defective vision.

14.5.2 SOLUTIONS

There are two potential solutions to the problem of water layers on retroreflective road markings. One is to increase the size of the retroreflective elements so that the water layer has to be deeper to cover them. The negative side of this solution is that the wear is more rapid, resulting in a faster reduction in road marking luminance. The other is to use a porous asphalt pavement where rain is frequent. This material is constructed so as to produce faster draining of water from the road surface.

As for road studs, one solution to the problems of rain and dirt would be to revert to the original design of "cat's eyes" in which a rubber housing deformed when a vehicle's wheel passed over it and thereby wiped the surface of the retroreflector.

One possible solution of the effect of rain or dirt on retroreflective road signs would be to illuminate them so they no longer rely on retroreflection alone to make them visible. Given the wide range of locations of such signs and the cost of bringing electricity to them, this course of action would be much more likely if sufficient power could be provided through a photovoltaic system. If the light pollution that can be caused by external illumination of road signs is a concern, then internally illuminated road signs are a solution. As for changeable message signs where it is necessary to adapt the luminous intensity of the elements of the sign to ambient conditions, there is advice available (Padmos et al. 1988) and the technology required to implement that advice is relatively simple.

There is one other aspect of road markings, signs, and signals that can cause problems for drivers but it has little to do with lighting. It is simply the number of markings, signs, and signals. There is no doubt that duplicating signs or repeating signs increases the likelihood that they will be noticed, but ultimately, there is a risk of cognitive overload. All signs are not of equal importance so some could be omitted where cognitive overload is likely. What this means is that some discipline is required by road engineers to avoid excessive use of markings and signs.

14.6 WHY CHANGE?

This discussion of the future of lighting for driving has ignored one major factor, why should anyone bother to make any of the changes suggested? After all, most drivers get from A to B without an accident and, apart from the old or ill, think nothing of driving at night. The answer is that there are three major influences that are focusing attention on various aspects of lighting for driving. The first but not necessarily the most potent is the political will to reduce road deaths. Figure 14.3 shows that progress has been made on this objective in several countries since 1995. Much of this progress has been due to improvements in the design of motor vehicles. Such features as air bags, crumple zones, child car seats, and better brakes have made the consequences of a collision less severe. One vehicle class where these improvements are not applicable, motorcycles, has shown a consistent increase in fatalities over the same period (ERSO 2007). Road design in the sense of separating vehicles from pedestrians and cyclists has also helped, as has more rigorous enforcement of driving laws. However, recognition of the fundamental conflict between visibility and glare for headlamps is growing, as is the capability to do something about it. The result is a renewed interest in what contribution better vehicle lighting might make to reducing road deaths and injuries.

Another factor influencing vehicle lighting is the desire of vehicle manufacturers to differentiate their product from others. The improvement in vehicle interior lighting over the last decade is mainly due to this. For many decades the diffusing dome light was the only form of interior lighting, but once one manufacturer started to provide more sophisticated interior lighting others soon followed. The same approach has now begun for vehicle forward lighting with the introduction of the swiveling headlamp in luxury vehicles. No doubt this and other elements of the adaptive forward lighting system will soon begin to appear in other vehicles alongside other forms of supplementary information such as infrared imaging systems.

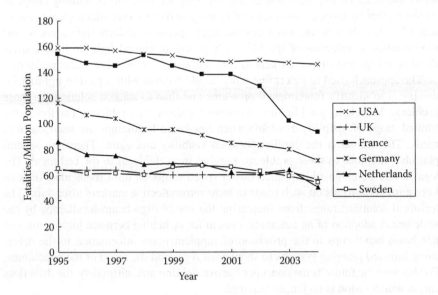

FIGURE 14.3 Trends in road fatalities per million population in the United States, the UK, France, Germany, The Netherlands, and Sweden from 1995 to 2004. The data sources used were the Fatality Analysis Reporting System for the United States (NHTSA 2006a) and European Road Safety Observatory for all the other countries (ERSO 2007).

Finally, road lighting is under two very different pressures. The first is that of costs. The increasing price of electricity in many countries is forcing the authorities responsible for road lighting to justify what they do. Typical of the questions being raised are why do they need to use so much light and why does it have to be operating at the same level throughout the night? Fortunately, the remote monitoring technology now becoming available makes the possibility of adjusting road lighting to match the traffic density a real one. The second pressure is that of road lighting's impact on the environment. This is being discussed at two levels, national and local. Nationally, the concern about global warming is forcing people in many countries to look at how electricity consumption might be reduced. Road lighting is a very conspicuous example of electricity consumption and, as such, is certainly under scrutiny. Locally, there is an increasing awareness of light pollution and road lighting is seen as a major contributor to it. Taken together, these pressures are likely to influence where, when, and how much road lighting is used in the future.

14.7 SUMMARY

This chapter seeks to envision what might be the future of lighting for driving, at two levels. One attempts to predict what is most likely to happen. The other attempts to set out what should happen. Over the next decade, what is most likely to happen to vehicle forward lighting is that it will become more sophisticated as elements of the adaptive forward lighting system trickle down from up-market vehicles to more modestly priced vehicles. Vehicle signal lighting will change little apart from the wider use of LEDs as a light source and the adoption of daytime running lamps as a requirement by more countries. Road lighting is likely to see much greater use of metal halide light sources, much less use of low-pressure sodium light sources, and the eventual introduction of the LED as a practical general lighting light source. Road markings and signs are likely to continue to grow in number and complexity.

The approach used to determine what should happen with regard to lighting for driving is to identify fundamental problems and then to suggest solutions to those problems. Vehicle forward lighting faces two fundamental problems. The first is the limited range of visibility available when low-beam headlamps are used on unlit roads. The second is the conflict between visibility and glare. There are several plausible solutions to these problems, ranging from the legal to the technical. The legal would be either to introduce slower speed limits on unlit roads after dark or to require pedestrians on such roads to wear retroreflective markers after dark. The technical solutions range from increasing the use of high-beam headlamps by the widespread adoption of an automatic system for switching between high-beam and low-beam headlamps to the provision of supplementary information to the driver using infrared imaging systems to show what is beyond the range of the headlamps. Further into the future is the concept of active lighting and, ultimately, the driverless car, in which vision is no longer required.

Vehicle signal lighting has only one problem, most of its limitations being more a matter of missed opportunities than problems. The problem faced by vehicle signal lighting is adaptation to prevailing conditions, something that is well within the capabilities of modern technology to solve. The missed opportunities are essentially forms of signal lighting that might be used to convey more information to other drivers. One type of missed opportunity relates to the use of lighting to ensure stable perception of movement in varying conditions. This could be done by using either contour lighting to outline the vehicle or lighting under the vehicle to provide a constant pattern of luminance contrasts as the car moves along the road. Vehicle lighting could also be used to provide more information about the speed of and distance from other vehicles by insisting that all signal lamps on vehicles other than motorcycles should be in pairs mounted on opposite sides of the vehicle. Another aspect of signal lighting that deserves attention is masking of the signal by an intervening vehicle. This limitation could be alleviated somewhat if the conventional rear signal assemblies were to be duplicated high up on the vehicle. However, the main opportunity that is missed with current vehicle signal lighting is to provide other drivers with additional information, such as the force with which the brakes have been applied. This could be done by linking the area, luminous intensity, or flash rate of the stop lamps to that force.

Road lighting, as currently practiced, does not have a problem in the sense that it fails to do what it sets out to do. Rather, the problems it does have are whether or not what it sets out to do is what it should be doing and if what it sets out to do represents the best use of resources. The problems that surround road lighting can be considered under three headings: technical, financial, and environmental. Recent advances in remote monitoring technology have the potential to reduce the financial and environmental problems of road lighting by dimming when less light is required. Unfortunately, asking when less light is required itself raises questions about where should road lighting be provided and what form should it take? To answer the question about where road lighting should be provided means identifying those areas where it will do most good. Examination of the accident data shows that the type of accidents most sensitive to the change of light level from day to night are those involving pedestrians. This implies that a rational approach would be to concentrate road lighting where pedestrians are most common and to use it at its highest level when pedestrians are most active.

The first step towards implementing such an approach would be to revise the road classification system giving greater emphasis to the level of pedestrian activity based on three types of road: residential roads, traffic routes, and conflict areas. In residential areas the lighting should be designed primarily for pedestrians and cyclists. On traffic routes the lighting should be designed to serve the interests of the driver. In conflict areas, vehicular traffic, pedestrians, and cyclists all come together so the lighting should be designed to give good visibility for all three user groups. The problem with lighting for all these three road classes is that there is little evidence about what is required for reducing accidents or improving driving performance. There needs to be more research in this area if road lighting is to be put on a rational footing.

Road markings, road studs, and road signs that rely on retroreflection of light from headlamps to be seen at night have a common problem. This is the decline in the luminances of the sign in the presence of a layer of water or dirt. Self-luminous road markings and road studs and illuminated road signs do not have this problem, but where light is used to convey the message, different luminous intensities are required by day and night. Technology to achieve this is available. Where the sign contains detail with inherent luminance contrast, the signs need only be illuminated after dark and in adverse weather but often they are illuminated continuously.

There are two potential solutions to the problem of water layers on retroreflective road markings. One is to increase the size of the retroreflective elements so that the water layer has to be deeper to cover them. The other is to use a porous asphalt road, which allows faster draining of water from the road surface. Retroreflective road studs could also be made more resistant to water by reverting to the original "cat's eyes" design where a flexible rubber housing wipes the retroreflector whenever a vehicle's wheel passes over. As for retroreflective road signs, one possible solution would be to illuminate them so they no longer rely on retroreflection alone to be visible, although there are also concerns that such illumination contributes to light pollution. This concern can be alleviated by using internally illuminated signs.

There is one other aspect of road markings, signs, and signals that can cause problems for drivers but it has little to do with lighting. It is simply the number of markings, signs, and signals. There is no doubt that duplication of signs increases the likelihood that they will be noticed, but there is a risk of cognitive overload. Some discipline is required by road engineers to avoid excessive use of markings, signs, and signals.

Given all these possibilities, it is necessary to ask why anyone should bother to make any of the changes suggested? After all, most drivers get from A to B without an accident and, apart from the old and ill, think nothing of driving at night. The answer is that there are three major influences that are focusing attention on various aspects of lighting for driving. The first is the political will to reduce road deaths and injuries. This is raising questions about the effectiveness of both vehicle and road lighting. The second is the desire of vehicle manufacturers to differentiate their product from others, the trend being towards more sophisticated vehicle lighting. The third is the financial and environmental pressures being applied to road lighting. The increasing price of electricity in many countries is forcing the authorities responsible for road lighting to justify what they do. In addition, concern about global warming is forcing people in many countries to look at how electricity consumption might be reduced. Road lighting is a very conspicuous example of electricity use and, as such, is under scrutiny. Taken together, these pressures are likely to influence where, when, and how much road lighting is used in the future.

References

Adrian, W. (1976) Method of calculating the required luminances in tunnel entrances, *Lighting Research and Technology*, 8, 103–106.

Adrian, W. (1982) Investigations on the required luminance in tunnel entrances, *Lighting Research and Technology*, 14, 151–159.

Adrian, W. (1987a) Visibility levels under night-time driving conditions, *Journal of the Illuminating Engineering Society*, 16, 3–12.

Adrian, W. (1987b) Adaptation luminance when approaching a tunnel in daytime, *Lighting Research and Technology*, 19, 73–79.

Adrian, W. (1989) Visibility of targets: Model for calculation, *Lighting Research and Technology*, 21, 181–188.

Adrian, W. (1991) Comparison between the CBE and CIE glare mark formula and earlier discomfort glare descriptions, *First International Symposium on Glare*, New York: Lighting Research Institute.

Adrian, W. (1995) Change of visual acuity with age, in W. Adrian (ed.) *Lighting for Aging Vision and Health*, New York: Lighting Research Institute.

Adrian, W., and Schreuder, D.A. (1970) A simple method for the appraisal of glare in street lighting, *Lighting Research and Technology*, 2, 61–73.

Adrian, W.K., Gibbons, R.G., and Thomas, L. (1993) Amendments in calculating STV: Influence of light reflected from the road surface on the target luminance, *Proceedings of the 2nd International Symposium on Visibility and Luminance in Roadway Lighting*, New York: Lighting Research Institute.

Akashi, Y., and Rea, M.S. (2002) Peripheral detection while driving under a mesopic light level, *Journal of the Illuminating Engineering Society*, 31, 85–94.

Akashi, Y., Rea, M.S., and Bullough, J.D. (2007) Driver decision making in response to peripheral moving targets under mesopic light levels, *Lighting Research and Technology*, 39, 53–67.

Akerboom, S.P., Kruysse, H.W., and La Heij, W. (1993) Rear light configurations: The removal of ambiguity by a third brake light, in E.G. Gale (ed.) *Vision in Vehicles IV*, Amsterdam: North Holland.

Alferdinck, J.W.A.M. (1996) Traffic safety aspects of high-intensity discharge headlamps; discomfort glare and direction indicator conspicuity, in A.G. Gale, I.D. Brown, C.M. Haslegrave, and S.P. Taylor (eds.) *Vision in Vehicles V,* Amsterdam: North-Holland.

Alferdinck, J.W.A.M., and Padmos, P. (1988) Car headlamps: Influence of dirt, age and poor aim on glare and illumination intensities, *Lighting Research and Technology,* 20, 195–198.

Alferdinck, J.W.A.M., and Varkevisser, J. (1991) *Discomfort Glare from D1 Headlamps of Different Size,* Report IZF 1991 C-21, Soesterberg, The Netherlands: TNO Institute for Perception.

Alm, H., and Nilsson, L. (2000) Incident warning systems and traffic safety: A comparison between the PORTICO and MELYSSA test site systems, *Transportation Human Factors*, 2, 77–93.

American Association of State Highways and Transportation Officials (AASHTO) (1984) *An Informational Guide to Roadway Lighting*, Washington, DC: AASHTO.

American Association of State Highway Transportation Officials (AASHTO) (2001) *A Policy on Geometric Design of Highways and Streets*, Washington, DC: AASHTO.

American Society for Testing and Materials (ASTM) (1996) *Standard Practice for Specifying Colors by the Munsell System*, D1535–1596, Philadelphia: ASTM.

Andersson, K., and Nilsson, G. (1981) *The Effects on Accidents of Compulsory Use of Running Lights During Daylight in Sweden*, VTI–Report 208A, Linkoping, Sweden: National Road and Traffic Research Institute.

Andre, J.T., and Owens, D.A. (1999) The Twilight Envelope: An Alternative Approach Toward Defining Roadway Visibility at Night, *Proceedings of the Fourth International Lighting Research Symposium: Vision at Low Light Levels*, Palo Alto, CA: Electric Power Research Institute.

Arnold, E.D. (2004) *Development of Guidelines for In-Roadway Warning Lights*, Richmond, VA: Commonwealth of Virginia.

Arora, H., Collard, D., Robbins, G., Welbourne, E.R., and White, J.G. (1994) *Effectiveness of Daytime Running Lights in Canada*, Report TP 12298 (E), Ottawa: Transport Canada.

Association Francoise de l'Eclairage (AFE) (2006) *Les Nuisances Dues a la Lumiere*, Paris: AFE.

Assum, T., Bjornskau, T., Fosser, S., and Sagberg, F. (1999) Risk compensation — the case of road lighting, *Accident Analysis and Prevention*, 31, 545–553.

Attwood, D.A. (1979) The effects of headlight glare on vehicle detection at dusk and dawn, *Human Factors*, 21, 35–45.

Ayama, M., and Ikeda, M. (1998) Brightness-to-luminance ratio of coloured light in the entire chromaticity diagram, *Color Research and Application*, 23, 274–287.

Bacelar, A. (2004) The contribution of vehicle lights in urban and peripheral urban environments, *Lighting Research and Technology*, 36, 69–78.

Bacelar, A., Cariou, J., and Hamard, M. (1999) Calculational visibility model for road lighting installations, *Lighting Research and Technology*, 31, 177–180.

Baddiley, C.J., and Webster, T. (2007) *Towards Understanding Skyglow*, Rugby, UK: Institution of Lighting Engineers.

Bajorski, P., Dhar, S., and Sandhu, D. (1996) Forward-lighting configurations for snowplows, *Transportation Research Record*, 1533, 59–66.

Balazsi, A.G., Rootman, J., Drance, S.M., Schulze, M., and Douglas, G.R. (1984) The effect of age on the nerve fiber population of the human optic nerve, *American Journal of Ophthalmology*, 97, 760–766.

Ball, K., Owsley, C., Sloane, M., Roeker, D., and Bruni, J. (1993) Visual attention problems as a predictor of vehicle crashes in older drivers, *Investigative Ophthalmology and Vision Science*, 34, 3110–3123.

Bergkvist, P. (2001) Daytime running lamps (DRLs) — a North American success story, *Proceedings of the 17th International Technical Conference on the Enhanced Safety of Vehicles*, Washington DC: Department of Transportation.

Berman, S.M. (1992) Energy efficiency consequences of scotopic sensitivity, *Journal of the Illuminating Engineering Society*, 21, 3–14.

Berman, S.M., Fein, G., Jewett, D.L., Saika, G., and Ashford, F. (1992) Spectral determinants of steady-state pupil size with a full field of view, *Journal of the Illuminating Engineering Society*, 21, 3–13.

Billmeyer Jr., F.W. (1987) Survey of color order systems, *Color Research and Application*, 12, 173–186.

Blackwell, H.R. (1959) Development and use of a quantitative method for specification of interior illumination levels on the basis of performance data, *Illuminating Engineering*, 54, 317–353.

Blackwell, H.R., and Blackwell, O.M. (1971) Visual performance data for 156 normal observers of various ages, *Journal of the Illumination Engineering Society*, 1, 3–13.

Blackwell, H.R., and Blackwell, O.M. (1977) A basic task performance assessment of roadway luminous environments, in *Measures of Road Lighting Effectiveness,* Berlin: Lichttechnische Gesellschaft LiTG.

Blackwell, H.R., Schwab, R.N., and Pritchard, B.S. (1964) Visibility and illumination variables in roadway visual tasks, *Illuminating Engineering*, 59, 277–308.

Blazer, P., and Dudli, H. (1993) Tunnel lighting: Method of calculating luminance of access zone L20, *Lighting Research and Technology*, 25, 25–30.

Bodmann, H.W., and Schmidt, H.J. (1989) Road surface reflection and road lighting: Field investigations, *Lighting Research and Technology*, 21, 159–170.

Boff, K.R., and Lincoln, J.E. (1988) *Engineering Data Compendium: Human Perception and Performance*, Wright–Patterson AFB, OH: Harry G. Armstrong Aerospace Medical Research Laboratory.

Bornstein, M.H., Kessen, W., and Weiskopf, S. (1976) The categories of hue in infancy, *Science*, 191, 201–202.

Bourdy, C., Chiron, A., Cottin, F., and Monot, A. (1988) Visibility at a tunnel entrance: Effect of temporal luminance variation, *Lighting Research and Technology*, 20, 199–200.

Bowers, A., Peli, E., Elgin, J., McGwin, G. Jr., and Owsley, C. (2005) On-road driving and moderate visual field loss, *Optometry and Vision Science*, 82, 657–667.

Box, P.C. (1981) Parking lot accident characteristics, *ITE Journal*, 51, 12–15.

Boyce, P.R. (1996) Illuminance selection based on visual performance — and other fairy stories, *Journal of the Illuminating Engineering Society*, 25, 41–49.

Boyce, P.R. (2003) *Human Factors in Lighting*, London: Taylor and Francis.

Boyce, P.R., and Fan, J. (1994) Drivers' headlight use: A lighting factoid, *Lighting Research and Technology*, 26, 23–27.

Boyce, P., Bierman, A., Carter, B., Hunter, C., Bullough, J., Figueiro, M., and Conway, K. (2000) *The Color Identification of Traffic Signals*, Troy, NY: Lighting Research Center.

Boyce, P.R., Eklund, N.H., Hamilton, B.J., and Bruno, L.D. (2000) Perceptions of safety at night in different lighting conditions, *Lighting Research and Technology*, 32, 79–91.

Boynton, R.M., and Clarke, F.J.J. (1964) Sources of entoptic scatter in the human eye, *Journal of the Optical Society of America*, 54, 110–119.

Boynton, R.M., and Gordon, J. (1965) Bezold–Brucke hue shift measured by a color naming technique, *Journal of the Optical Society of America*, 55, 78–86.

Boynton. R.M., and Purl, K.F. (1989) Categorical colour perception under low pressure sodium lighting with small amounts of added incandescent illumination, *Lighting Research and Technology*, 21, 23–27.

Brandston, H.M., Peterson, A.J. Jr., Simonson, E.K., and Boyce, P.R. (2000) A white-LED post-top luminaire for rural applications, *Proceedings of the Illuminating Engineering Society of North America Annual Conference, Washington, DC,* New York: IESNA.

British Standards Institution (2003a) BS EN 13201-2:2003, *Road Lighting — Part 2: Performance Requirements,* London: BSI.

British Standards Institution (2003b) BS 5489-1:2003, *Code of Practice for the Design of Road Lighting — Part 1: Lighting of Roads and Public Amenity Areas,* London: BSI.

Brons, J., Bullough, J., and Rea, M.S. (2007) Thinking inside the box, *Lighting Journal*, 72, 27–34.

Brons, J.A., Bullough, J.D., and Rea, M.S. (2008) Outdoor site-lighting performance (OSP): A comprehensive and quantitative framework for assessing light pollution, *Lighting Research and Technology*, 40, 201–224.

Brooks, J.O., Tyrrell, R.A., and Frank, T.A. (2005) The effect of severe visual challenges on steering performance in visually healthy young drivers, *Optometry and Vision Science*, 82, 689–697.

Brown, W.R.J. (1951) The influence of luminance level on visual sensitivity to color differences, *Journal of the Optical Society of America*, 41, 684–688.

Bruneau, J-F. and Morin, D. (2005) Standard and non-standard roadway lighting compared with darkness at rural intersections, *Transportation Research Record*, 1918, 116–122.

Brusque, C., Paulmier, G., and Carta, V. (1999) Study of the influence of background complexity on the detection of pedestrians in urban sites, *Proceedings of the CIE, 24th Session, Warsaw*, Vienna: CIE.

Buck, J.A., McGowan, T.K., and McNelis, J.F. (1975) Roadway visibility as a function of light source color, *Journal of the Illuminating Engineering Society*, 5, 20–25.

Bullough, J.D. (2002a) Interpreting outdoor luminaire cutoff classification, *Lighting Design and Application*, 32, 44–46.

Bullough, J.D. (2002b) Modeling peripheral visibility under headlamp illumination, *Proceedings of the TRB 16th Biennial Symposium on Visibility and Simulation, Iowa City, IA*, Washington DC: Transportation Research Board.

Bullough, J., and Rea, M.S. (1997) A simple model of forward visibility for snow plow operators through snow and fog at night, *Transportation Research Record*, 1585, 19–24.

Bullough, J., and Rea, M.S. (2000) Simulated driving performance and peripheral detection at mesopic and low photopic light levels, *Lighting Research and Technology*, 32, 194–198.

Bullough, J.D., and Rea, M.S. (2001a) Forward vehicle lighting and inclement weather conditions, *Proceedings of the Symposium Progress in Automobile Lighting 2001*. Munich, Germany: Technical University of Darmstadt.

Bullough, J.D., and Rea, M.S. (2001b) Driving in snow: Effects of headlamp color at mesopic and photopic light levels, *Lighting Technology Developments for Automobiles*, Warrendale, PA: Society of Automotive Engineers.

Bullough, J.D., Boyce, P.R., Bierman, A., Conway, K.M., Huang, K., O'Rourke, C.P., Hunter, C.M., and Nakata, A. (2000) Response to simulated traffic signals using light emitting diode and incandescent sources, *Transportation Research Record*, 1724, 39–46.

Bullough, J.D., Boyce, P.R., Bierman, A., Hunter, C.M., Conway, K.M., Nakata, A., and Figueiro, M.G. (2001a) Traffic signal luminance and visual discomfort at night, *Transportation Research Record*, 1754, 42–47.

Bullough, J.D., Rea, M.S., Pysar, R.M., Nakhla, H.K., and Amsler, D.E. (2001b) Rear lighting configurations for winter maintenance vehicles, *Proceedings of the IESNA Annual Conference, Ottawa*, New York: IESNA.

Bullough, J.D., Zu, F., and Van Derlofske, J. (2002) *Discomfort and Disability Glare from Halogen and HID Headlamp Systems*, SAE Paper 2002-01-1-0010, Warrendale, PA: Society of Automotive Engineers.

Bullough, J.D., Van Derlofske, J., Fay, C.R., and Dee, P. (2003) *Discomfort Glare from Headlamps: Interactions among Spectrum, Control of Gaze and Background Light Level*, SAE Paper 2003-01-0296, Warrendale, PA: Society of Automotive Engineers.

Bullough, J.D., Yuan, Z., and Rea, M.S. (2007a) Perceived brightness of incandescent and LED aviation signal lights, *Aviation, Space and Environmental Medicine*, 78, 893–900.

Bullough, J.D., Van Derlofske, J., and Kleinkes, M. (2007b) *Rear Signal Lighting: From Research to Standards, Now and in the Future*, SAE Paper 2007-01-1229, Warrendale, PA: Society of Automotive Engineers.

Bullough, J.D., Brons, J.A., Qi, R., and Rea, M.S. (2008) Predicting discomfort glare from outdoor lighting installations, *Lighting Research and Technology*, 40, 225–242.

Burg, A. (1967) *The Relationship Between Vision Test Scores and Driving Records: General Findings*, Report 67-24, Los Angeles, CA: Department of Engineering, University of California.

Burg, A. (1971) Vision and driving: A report on research, *Human Factors*, 13, 79–87.

Burghout, F. (1979) On the relationship between reflection properties, composition and texture of road surfaces, *Proceedings of the CIE, 19th Session, Kyoto, Japan*, Vienna: CIE.

Caird, J.K., and Hancock, P.A. (2002) Left turn and gap acceptance crashes, in R.E. Dewar and P.L. Olson (eds.) *Human Factors in Traffic Safety*, Tucson, AZ: Lawyers and Judges Publishing Company.

Cairney, P. and Catchpole, J. (1996) Patterns of perceptual failures at intersections of arterial roads and local streets, in A.G. Gale, I.D. Brown, C.M. Haslegrave, and S.P. Taylor (eds.) *Vision in Vehicles-V*, Amsterdam: North-Holland.

California Energy Commission (CEC) (2005) *Title 24, Part 6 of the California Code of Regulations, California's Energy Efficiency Standard for Residential and Non-Residential Buildings*, Sacramento, CA: CEC.

Casson, E.J., and Racette, L. (2000) Vision standards for driving in Canada and the United States: A review for the Canadian Ophthalmological Society, *Canadian Journal of Ophthalmology*, 35, 192–203.

Cavallo, V., Colomb, M., and Dure, J. (2001) Distance perception of vehicle rear lights in fog, *Human Factors*, 43, 442–451.

Chandler, D. (1949) *The Rise of the Gas Industry in Britain*, London: Kelly and Kelly.

Chandra, D., Sivak, M., Flannagan, M.J., Sato, T., and Traube, E.C. (1992) *Reaction Times to Body-Colour Brake Lamps*, UMTRI-92-15, Ann Arbor, MI: University of Michigan Transportation Research Institute.

Chang, Y-M., and Wang, L.L. (2007) *The Visual Power of the Dashboard of a Passenger Car by Applying Eye-Tracking Theory*, SAE Paper 2007-01-0425, Warrendale, PA: Society of Automotive Engineers.

Chapparo, A., Wood, J.M., and Carberry, T. (2005) Effect of age and auditory and visual dual tasks on closed-road driving performance, *Optometry and Vision Science*, 82, 747–754.

Charlton, S.G. (2007) The role of attention in horizontal curves: A comparison of advance warnings, delineation and road marking treatments, *Accident Analysis and Prevention*, 39, 873–885.

Charman, W.N. (1997) Vision and driving — a literature review and commentary, *Ophthalmic and Physiological Optics* 57, 371–391.

Chartered Institution of Building Services Engineers (1992) *Lighting Guide 6: The Outdoor Environment*, London: CIBSE.

Clay, O.J., Wadley, V.G., Edwards, J.D., Roth, D.L, Roenker, D.L., and Ball, K.K. (2005) Cumulative meta-analysis of the relationship between useful field of view and driving performance in older adults: Current and future implications, *Optometry and Vision Science*, 82, 724–731.

Coaton, J.R., and Marsden, A.M. (1996) *Lamps and Lighting*, 4th Edition, London: Butterworth-Heinemann.

Cobb, J. (1990) *Roadside Survey of Vehicle Lighting 1989,* Research Report 290, Crowthorne, UK: Transport and Road Research Laboratory.

Cobb, J. (1992) *Daytime Conspicuity Lights*, Report WP/RUB/14), Crowthorne, UK: Transport Research Laboratory.

Codling, P.J. (1971) *Thick Fog and its Effects on Traffic Flow and Accidents*, RRL Report LR 397, Crowthorne, UK: Road Research Laboratory.

Cole, B.L., and Brown, B. (1966) Optimum intensity of red road-traffic signal lights for normal and protanopic observers, *Journal of the Optical Society of America*, 56, 516–522.

Collins, J.J., and Hall, R.D. (1992) Legibility and readability of light reflecting matrix variable message road signs, *Lighting Research and Technology*, 24, 143–148.

Commission Internationale de l'Eclairage (1976) *Calculation and Measurement of Luminance and Illuminance in Road Lighting*, CIE Publication 30, Vienna: CIE.

Commission Internationale de l'Eclairage (CIE) (1978) *Light as a True Visual Quantity*, CIE Publication 41, Vienna: CIE.

Commission Internationale de l'Eclairage (CIE) (1979) *Road Lighting for Wet Conditions*, CIE Publication 47, Vienna: CIE.

Commission Internationale de l'Eclairage (CIE) (1983) *The Basis of Physical Photometry*, CIE Publication 18.2, Vienna: CIE.

Commission Internationale de l'Eclairage (CIE) (1984) *Road Surfaces and Lighting*, CIE Publication 66, Vienna: CIE.

Commission Internationale de l'Eclairage (CIE) (1986) *Colorimetry*, CIE Publication 15.2, Vienna: CIE.

Commission Internationale de l'Eclairage (CIE) (1987) *Guide to the Properties and Uses of Retro-reflectors at Night*, CIE Publication 72, Vienna: CIE.

Commission Internationale de l'Eclairage (CIE) (1989) *Mesopic Photometry: History, Special Problems and Practical Solutions*, CIE Publication 81, Vienna: CIE.

Commission Internationale de l'Eclairage (CIE) (1990a), *CIE 1988 2° Spectral Luminous Efficiency Function for Photopic Vision*, CIE Publication No 86,Vienna: CIE.

Commission Internationale de l'Eclairage (CIE) (1990b) *Guide for the Lighting of Road Tunnels and Underpasses*, CIE Publication No. 88, Vienna: CIE.

Commission Internationale de l'Eclairage (CIE) (1992a) *Fundamentals of the Visual Task of Night Driving*, CIE Publication 100, Vienna: CIE.

Commission Internationale de l'Eclairage (CIE) (1992b) *Road Lighting as an Accident Countermeasure,* CIE Publication 93, Vienna: CIE.

Commission Internationale de l' Eclairage (CIE) (1993) *Daytime Running Lights (DRL),* Report CIE 104, Vienna: CIE.

Commission Internationale de l'Eclairage (CIE) (1994a) *Review of the Official Recommendations of the CIE for the Colors of Signal Lights*, CIE Technical Report 107, Vienna: CIE.

Commission Internationale de l'Eclairage (CIE) (1994b) *Variable Message Signs*, CIE Publication 111–1994, Vienna: CIE.

Commission Internationale de l'Eclairage (CIE) (1995a) *Method of Measuring and Specifying Color Rendering Properties of Light Sources*, CIE Publication 13.3, Vienna: CIE.

Commission Internationale de l'Eclairage (CIE) (1995b) *Recommendations for the Lighting of Roads for Motor and Pedestrian Traffic,* CIE Technical Report 115, Vienna: CIE.

Commission Internationale de l'Eclairage (CIE) (1997) *Guidelines for Minimizing Sky Glow*, CIE Publication 126, Vienna: CIE.

Commission Internationale de l'Eclairage (1999) *Road Surface and Road Marking Reflection Characteristics*, CIE Publication 13x-1999, Vienna: CIE.

Commission Internationale de l'Eclairage (CIE) (2002) *CIE Collection on Glare*, CIE Publication 146-2002, Vienna: CIE.

Commission Internationale de l'Eclairage (CIE) (2003) *Guide on the Limitation of the Effects of Light Trespass from Outdoor Lighting Installations*, CIE Publication 150, Vienna: CIE.

Commission Internationale de l'Eclairage (CIE) (2007) *Road Transport Lighting for Developing Countries,* CIE Publication 180, Vienna: CIE.

Crawford, B.H. (1949) The scotopic visibility function, *Proceedings of the Physical Society: Section B*, 62, 321.

Crawford, B.H. (1972) The Stiles–Crawford effects and their significance in vision, in D. Jameson and L.M. Hurvich (eds.) *Handbook of Sensory Physiology, Vol. VII/4 Visual Psychophysics*, Berlin: Springer-Verlag.

Croft, T.A. (1971) Failure of visual estimation of motion under strobe, *Nature*, 231, 397.

Crosley, J., and Allen, M.J. (1966) Automobile brake light effectiveness: An evaluation of high placement and accelerator switching, *American Journal of Optometry and Archives of the American Academy of Optometry*, 43, 299–304.

Crundall, P., and Underwood, G. (1998) Effect of experience and processing demand on visual information acquisition in drivers, *Ergonomics*, 41, 448–458.

Davidse, R., van Driel, C., and Goldenbeld, C. (2004) *The Effect of Altered Road Marking on Speed and Lateral Position: A Meta-Analysis*, Leidschendem, The Netherlands: SWOV.

De Boer, J.B. (1951) Fundamental experiments on visibility and admissible glare in road lighting, *Proceedings of the CIE, 12th Session, Stockholm*, Paris: CIE.

De Boer, J.B. (1974) Modern light sources for highways, *Journal of the Illuminating Engineering Society*, 3, 142–152.

De Boer, J.B., and Westermann, H.O. (1964a) Characterisation and classification of road surfaces from the point of view of luminance in public lighting, *Lux*, 30, 385.

De Boer, J.B., and Westermann, H.O. (1964b) The discrimination of road surfaces depending on the reflection properties and its meaning for road lighting, *Lichttechnik*, 16, 487.

De Boer, J. B., Onate, V., and Oostrijk, A. (1952) *Practical Methods for Measuring and Calculating the Luminance of Road Surfaces*, Philips Research Reports, 7, 54–76.

De Boer, J.B., Burghout, F., and van Heemskerck Veekens, J.F.T. (1959) Appraisal of the quality of public lighting based on road surface luminance and glare, *Proceedings of the CIE, 14th Session, Brussels*, Vienna: CIE.

De Lange, H. (1958) Research into the dynamic nature of the human fovea cortex systems with intermittent and modulated light. 1. Attenuation characteristics with white and colored lights, *Journal of the Optical Society of America*, 48, 777–789.

Deans, R.L., Miller, A.R., Murrill, J.K., Sanders, J.R., Turley, T.C., Lambert, J.H., Bridewell, T.A., and Cottrell, B.H. (2003) Screening needs for roadway lighting by exposure assessment and site parameters, in M.H. Jones, B.E. Tawney, and K. Preston White Jr (eds.) *Proceedings of the 2003 IEEE Systems and Information Engineering Design Symposium*, Charlottesville, VA: Department of System and Information Engineering, University of Virginia.

Dee, P. (2003) *The Effect of Spectrum on Discomfort Glare*, MS Thesis, Troy, NY: Rensselaer Polytechnic Institute.

Department for Transport (DfT) (2005a) *Traffic Sign Manual*, DfT website accessed 25/6/2007.

Department for Transport (DfT) (2005b) *The Design of Pedestrian Crossings*, Local Transport Note 2/95, London: Department for Transport.

Department for Transport (DfT) (2005c) *Road Casualties Great Britain, 2004*, London: DfT.

Department for Transport (DfT) (2007a) *Interim Advice Note 89/07 Appraisal of New and Replacement Road Lighting on the Strategic Motorway Network and All Purpose Trunk Road Network*, London: DfT.

Department for Transport (DfT) (2007b) *Highways Economics Note No 1, 2005 Valuation of the Benefits of Prevention of Road Accidents and Casualties*, London: DfT.

Department of Trade and Industry (DTI) (2006) *Digest of UK Energy Statistics, 2006*, London: DTI.

Desimone, R. (1991) Face-selective cells in the temporal cortex of monkeys, *Journal of Cognitive Neuroscience*, 3, 1–8.

Devaney, K.O., and Johnson, H.A. (1980) Neuron loss in the ageing visual cortex of man, *Journal of Gerontology*, 35, 836–841.

Drasdo, N. (1977) The neural representation of visual space, *Nature*, 266, 554–556.

Drasdo, N., and Haggerty, C.M. (1984) A comparison of the British number plate and Snellen vision test for car drivers, *Ophthalmic and Physiological Optics*, 1, 39–54.

Duke-Elder, W.S. (1944) *Textbook of Ophthalmology, Vol. 1.*, St Louis, MO: C.V. Mosby & Co.

Dumont, E., and Paumier, J.-L. (2007) Are standard r-tables still representative of the properties of road surfaces in France? *Proceedings of the CIE, 26th Session, Beijing*, Vienna: CIE.

Dunbar, C. (1938) Necessary values of brightness contrast in artificially lighted streets, *Transactions of the Illuminating Engineering Society (London)*, 3, 187–195.

Eastman, A.A., and McNelis, J.F. (1963) An evaluation of sodium, mercury and filament lighting for roadways, *Illuminating Engineering*, 58, 28–34.

Edwards, C.S., and Gibbons, R.B. (2007) *The Relationship of Vertical Illuminance to Pedestrian Visibility in Crosswalks*, TRB Visibility Symposium, College Station, TX: Transportation Research Board.

Eklund, N.H. (1999) Exit sign recognition for color normal and color deficient observers, *Journal of the Illuminating Engineering Society*, 28, 71–81.

Eklund, N.H., Rea, M.S., and Bullough, J. (1997) Survey of snowplow operators about forward lighting and visibility during nighttime operations, *Transportation Research Record*, 1585, 25–29.

Ekrias, A., Eloholma, M., and Halonen, L. (2007) Analysis of road lighting quantity and quality in varying weather conditions, *Leukos*, 4, 89–98.

Electric Power Research Institute (EPRI) (2005) *Real World Background Luminance for Objects Viewed by Night Drivers*, Palo Alto, CA: EPRI.

Elexon Limited, 2005, *BSCP520 Appendices for BSC Procedure: Unmetered Supplies Registered in SMRS*, Version 3.

Ellis, R.F., Amos, S., and Kumar, A. (2003) *Illumination Guidelines for Nighttime Highway Work*, NCHRP Report 498, Washington, DC: Transportation Research Board.

Eloholma, M., and Halonen, L. (2006) New model for mesopic photometry and its application to roadway lighting, *Leukos*, 2, 263–293.

Eloholma, M., Halonen, L., and Ketomaki, J. (1999) The effects of light spectrum on visual performance at mesopic light levels, *Proceedings of the CIE Symposium: 75 Years of CIE Photometry*, Vienna: CIE.

Elvik, R. (1993) The effects on accidents of compulsory use of daytime running lights for cars in Norway, *Accident Analysis and Prevention,* 25, 383–398.

Elvik, R. (1995) Meta-analysis of evaluations of public lighting as accident countermeasure, *Transportation Research Record*, 1485, 112–123.

Elvik, R. (1996) A meta-analysis of studies concerning the safety effects of daytime running lights on cars, *Accident Analysis and Prevention,* 28, 685–694.

Erbay, A. (1974) A new method for the characterisation of the reflection properties of road surfaces, *Lichttechnik*, 26, 239.

European Committee for Standardization (CEN) (2006) *European Standard: Traffic Control Equipment — Signal Heads*, EN 12368, Brussels: CEN.

European Road Safety Observatory (ERSO) (2007) *Traffic Safety Basic Facts, 2006,* Brussels: ERSO.

Falkmer, T., and Gregersen, N.P. (2005) A comparison of eye movement behaviour in experienced and inexperienced drivers in real traffic environment, *Optometry and Visual Science,* 82, 732–739.

Farmer, C.M., and Williams, A.F. (2002) Effects of daytime running lights on multiple vehicle daylight crashes in the United States, *Accident Analysis and Prevention,* 34, 197–203.

Federal Highways Administration (FHWA) (2003) *Manual on Uniform Traffic Control Devices,* Washington, DC: FHWA.

Feeny-Burns, L., Burns, R.P., and Gao, C.L. (1990) Age-related macular changes in humans over 90 years old, *American Journal of Ophthalmology,* 109, 265–278.

Ferguson S.A., Preusser, D.F., Lund, A.K., Zador, P.L., and Ulmer, R.G. (1995) Daylight saving time and motor vehicle crashes: The reduction in pedestrian and vehicle occupant fatalities, *American Journal of Public Health*, 85, 92–95.

Fildes, B.N., Corben, B., Kent, S., Oxley, J., Le, T.M., and Ryan, P. (1994) *Older Road User Crashes,* Clayton, Australia: Monash University Accident Research Centre.

Fisher, A.J., and Hall, R.R. (1976) Road luminances based on detection of change of visual angle, *Lighting Research and Technology,* 8, 187–194.

Flannagan, M.J. (2001) *The Safety Potential of Current and Improved Front Fog Lamps,* UMTRI-2001-40, Ann Arbor, MI: University of Michigan Transportation Research Institute.

Flannagan, M., Sivak, M., Ersing, M., and Simmon, C.J. (1989) *Effect of Wavelength on Discomfort Glare for Monochromatic Sources,* UMTRI-89-30, Ann Arbor, MI: University of Michigan Transportation Institute.

Folks, W.R., and Kreysar, D. (2000) Front fog lamp performance, *Human Factors in 2000, Driving, Lighting, Seating Comfort and Harmony in Vehicle Systems*, SP-1539, Warrendale PA: SAE.

Fontaine, H. (2003) Age des conducteurs de voiture et accidents de la route: Quel risque pour les seniors, *Rescherche-Transports-Securité*, 79–80, 107–120.

Forbes, T.W. (1972) Visibility and legibility of highway signs, in T.W. Forbes (ed.) *Human Factors in Highway Safety Traffic Research*, New York: Wiley Interscience.

Fotios, S.A., and Cheal, C. (2007) Lighting for subsidiary streets: Investigations of lamps of different SPD, Part 2 — Brightness, *Lighting Research and Technology*, 39, 215–232.

Fotios, S., Cheal, C., and Boyce, P.R. (2005) Light source spectrum, brightness perception and visual performance in pedestrian environments: A review, *Lighting Research and Technology*, 37, 271–294.

Frederiksen, E., and Rotne, N. (1978) *Calculation of Visibility in Road Lighting,* Report 17, Lyngby, Denmark: The Danish Illuminating Engineering Laboratory.

Frederiksen, E., and Sorensen, K. (1976) Reflection classification of dry and wet road surfaces, *Lighting Research and Technology*, 8, 175–186.

Freedman, M., Janoff, M.S., Kuth, B.W., and McCunney, W. (1975) *Fixed Illumination for Pedestrian Protection*, Report FHWA-RD-76-8, Washington, DC: Federal Highways Administration.

Freedman, M., Zador, P., and Staplin, L. (1993) Effects of reduced transmittance film on automobile rear window visibility, *Human Factors*, 35, 535–550.

Freeman, M., Mitchel, J., and Coe, G.A. (2003) *Safety Performance of Traffic Management at Major Motorway Road Works,* TRL Report 595, Crowthorne, UK: Transport Research Laboratory.

Freeman, E., Munoz, B., Turano, K., and West, S.K. (2005) Vision loss and time to driving cessation in older adults, *Optometry and Vision Science,* 82, 765–775.

Freeman, E.E., Munoz, B., Turano, K.A., and West, S.K. (2006) Measures of visual function and their association with driving modification in older adults, *Investigative Ophthalmology and Visual Science*, 47, 514–520.

Fry, G.A., and King, V.M. (1975) The pupillary response and discomfort glare, *Journal of the Illuminating Engineering Society*, 4, 307–324.

Gallagher, V.P., and Meguire, P.G. (1975) Contrast requirements of urban driving, *Transportation Research Board, Report 156: Driver Visual Needs in Night Driving*, Washington, DC: TRB.

Gallagher, V.P., Janoff, M.S., and Farber, E. (1974) Interaction between fixed and vehicular illumination systems on city streets, *Journal of the Illuminating Engineering Society*, 4, 3–10.

Garstang, R.H. (1986) Model for artificial night sky illumination, *Publications of the Astronomical Society of the Pacific*, 98, 364–375.

Gibson, J.J. (1950) *The Perception of the Visual World*, Boston, MA: Houghton-Mifflin.

Gibson, K.S., and Tyndell, E.P.T. (1923) Visibility of radiant energy, *Bulletin of the Bureau of Standards*, 19, 131–191.

Gilkes, M.J. (1988) The basis of the medical recommendations for driver's visual standards in the United Kingdom, in A.G. Gale, M.H. Freeman, C.M. Haslegrave, P. Smith, and S.P. Taylor (eds.) *Vision in Vehicles-II*, Amsterdam: North Holland.

Girasole, T., Roze, C., Maheu, B., Grehan, G., and Menard, J. (1998) Visibility distances in a foggy atmosphere: Comparison between lighting installations by Monte Carlo simulation, *Lighting Research and Technology*, 30, 29–36.

Goodman, T., Forbes, A., Walkey, H., Eloholma, M., Halonen, L., Alferdinck, J., Freiding, A., Bodrogi, P., Varady, G., and Szalmas, A. (2007) Mesopic visual efficiency IV: A model with relevance to nighttime driving and other applications, *Lighting Research and Technology*, 39, 365–392.

Graham, C.H., and Kemp, E.H. (1938) Brightness discrimination as a function of the duration of the increment in intensity, *Journal of General Physiology*, 21, 635–650.

Green, J., and Hargroves, R.A. (1979) A mobile laboratory for dynamic road lighting measurement, *Lighting Research and Technology*, 11, 197–203.

Groupe de Travail Bruxelles (GTB) (1999) *Rationale of Harmonized Dipped (Low) Beam Pattern*, Report No. C.E.-3160, Geneva, Switzerland: GTB.

Guler, O., and Onaygil, S. (2003) The effect of luminance uniformity on visibility level in road lighting, *Lighting Research and Technology*, 35, 199–215.

Guler, O., Onaygil, S., and Erkin, E. (2005) The effect of vehicle headlights on visibility level in road lighting, *Proceedings of CIE Midterm Meeting, Leon*, Vienna: CIE.

Guo, L., Eloholma, M., and Halonen, L. (2007) Lighting control strategies for telemanagement road lighting control systems, *Leukos*, 4, 157–171.

Hakamies-Blomqvist, L. (1994) *Older Drivers in Finland: Traffic Safety and Behaviour*, Report 40/1994, Helsinki, Finland: Liikenneturva.

Hakamies-Blomqvist, L., and Wahlstrom, B. (1998) Why do older drivers give up driving? *Accident Analysis and Prevention*, 30, 305–312.

Hakamies-Blomqvist, L., Johansson, K., and Lundberg, C. (1996) Medical screening in a comparative Finnish–Swedish evaluation study, *Journal of the American Geriatrics Society*, 44, 650–653.

Hakamies-Blomqvist, L., Raitanen, T., and O'Neill, D. (2002) Driver ageing does not cause higher accident rates per km, *Transport Research*, Part F5, 271–274.

Hale, J. (1989) *Snowplow Lighting Study: Final Report*, Report No MN/RD-89/03, St. Paul, MN: Minnesota Department of Transportation.

Hall, R.R., and Fisher, A.J. (1977) Measures of visibility and visual performance in road lighting, *Measures of Road Lighting Effectiveness*, Berlin: Lichttechnische Gesellschaft LiTG.

Hallet, P.E. (1963) Spatial summation, *Vision Research*, 3, 9–24.

Hamm, M. (2002) *Adaptive Lighting Functions History and Future — Performance Investigations and Field Test for User's Acceptance*, SAE Paper 2002-01-0526, Warrendale, PA: Society of Automotive Engineers.

Hankey, J.M., Kiefer, R.J., and Gibbons, R.B. (2005) *Quantifying the Pedestrian Detection Benefits of the General Motors Night Vision System*, SAE Technical Paper 2005-01-0443, Warrendale, PA: Society of Automotive Engineers.

Hansen, E.R. and Larsen, J.S. (1979) Reflection factors for pedestrian clothing, *Lighting Research and Technology*, 11, 154–157.

Hare, C.T., and Hemion, R.H. (1968) *High Beam Usage on US Highways*, Report QR-666, Washington DC: Bureau of Roads.

Hargroves, R.A. (1981) Road lighting — as calculated and as in service, *Lighting Research and Technology*, 13, 130–136.

Hargroves, R.A. (2001) Lighting for pleasantness outdoors, *Proceedings of the CIBSE National Conference*, London: Chartered Institution of Building Services Engineers.

Hargroves, R.A., and Scott, P.P. (1979) Measurements of road lighting and accidents — the results, *Public Lighting*, 44, 213–221.

Hartmann, E., Finsterwalder, J., and Muller, M. (1986) Kinetic luminance measurement and assessment of road tunnels, *Lighting Research and Technology*, 18, 28–36.

Hasson, P., Lutkevich, P., Ananthanarayanan, B., Watson, P., and Knoblauch, R. (2002) Field test for lighting to improve safety at pedestrian crosswalks, *Proceedings of the 16th Biennial Symposium on Visibility and Simulation*, Washington, DC: Transportation Research Board.

Hawkins, R.K. (1988) Motorway traffic behaviour in reduced visibility conditions, in A.G. Gale, M.H. Freeman, C.M. Haslegrave, P. Smith, and S.P. Taylor (eds.) *Vision in Vehicles-II*, Amsterdam: North Holland.

He, Y., Rea, M.S., Bierman, A., and Bullough, J. (1997) Evaluating light source efficacy under mesopic conditions using reaction times, *Journal of the Illuminating Engineering Society*, 26, 125–138.

Hellier-Symons, R.D., and Irving, A. (1981) *Masking of Brake Lights by High-Intensity Rear Lights in Fog*, TRRL Report 998, Crowthorne, UK: Transportation and Road Research Laboratory.

Helmers, G., and Rumar, K. (1975) High beam intensity and obstacle visibility, *Lighting Research and Technology*, 7, 38–42.

Hentschel, H.J. (1971) A physiological appraisal of the revealing power of street lighting installations for large composite targets, *Lighting Research and Technology*, 3, 268–273.

Higgins, K.E., and Bailey, I.L. (2000) Visual disorders and performance of specific tasks requiring vision, in B. Silverstone, M.A. Lang, B.P. Rosenthal, and E.E. Faye (eds.) *The Lighthouse Handbook on Vision Impairment and Vision Rehabilitation*, New York: Oxford University Press.

Higgins, K.E., Wood, J.M., and Tait, A. (1996) Closed road driving performance: The effect of degradation of visual acuity, *Vision Science and Its Applications*, Washington, DC: Optical Society of America.

Hills, B.L. (1975a) Visibility under night driving conditions, Part 1 Laboratory background and theoretical considerations, *Lighting Research and Technology*, 7, 179–184.

Hills, B.L. (1975b) Visibility under night driving conditions, Part 2 Field measurements using disc obstacles and a pedestrian dummy, *Lighting Research and Technology*, 7, 251–258.

Hills, B.L. (1976) Visibility under night driving conditions, Part 3 Derivation of (ΔL, A) characteristics and factors in their application, *Lighting Research and Technology*, 8, 11–26.

Hills, B.L. (1980) Vision, visibility and perception in driving, *Perception*, 9, 183–216.

Hilz, R., and Cavonius, C.R. (1974) Functional organization of the peripheral retina: Sensitivity to periodic stimuli, *Vision Research*, 14, 1333–1337.

Holladay, L.L. (1926) The fundamentals of glare and visibility, *Journal of the Optical Society of America*, 12, 271–319.

Holz, M., and Weidel, E. (1998) Night vision enhancement system using diode laser headlamps, *Electronics for Trucks and Buses,* Report SP-1401, Warrendale, PA: Society of Automotive Engineers.

Hopkinson, R.G. (1940) Discomfort glare in lighted streets, *Transactions of the Illuminating Engineering Society (London)*, 5, 1–29.

Horowitz, A.D. (1994) Human factors issues in advanced rear signaling systems, *Proceedings of the Fourteenth International Technical Conference on Enhanced Safety of Vehicles*, Washington, DC: Department of Transportation.

Huang, H., Zegeer, C., and Nassi, R. (2000) *Innovative Treatments at Unsignalized Pedestrian Crossing Locations*, ITE Annual Meeting Compendium, Washington, DC; Institute of Transportation Engineers.

Hughes, P.K., and Cole, B.L. (1984) Search attention conspicuity of road traffic control devices, *Australian Road Research*, 14, 1–9.

Huhn, W., Ripperger, J., and Befelein, C. (1997) *Rear Light Redundancy and Optimized Hazard Warning Signal — New Safety Functions for Vehicles*, SAE Technical Paper 970656, Warrendale, PA: Society of Automotive Engineers.

Hunter, W.W., Stutts, J.C., Pein, W.E., and Cox, C.C. (1996) *Pedestrian and Bicycle Crash Types of the Early 1990s,* Report FHWA-RD-95-163, Washington DC: Federal Highways Administration.

Hutt, D.L., Bissonnette, L.R., St. Germain, D., and Oman, J. (1992) Extinction of visible and infrared beams by falling snow, *Applied Optics*, 31, 5121–5132.

Ikeda, M., and Shimozono, H. (1981) Mesopic luminous efficiency functions, *Journal of the Optical Society of America*, 71, 280–284.

Illuminating Engineering Society of North America (IESNA) (1980) Roadway Lighting Subcommittee, *Visual Comfort: The CBE Recommendation*, New York: IESNA.

Illuminating Engineering Society of North America (IESNA) (1998) Recommended Practice RP-20-98: *Lighting for Parking Facilities*, New York: IESNA.

Illuminating Engineering Society of North America (IESNA) (1999) Recommended Practice RP-33-99 *Lighting the Exterior Environment,* New York: IESNA.

Illuminating Engineering Society of North America (IESNA) (2000) *The IESNA Lighting Handbook, 9th Edition,* New York: IESNA.

Illuminating Engineering Society of North America (IESNA) (2001) Recommended Practice RP-19-01 *Roadway Sign Lighting,* New York: IESNA.

Illuminating Engineering Society of North America (IESNA) (2005a) Recommended Practice RP-8-00 *Roadway Lighting,* New York: IESNA.

Illuminating Engineering Society of North America (2005b) Recommended Practice RP-22-05 *IESNA Recommended Practice for Tunnel Lighting,* New York: IESNA.

Illuminating Engineering Society of North America (IESNA) (2006) Technical Memorandum TM-12-06 *Spectral Effects of Lighting on Visual Performance at Mesopic Light Levels,* New York: IESNA.

Illuminating Engineering Society of North America (IESNA) (2007) Technical Memorandum TM-15-07 *Luminaire Classification System for Outdoor Luminaires,* New York: IESNA.

Institute of Transportation Engineers (ITE) (1985) *Vehicle Traffic Control Signal Heads: A Standard of the Institute of Transportation Engineers,* Washington, DC: ITE.

Institute of Transportation Engineers (ITE) (2005) *Vehicle Traffic Control Signal Heads Part 2: Light Emitting Diode Circular Signal Supplement,* Washington, DC: ITE.

Institution of Lighting Engineers (ILE) (1995) *Lighting the Environment, a Guide to Good Urban Lighting,* Rugby, UK: ILE.

Institution of Lighting Engineers (ILE) (2000) *Guidance Notes for the Reduction of Light Pollution,* Rugby, UK: ILE.

Institution of Lighting Engineers (ILE) (2007) *Technical Report 28: Measurement of Road Lighting Performance on Site,* Rugby, UK: ILE.

International Dark-Sky Association (IDA) (2002) *Outdoor Lighting Code Handbook,* Version 1.14, http://www.darksky.org (accessed March 21st, 2007).

Isebrands, H., Hallmark, S., Hans, Z., McDonald, T., Preston, H., and Storm, R. (2004) *Safety Impacts of Street Lighting at Isolated Rural Intersections, Part II, Year 1 Report,* St. Paul, MN: Minnesota Department of Transportation.

Jalink, C.J. (2002) *Distributive Lighting Systems for Interior Applications,* SAE Paper 2002-01-0979, Warrendale, PA: Society of Automotive Engineers.

Janoff, M.S. (1990) The effect of visibility on driver performance: A dynamic experiment, *Journal of the Illuminating Engineering Society,* 19, 57–63.

Janoff, M.S. (1992a) The relationship between visibility level and subjective ratings of visibility, *Journal of the Illuminating Engineering Society,* 21, 98–107.

Janoff, M.S. (1992b) Effect of headlights on small target visibility, *Journal of the Illuminating Engineering Society,* 21, 46–53.

Janoff, M.S., and Havard, J.A. (1997) The effect of lamp color on visibility of small targets, *Journal of the Illuminating Engineering Society,* 26, 173–181.

Janoff, M.S., Freedman, M., and Koth, B. (1977) Driver and pedestrian behavior — the effect of specialized crosswalk illumination, *Journal of the Illuminating Engineering Society,* 6, 202–208.

Janssen, W.H., Michon, J.A., and Lewis, O.H. (1976) The perception of lead vehicle movement in darkness, *Accident Analysis and Prevention,* 8, 151–166.

Johnson, C., and Keltner, J. (1983) Incidence of field loss in 20,000 eyes and its relationship to driving performance, *Archives of Ophthalmology.* 101, 371–375.

Jones, H.V., and Heimstra, N.W. (1964) Ability of drivers to make critical passing judgments, *Journal of Engineering Psychology,* 3, 117–122.

Judd, D.B. (1951) Report of the US Secretariat Committee on Colorimetry and Artificial Daylight, *Proceeding of the CIE 12th Session, Stockholm,* Vienna: CIE.

Kahane, C.J., and Hertz, E. (1998) *The Long Term Effectiveness of Center High-Mounted Stop Lamps in Passenger Cars and Trucks*, DOT HS 808 696, Washington DC: National Highway Traffic Safety Administration.

Kaiser, P.K., and Boynton, R.M. (1996) *Human Color Vision*, Washington DC: Optical Society of America.

Keith, D.M. (2000) Roadway lighting design for optimization of UPD, STV and uplight, *Journal of the Illuminating Engineering Society*, 29, 15–23.

Kelly, D.H. (1959) Effects of sharp edges in a flickering field, *Journal of the Optical Society of America*, 49, 730–732.

Kelly D.H. (1961) Visual response to time-dependent stimuli, 1. Amplitude sensitivity measurements, *Journal of the Optical Society of America*, 51, 422–429.

Kelly, K.L., and Judd, D.B. (1965) *The ISCC-NBS Centroid Color Charts*, Washington DC: National Bureau of Standards.

Kirkpatrick, M., Baker, C.C., and Heasly, C.C. (1987) *A Study of Daytime Running Light Design Factors. Final Report*, Report DOT/HS 807–193, Washington, DC: Department of Transportation.

Kline, D. (1995) Visual Requirements for the Aging Driver, in W. Adrian (ed.) *Lighting for Aging Vision and Health*, New York: Lighting Research Institute.

Knoblauch, K., Saunders, F., Kusada, M., Hynes, R., Podgor, M., Higgins, K.E., and de Monasterio, F.M. (1987) Age and illuminance effects in the Farnsworth-Munsell 100 hue test scores, *Applied Optics*, 26, 1441–1448.

Kocifaj, M. (2007) Light pollution model for cloudy and cloudless night skies with ground based light sources, *Applied Optics*, 46, 3013–3022.

Kocmond, W.C., and Perchonok, K. (1970) *Highway Fog*, National Cooperative Highway Research Program, Report 171, Washington, DC: The National Research Council.

Konyukhov, V.V., Koroleva, Y.E., Novakovsky, L.G., and Novikova, L.A. (2006) Motorcycle headlamps featuring a light beam stabilization dynamic adjuster during turning, *Light and Engineering*, 14, 50–61.

Koornstra, M.J. (1993) *Daytime Running Lights: Its Safety Revisited*, Report D-93-25, Leidschendam, The Netherlands: SWOV Institute for Road Safety Research.

Kosnik, W., Winslow, L., Kline, D., Rasinski, K., and Sekular, R. (1988) Visual changes throughout adulthood, *Journal of Gerontology: Psychological Sciences*, 43, 63–70.

Koth, B.W., McCunney, W.D., Duerk, C.P., Janoff, M.S., and Freedman, M. (1978) *Vehicle Fog Lighting: An Analytical Evaluation*, Report DOT HS-803-442, Washington, DC; National Highway Traffic Safety Administration.

Kraus, A., Benter, N., and Boerner, H. (2007) *OLED Technology and Its Possible Use in Automotive Applications*, SAE Paper 2007-01-1230, Warrendale, PA: Society of Automotive Engineers.

Kwong, R.C., Michaiski, L., Nugent, M., Rajan, K., Ngo, T., Brown, J.J., Lamansky, S., Djurovich, P., Murphy, D., Abdel-Razzaq, F., Brooks, J., Thompson, M.E., Adachi, C., Baldo, M., and Forrest, S.R. (2001) Recent advances in organic light emitting diodes, *Proceedings of the 9th International Symposium on the Science and Technology of Light Sources*, Ithica, NY: Cornell University Press.

Lamm, R., Kloeckner, J.H., and Choueiri, E.M. (1985) *Freeway Lighting and Traffic Safety: A Long Term Investigation*, TRB-1985-17, 64th TRB Annual Meeting, Washington DC, Transportation Research Board.

Land, M.F., and Horwood, J. (1995) Which part of the road guides steering, *Nature (London)*, 377, 339–340.

Langford, J., Methorst, R., and Hakamies-Blumqvist, L. (2006) Older drivers do not have a high crash risk — a replication of low mileage bias, *Accident Analysis and Prevention*, 38, 574–578.

Lansdown, T., Brook-Carter, N., and Kersloot, T. (2004) Distraction from multiple in-vehicle secondary tasks: Visual performance and mental workload implications, *Ergonomics*, 47, 91–104.

Lecocq, J. (1993) VL application for hemispherical multi-faceted targets, *Proceedings of the 2nd International Symposium on Visibility and Luminance in Roadway Lighting, Orlando, FL*, Cleveland OH: Lighting Research Office.

Lecocq, J. (1994) Visibility and lighting of wet road surfaces, *Lighting Research and Technology*, 26, 75–87.

Leibowitz, H.W., and Owens, D.A. (1975) Night myopia and the intermediate dark focus of accommodation, *Journal of the Optical Society of America*, 65, 1121–1128.

Leibowitz, H.W., and Owens, D.A. (1977) Nighttime driving accidents and selective visual degradation, *Science*, 197, 422–423.

Lestina, D.C., Miller, T.R., Knoblauch, R., and Nitzburg, M. (1999) Benefits and costs of ultraviolet fluorescent lighting, *Proceedings of the Association for the Advancement of Automotive Medicine*, Barcelona, Spain.

Lewis, A.L. (1999) Visual performance as a function of spectral power distribution of light sources at luminances used for general outdoor lighting, *Journal of the Illuminating Engineering Society*, 28, 37–42.

Lighting Research Center (LRC) (1996) *Evaluating Interior Lighting Schemes: General Motors Automobile Study*, Troy, NY: Lighting Research Center.

Lighting Research Center (LRC) (2007) *Lighting Answers: Light Pollution*, Troy, NY: Lighting Research Center.

Lin, Y., Chen, W., Chen, D., and Shao, H. (2004) The effect of spectrum on visual field in road lighting, *Building and Environment*, 39, 433–439.

Lingard, R., and Rea, M.S. (2002) Off-axis detection at mesopic light levels in a driving context, *Journal of the Illuminating Engineering Society*, 31, 33–39.

Liu, A. (1998) What the driver's eyes tells the car's brain, in G. Underwood (ed.) *Eye Guidance in Reading and Scene Perception*, Amsterdam: Elsevier.

Livingston, M.S., and Hubel, D.H. (1981) Effects of sleep and arousal on the processing of visual information in the cat, *Nature*, 291, 554–561.

Luoma, J., Flannagan, M.J., Sivak, M., Aoki, M., and Traube, E.C. (1995a) *Effects of Turn-Signal Color on Reaction Times to Brake Signals*, UMTRI-95-5, Ann Arbor, MI: University of Michigan Transportation Research Institute.

Luoma, J., Schumann, J., and Traube, E.C. (1995b) *Effect of Retroreflector Positioning on Nighttime Recognition of Pedestrians*, UMTRI-95-18, Ann Arbor, MI: University of Michigan Transportation Research Institute.

Luoma, J., Sivak, M., and Flannagan, M.J. (2006) Effects of dedicated stop-lamps on nighttime rear-end collisions, *Leukos*, 3, 159–165.

Lynes, J.A. (1971) Lightness, colour and constancy in lighting design, *Lighting Research and Technology*, 3, 24–42.

Lynes, J.A. (1977) Discomfort glare and visual distraction, *Lighting Research and Technology*, 9, 51–52.

MacAdam, D.L. (1942) Visual sensitivity to color differences in daylight, *Journal of the Optical Society of America*, 32, 247–274.

Mace, D.J., Hosletter, R.S., Pollack, L.E., and Sweig, W.D. (1986) *Minimal Luminance Requirements for Official Highway Signs*, FHWA–RD-86-151, Washington, DC: Federal Highways Administration.

Mace, D., Garvey, P., Porter, R.J., Schwab, R., and Adrian, W. (2001) *Countermeasures for Reducing the Effects of Headlight Glare*, Washington, DC: AAA Foundation for Traffic Safety.

Mandlebaum, J., and Sloan, L.L. (1947) Peripheral visual acuity, *American Journal of Ophthalmology*, 30, 581–588.

Marshall, J., Grindle, J., Ansell, P.L., and Borwein, B. (1979) Convolution in human rods: An aging process, *British Journal of Ophthalmology*, 63, 181–187.

Martin, A. (2006) *Factors Influencing Pedestrian Safety: A Literature Review*, TRL Report PPR241, Crowthorne, UK: Transport Research Laboratory.

Massart, P. (1973) *Definition of Synthetic Surfaces in Road Lighting*, Thesis, University of Liege, Belgium.

Massie, D.L., and Campbell, K.L. (1993) *Analysis of Accident Rate by Age, Gender and Time of Day Based on the 1990 Nationwide Personal Transportation Survey*, UMTRI-93-7, Ann Arbor, MI: University of Michigan Transportation Research Institute.

Maycock, G., Lockwood, G.R., and Lester, J.F. (1991) *The Accident Liability of Drivers*, TRRL Report 315, Crowthorne, UK: Transport and Road Research Laboratory.

McLean, D. (2006) Adaptive roadway lighting, *Journal of the IMSA*, 41, 54–58.

McColgan, M.W., Van Derlofske, J., Bullough, J.D., and Shakir, I. (2002) *Subjective Color Preferences of Common Road Sign Materials under Headlamp Bulb Illumination*, SAE Paper 2002-01-0261, Warrendale, PA: Society of Automotive Engineers.

McGrath, C., and Morrison, J.D. (1981) The effects of age on spatial frequency perception in human subjects, *Quarterly Journal of Experimental Physiology*, 66, 253–261.

McGwin, G. Jr., and Brown, D.B. (1999) Characteristics of traffic crashes among young, middle-aged and older drivers, *Accident Analysis and Prevention*, 31, 181–198.

McKnight, A.J., Shinar, D., and Hilburn, B. (1991) The visual and driving performance of monocular and binocular heavy duty truck drivers, *Accident Analysis and Prevention*, 23, 225–237.

McNally, D. (1994) *The Vanishing Universe*, Cambridge, UK: Cambridge University Press.

McSweeney, D.J. (2007) The future of lighting, *Lighting Journal*, 72, 22–28.

Middleton, W.E.K. (1952) *Vision Through the Atmosphere*, Toronto: University Press.

Miller, J.W. (1958) Study of visual acuity during the ocular pursuit of moving test objects. II Effects of direction of movement, relative movement and illumination, *Journal of the Optical Society of America*, 48, 803–808.

Mizon, B. (2002) *Light Pollution, Responses and Remedies*, London: Springer.

Monahan, D.R. (1995) Safety considerations in parking facilities, *Proceedings of the International Parking Conference and Exposition*, Fredericksburg, VA: International Parking Institute.

Moon, P., and Hunt, R.M. (1938) Reflection characteristics of road surfaces, *Journal of the Franklin Institute*, 225, 1–21.

Moore, R.L. (1952) *Rear Lights of Motor Vehicles and Pedal Cycles*, Road Research Technical Paper 25, London: His Majesty's Stationery Office, London.

Moore, D.W., and Rumar, K. (1999) *Historical Development and Current Effectiveness of Rear Lighting Systems*, UMTRI-99-31, Ann Arbor, MI: University of Michigan Transportation Research Institute.

Morgan, C. (2001) *The Effectiveness of Retro-reflective Tape on Heavy Trailers*, NHTSA Technical Report DOT HS 809 222, Washington DC: National Highway Traffic Safety Administration.

Mortimer, R.G. (1977) A decade of research in rear lighting: What have we learned, *Proceedings of the 21st Conference of the American Association for Automotive Medicine*, Morton Grove, IL: AAAM.

Mortimer, R.G. (1981) *Field Test Evaluation of Rear Lighting Deceleration Signals*, Safety Research Report 81-1, Champaign, IL: University of Illinois.

Mortimer, R.G., and Becker, J.M. (1973) *Development of a Computer Simulation to Predict the Visibility Distances Provided by Headlamp Beams*, Report UM-HSRI-IAF-73-15, Ann Arbor, MI: University of Michigan.

Mortimer, R.G., and Jorgeson, C.M. (1974) *Eye Fixations of Drivers in Night Driving with Three Headlight Beams*, Report UM-HSRI-74-17, Ann Arbor, MI: University of Michigan.

Murata, Y. (1987) Light absorption characteristics of the lens capsule, *Ophthalmic Research*, 19, 107–112.

Murray, J., Feather, J., and Carden, D. (2002) Dynamic discomfort glare and driver fatigue, *Lighting Journal*, 67, 20–23.

Narendran, N., Vasconez, S., Boyce, P., and Eklund, N. (2000) Just-perceivable color difference between similar light sources in display lighting applications, *Journal of the Illuminating Engineering Society*, 29, 78–82.

Narendran, N., Bullough, J.D., Maliyagoda, N., and Bierma, A. (2001) What is useful life for white light LEDs? *Journal of the Illuminating Engineering Society*, 30, 57–67.

Narisada, K. (1995) Perception in complex fields under road lighting conditions, *Lighting Research and Technology*, 27, 123–131.

Narisada, K., and Karasawa, Y. (2001) Re-consideration of the revealing power on the basis of visibility level, *Proceedings of the International Lighting Congress, Istanbul*, 2, 473–480.

Narisada, K., and Yoshikawa, K. (1974) Tunnel entrance lighting — effect of fixation point and other factors on the determination of requirements, *Lighting Research and Technology*, 6, 9–18.

Narisada, K., Karasawa, Y., and Shirao, K. (2003) Design parameters of road lighting and revealing power, *Proceedings of the CIE, 25th Session, San Diego*, Vienna: CIE.

Nathan, J., Henry, G., and Cole, B. (1964) Recognition of colored traffic light signals by normal and color vision defective observers, *Journal of the Optical Society of America*, 54, 1041–1045.

National Bureau of Standards (1976) *Color: Universal Language and Dictionary of Names*, Special Publication 440, Washington, DC: National Bureau of Standards.

National Center for Statistics and Analysis (NCSA) (2004) *Traffic Safety Facts, 2004 Data, Rural/Urban Comparison*, Washington, DC: NCSA.

National Highway Traffic Safety Administration (NHTSA) (2006a) *Fatality Analysis Reporting System*, Washington, DC: US Department of Transportation.

National Highway Traffic Safety Administration (NHTSA) (2006b) *Traffic Safety Facts 2005*, Washington, DC: US Department of Transportation.

National Police Agency (Japan) (1986) *Specification for Metal Vehicle Traffic Control Signals*, Tokyo: The National Police Agency.

Neitz, J., Carroll, J., and Neitz, M. (2001) Color vision: Almost reason enough for having eyes, *Optics and Photonics News*, January, 26–33.

Neubauer, O., Harrer, S., Marre, M., and Verriest, G. (1978) Colour vision and traffic, in G. Verriest (ed.) *Modern Problems in Ophthalmology*, Basel, Switzerland: Karger.

Nickerson, D. (1957) Horticultural color chart names with a Munsell key, *Journal of the Optical Society of America*, 47, 619–621.

Nilsson, L., and Alm, H. (1996) Effects of a vision enhancement system on drivers' ability to drive safely in fog, in A.G. Gale, I.D. Brown, C.M. Haslegrave, and S.P. Taylor (eds.) *Vision in Vehicles V*, Amsterdam: North-Holland.

Novellas, F., and Perrier, J. (1985) New lighting method for road tunnels, *CIE Journal*, 4, 58–70.

Olson, P.L., and Sivak, M. (1983) Comparison of headlamp visibility distance and stopping distance, *Perceptual and Motor Skills*, 57, 1177–1178.

Olson, P.L., Cleveland, D.E., Fancher, P.S., and Schneider, L.W. (1984) *Parameters Affecting Stopping Sight Distances*, Report UMTRI-84-15, Ann Arbor, MI: University of Michigan Transportation Research Institute.

Olson, P.L., Battle, D.S., and Aoki, T. (1989) *The Detection Distance of Highway Signs as a Function of Color and Photometric Properties*, Report UMTRI-89-36, Ann Arbor, MI: University of Michigan Transportation Research Institute.

Organization for Economic Co-operation and Development (OECD) (1998) *Safety of Vulnerable Road Users*, Paris: OECD.

Owens, D.A., and Tyrrell, R.A. (1999) Effects of luminance, blur and age on nighttime visual guidance: A test of the selective degradation hypothesis, *Journal of Experimental Psychology: Applied*, 5, 1–14.

Owsley, C., Sekular, R., and Siemsen, D. (1983) Contrast sensitivity throughout adulthood, *Vision Research*, 23, 689–699.

Oya, H., Ando, K., and Kanoshima, H. (2002) A research on interrelation between illuminance at intersections and reduction in traffic accidents, *Journal of Light and Visual Environment*, 26, 29–34.

Padmos, P., and Alferdinck, J.W.A.M. (1983a) *Glare in Tunnel Entrances II: The Effect of Stray Light on the Windscreen*, IZF 1983 C-10. Soesterberg, The Netherlands: IZF-TNO.

Padmos, P., and Alferdinck, J.W.A.M. (1983b) *Glare in Tunnel Entrances III: The Effect of Atmospheric Stray Light*, IZF 1983 C-9. Soesterberg, The Netherlands: IZF-TNO.

Padmos, P., Van den Brink, T.D.J., Alferdinck, J.W.A.M., and Folles, E. (1988) Matrix signs for motorways: System design and optimum photometric features, *Lighting Research and Technology*, 20, 55–60.

Painter, K. (1999) The social history of street lighting (part 1), *Lighting Journal*, 64, 14–24.

Painter, K. (2000) The social history of street lighting (part 2), *Lighting Journal*, 65, 24–30.

Painter, K.A., and Farrington, D.P. (2001) The financial benefits of improved street lighting based on crime reduction, *Lighting Research and Technology*, 33, 3–12.

Palmer, D.A. (1968) Standard observer for large-field photometry at any level, *Journal of the Optical Society of America*, 58, 1296–1299.

Peli, E. (1990) Contrast in complex images, *Journal of the Optical Society of America A*, 7, 2032–2040.

Peli, E., and Peli, D. (2002) *Driving with Confidence: A Practical Guide to Driving with Low Vision*, Hackensack, NJ: World Scientific Publishing.

Perel, M., Olson, P.L., Sivak, M., and Medlin J.W. Jr., (1983) *Motor Vehicle Forward Lighting*, SAE Technical Paper 830567, Warrendale, PA: Society of Automotive Engineers.

Plainis, S., Murray, I.J., and Charman, W.N. (2005) The role of retinal adaptation in night driving, *Optometry and Vision Science*, 82, 682–688.

Polus, A., and Katz, A. (1978) An analysis of nighttime pedestrian accidents at specially illuminated crosswalks, *Accident Analysis and Prevention*, 10, 223–228.

Ponziani, R.L. (2006) *Electronic Intelligent Turn Signal System*, SAE Paper 2006-01-0714, Warrendale, PA: Society of Automotive Engineers.

Purves, D., and Beau Lotto, R. (2003) *Why We See What We Do: An Empirical Theory of Vision*, Sunderland MA: Sinauer Associates Inc.

Racette, L., and Casson, E.J. (2005) The impact of visual field loss on driving performance: Evidence from on-road assessments, *Optometry and Vision Science*, 82, 668–674.

Range H.D. (1972) A simplified method for the characterisation of road surfaces for lighting, *Lichttechnik*, 24, 608.

Raynham, P. (2004) An examination of the fundamentals of roads lighting for pedestrians and drivers, *Lighting Research and Technology*, 36, 307–316.

Rea, M.S. (2001) The road not taken, *Lighting Journal*, 66, 18–25.

Rea, M.S., and Bullough, J.D. (2007) Move to a unified system of photometry, *Lighting Research and Technology*, 39, 393–408.

Rea, M.S., and Ouellette, M.J. (1991) Relative visual performance: A basis for application, *Lighting Research and Technology*, 23, 135–144.

Rea, M.S., Bierman, A., McGowan, T., Dickey, F., and Havard, J. (1997) A field study comparing the effectiveness of metal halide and high pressure sodium illuminants under mesopic conditions, *Proceedings of the CIE Symposium on Visual Scales: Photometric and Colourimetric Aspects, Teddington, UK*, Vienna: CIE.

Rea, M.S., Bullough, J.D., Freyssinier-Nova, J-P., and Bierman, A. (2004) A proposed unified system of photometry, *Lighting Research and Technology*, 36, 85–111.

Reed, M.P., and Flannagan, M.J. (2003) *Geometric Visibility of Mirror-Mounted Turn Signals*, UMTRI-2003-18, Ann Arbor, MI: University of Michigan Transportation Research Institute.

Regan, D.M., and Beverley, K.I. (1982) How do we avoid confounding the direction we are looking with the direction we are moving? *Science*, 213, 194–196.

Rich, C., and Longcore, T. (2006) *Ecological Consequences of Artificial Night Lighting*, Washington DC: Island Press.

Richter, M., and Witt, K. (1986) The story of the DIN color system, *Color Research and Application*, 11, 138–145.

Robertson, A.R. (1977) The CIE 1976 color difference formulae, *Color Research and Application*, 2, 7–11.

Roch, J., and Smiatek, G. (1972), The q_s-method for the characterisation of road surfaces for lighting, *Lichttechnik*, 24, 329.

Rockwell, T.H., and Safford, R.R. (1968) *An Evaluation of Automotive Rear Signal System Characteristics in Night Driving*, Report EES-272 B, Columbus, OH: The Ohio State University.

Rogers, J.G. (1972) Peripheral contrast thresholds for moving images, *Human Factors*, 14, 199–205.

Roper, V.J., and Howard, E.A. (1938) Seeing with motor car headlamps, *Transactions of the Illuminating Engineering Society (London)*, 33, 417–438.

Rosenhahn, E.O., and Hamm, M. (2001) Measurements and ratings of HID headlamp impact on traffic safety aspects, *Lighting Technology Developments for Automobiles*, SAE Report, SP1595, Warrendale, PA: Society of Automotive Engineers.

Rumar, K. (1990) The basic driver error: Late detection, *Ergonomics*, 33, 1281–1290.

Rumar, K. (1998) *Vehicle Lighting and the Aging Population*, UMTRI-98-9, Ann Arbor, MI: University of Michigan Transportation Research Institute.

Rumar, K. (2000) *Relative Merits of the US and ECE High Beam Maximum Intensities and of Two and Four Headlamp Systems*, UMTRI 2000-41, Ann Arbor, MI: University of Michigan Transportation Institute.

Rumar, K. (2003) *Functional Requirements for Daytime Running Lights*, UMTRI-2003-11, Ann Arbor, MI: University of Michigan Transportation Research Institute.

Rumar, K., and Marsh II, D.K. (1998) *Lane Marking in Night Driving: A Review of Past Research and of the Present Situation*, UMTRI-98-50, Ann Arbor, MI: University of Michigan Transportation Research Institute.

Rutley, K.S., and Mace, D.G.W. (1969) *An Evaluation of a Brakelight Display Which Indicates the Severity of Braking*, LR-287, Crowthorne, UK: Transport and Road Research Laboratory.

Sabey, B.E., and Staughton, E.C. (1975) Interacting roles of road environment, vehicle and road user in accidents, *Proceedings of the 5th Conference of the International Association for Accident and Traffic Medicine*, London.

Sagawa, K., and Takahashi, Y. (2001) Spectral luminous efficiency as a function of age, *Journal of the Optical Society of America*, 18, 2659–2667.

Sagawa, K., and Takeichi, K. (1992) System of mesopic photometry for evaluating lights in terms of comparative brightness relationships, *Journal of the Optical Society of America A*, 9, 1240–1246.

Sayer, J.R., Mefford, M.L., Flannagan, M.J., and Sivak, M. (1996) *Reaction Time to Center High-Mounted Stop Lamps: Effects of Context, Aspect Ratio, Intensity and Ambient Illumination*, UMTRI-96-3, Ann Arbor, MI: University of Michigan Transportation Research Institute.

Sayer, J.R., Mefford, M.L., and Blower, D. (2001) *The Effects of Rear-Window Transmittance and Back-Up Lamp Intensity on Backing Behavior*, UMTRI-2001-6, Ann Arbor, MI: University of Michigan Transportation Research Institute.

Schieber, F. (1994) *Recent Developments in Vision, Aging and Driving: 1988–1994*, UMTRI-94-26, Ann Arbor, MI: University of Michigan Transportation Research Center.

Schivelbusch, W. (1988) *Disenchanted Night: The Industrialization of Light in the Nineteenth Century*, Oxford, UK: Berg Publishing.

Schmidt-Clausen, H.J. (1985) Optimum luminances and areas of rear-position lamps and stop lamps, *Proceedings of the 10th International Technical Conference on Experimental Safety Vehicles*, Washington, DC: National Highway Traffic Safety Administration.

Schmidt-Clausen, H.J., and Bindels, J.H. (1974) Assessment of discomfort glare in motor vehicle lighting, *Lighting Research and Technology*, 6, 79–88.

Schmidt-Clausen, H.J., and Finsterer, H. (1989) Large scale experiment about improving the night-time conspicuity of trucks, *Proceedings of the 12th International Technical Conference on Experimental Safety Vehicles*, Washington, DC: National Highway Traffic and Safety Administration.

Schoettle, B., Sivak, M., and Flannagan, M.J. (2002) *High-Beam and Low-Beam Headlighting Patterns in the US and Europe at the Turn of the Millenium*, SAE paper 2002-01-0262, Warrendale, PA: Society of Automotive Engineers.

Schreuder, D.A. (1964) *The Lighting of Vehicular Traffic Tunnels*, Thesis, Centrex: Techische Hochschule, Eindhoven.

Schreuder, D.A. (1976) *White or Yellow Light for Vehicle Head-Lamps*, Voorburg, The Netherlands: Institute for Road Safety Research.

Schreuder, D.A. (1993) Energy saving in tunnel entrance lighting, *Proceedings of Right Light, the Second European Conference on Energy-Efficient Lighting*, Arnhem, The Netherlands: Netherlands Institute of Illuminating Engineering.

Schreuder, D.A. (1998) *Road Lighting for Safety*, London: Thomas Telford.

Schumann, J., Sivak, M., Flannagan, M.J., and Schoettle, B. (2003) *Conspicuity of Mirror-Mounted Turn Signals*, UMTRI-2003-26, Ann Arbor, MI: University of Michigan Transportation Research Institute.

Schwab, R.N., and Mace, D.J. (1987) Luminance measurements for signs with complex backgrounds, *Proceedings of the CIE, 21st Session, Venice*, Vienna: CIE.

Scott, P. P. (1980) *The Relationship between Road Lighting Quality and Accident Frequency*, TRRL Laboratory Report 929, Crowthorne, UK: Transport and Road Research Laboratory.

Sekular, R., and Blake, R. (1994) *Perception*, New York: McGraw-Hill.

Shlaer, S. (1937) The relation between visual acuity and illumination, *Journal of General Physiology*, 21, 165–168.

Silverstone, B., Lang, M.A., Rosenthal, B.P., and Faye, E.E. (2000) *The Lighthouse Handbook on Vision Impairment and Vision Rehabilitation,* New York: Oxford University Press.

Simons, R.H., and Bean, A.R. (2000) *Lighting Engineering*, London: Butterworth-Heinemann.

Sivak, M. (2002) How common sense fails us on the road: Contribution of bounded rationality to the annual worldwide toll of one million traffic fatalities, *Transportation Research, Part 5*, 259–269.

Sivak, M., and Flannagan, M.J. (1993) A fast rise brake light as a collision-prevention device, *Ergonomics*, 36, 391–395.

Sivak, M., and Olson, P.L. (1985) Optimal and minimal luminance characteristics for retro-reflective highway signs, *Transportation Research Record*, 1027, 53–57.

Sivak, M., Olson, P., and Pastalan, L. (1981) Effect of driver's age on nighttime legibility of highway signs, *Human Factors*, 23, 59–64.

Sivak, M., Simmons, C.J., and Flannagan, M. (1990) Effect of headlamp area on discomfort glare, *Lighting Research and Technology*, 22, 49–52.

Sivak, M., Flannagan, M.J., Ensing, M., and Simmons, C.J. (1991) Discomfort glare is task dependent, *International Journal of Vehicle Design*, 12, 152–159.

Sivak, M., Flannagan, M., and Gellatly, W. (1993) Influence of truck driver eye position on effectiveness of retro-reflective traffic signs, *Lighting Research and Technology*, 25, 31–36.

Sivak, M., Campbell, K., Scheider, L., Sprague, J., Streff, F., and Waller, P.F. (1995) The safety and mobility of older drivers: What we know and promising research issues, *UMTRI Research Reviews*, 26, 1–24.

Sivak, M, Flannagan, M.J., Traube, E.C., Hashimoto, H., and Kojima, S. (1996) *Fog Lamps: Frequency of Installation and Nature of Use*, UMTRI-96-31, Ann Arbor, MI: University of Michigan Transportation Research Institute.

Sivak, M., Flannagan, M.J., Traube, E.C., and Kojima, S. (1998) Automobile rear signal lamps: Effects of realistic levels of dirt on light output, *Lighting Research and Technology*, 30, 24–28.

Sivak, M., Flannagan, M.J., Miyokawa, T., and Traube, E.C. (1999) *Color Identification in the Visual Periphery: Consequences for Color Coding of Vehicle Signals*, UMTRI-99-20, Ann Arbor, MI: University of Michigan Transportation Research Institute.

Sivak, M., Flannagan, M.J., and Miyokawa, T. (2001) A first look at visually aimable and internationally harmonized low-beam headlamps, *Journal of the Illuminating Engineering Society*, 30, 26–33.

Sivak, M., Flannagan, M.J., Schoettle, B., and Mefford, M.L. (2002) *Driving Performance With and Preference For HID Headlamps*, UMTRI 2002-3, Ann Arbor, MI: University of Michigan Transportation Institute.

Sivak, M., Schoettle, B., Flannagan, M.J., and Minoda, T. (2004) *Optimal Strategies for Adaptive Curve Lighting*, UMTRI 2004-22, Ann Arbor, MI: University of Michigan Transportation Institute.

Sivak, M., Schoettle, B., and Flannagan, M.J. (2006a) Mercury-free HID headlamps: Glare and colour rendering, *Lighting Research and Technology*, 38, 33–40.

Sivak, M., Schoettle, B., and Flannagan, M.J. (2006b) *Mirror-Mounted Turn Signals and Traffic Safety*, UMTRI-2006-23, Ann Arbor, MI: University of Michigan Transportation Research Institute.

Sivak, M., Schoettle, B., Flannagan, M.J., and Minoda, T. (2006c) Effectiveness of clear-lens turn signals in direct sunlight, *Leukos*, 2, 199–209.

Sliney, D. Fast, P., and Ricksand, A. (1995) Optical radiation hazards analysis of ultraviolet headlamps, *Applied Optics*, 34, 4912–4922.

Smith, F.C. (1938) Reflection factors and revealing power, *Transactions of the Illuminating Engineering Society (London)*, 3, 196–200.

Society of Automotive Engineers (SAE) (1990) *Lighting Committee Summary of DRL Tests*, Warrendale, PA: SAE.

Society of Automotive Engineers (SAE) (1995) *Harmonized Vehicle Headlamp Performance Requirements, SAE J1735*, Warrendale, PA: SAE.

Society of Automotive Engineers (SAE) (2001) *Ground Vehicle Lighting Standards Manual, SAE HS-34*, Warrendale, PA: SAE.

Society of Light and Lighting (SLL) (2006) *Factfile 2: Car Park Lighting — Dilemma Solved*, London: CIBSE.

Sorensen, K. (1974) *Report 7: Description and Classification of Light Reflection Properties of Road Surfaces*, Lyngby, Denmark: The Danish Illuminating Engineering Laboratory.

Sorensen, K. (1975) *Report 10: Road Surface Reflection Data*, Lyngby, Denmark: The Danish Illuminating Engineering Laboratory.

Sorensen, K. (1977) Road lighting for wet conditions, *Measures of Road Lighting Effectiveness*, Berlin: Lichttechnische Gesellschaft LiTG.

Stahl, F. (2004) *Kongruenz des blickverlaufs, bei virtuellen und raelen autofahrten — Validierung eines nachtfahrsimulators*, Diplomarbeit, Ilmenau, Germany: Ilmenau Technische Universitat.

Stanton, N.A., and Pinto, M. (2000) Behavioural compensation by drivers of a simulator when using a vision enhancement system, *Ergonomics*, 43, 1359–1370.

Stiles, W.S. (1930) The scattering theory of the effect of glare on the brightness difference threshold, *Proceedings of the Royal Society London*, 105B, 131–146.

Stiles, W.S., and Crawford, B.H. (1937) The effects of a glaring light source on extrafoveal vision, *Proceedings of the Royal Society London*, 112B, 255–280.

Sullivan, J., and Flannagan, M.J. (2001) *Reaction Time to Clear-Lens Turn Signals Under Sun-Loaded Conditions*, UMTRI-2001-30, Ann Arbor, MI: University of Michigan Transportation Research Institute.

Sullivan, J.M., and Flannagan, M.J. (2002) The role of ambient light level in fatal crashes: Inferences from daylight saving time transitions, *Accident Analysis and Prevention*, 34, 487–498.

Sullivan, J.M., and Flannagan, M.J. (2004) *Visibility and Rear-End Collisions Involving Light Vehicles and Trucks,* UMTRI-2004-14, Ann Arbor, MI: University of Michigan Transportation Research Institute.

Sullivan, J.M., and Flannagan, M.J. (2007) Determining the potential safety benefit of improved lighting in three pedestrian crash scenarios, *Accident Analysis and Prevention*, 39, 638–647.

Sullivan, J.M., Adachi, G., Mefford, M.L., and Flannagan, M.J. (2004a) High-beam headlamp usage on unlighted rural roadways, *Lighting Research and Technology*, 36, 59–67.

Sullivan, J.M., Bargman, J., Adachi, G., and Schoettle, B. (2004b) *Driver Performance and Workload Using a Night Vision System*, UMTRI 2004-8, Ann Arbor, MI: University of Michigan Transportation Institute.

Sumner, R., Baguley, C., and Burton, J. (1977) *Driving in Fog on the M4.* Supplementary Report 281, Crowthorne, UK: Transport Research Laboratory.

Tanner, J.C., and Harris, A.J. (1956) Comparison of accidents in daylight and in darkness, *International Road Safety and Traffic Review*, 4, 11–14, 39.

Tansley, B.W., and Boynton, R.M. (1978) Chromatic border perception: The role of red- and green-sensitive cones, *Vision Research*, 18, 683–697.

Taylor, G.W., and Ng, W.K. (1981) *Measurement of Effectiveness of Rear-Turn-Signal Systems in Reducing Vehicle Accidents from an Analysis of Actual Accident Data*, SAE Technical Paper Series 810192, Warrendale. PA: Society of Automotive Engineers.

Teichner, W., and Krebs, M. (1972) The laws of simple reaction time, *Psychological Review*, 79, 344–358.

Tenkink, E. (1988) Lane keeping and speed choice with restricted sight, in T. Rothengatter and R. de Bruin (eds.) *Road User Behaviour: Theory and Research*, Assen/Maastricht, The Netherlands: Van Gorcum.

Theeuwes, J., and Alferdinck, J.W.A.M. (1996) *The Relation Between Discomfort Glare and Driver Behavior,* Report DOT HS 808 452, Washington, DC: US Department of Transportation.

Theeuwes, J., and Riemersma, J. (1995) Daytime running lights as a vehicle collision countermeasure: The Swedish evidence reconsidered, *Accident Analysis and Prevention*, 27, 633–642.

Thompson, P.A. (2003) *Daytime Running Lamps (DRLs) for Pedestrian Protection*, SAE Technical Paper 2003-01-2072, Warrendale, PA: Society of Automotive Engineers.

Tregenza, P.R. (1987) Daylight availability at low solar altitudes, *Proceedings of the CIE, Venice*, Vienna: CIE.

Trezona, P.W. (1991) A system of mesopic photometry, *Color Research and Application*, 16, 202–216.

Troutbeck, R., and Wood, J.M. (1994) Effect of restriction of view on driving performance, *Journal of Transportation Engineering*, 120, 737–752.

Tsimhoni, O., Flannagan, M.J., and Minoda, T. (2005) *Pedestrian Detection with Night Vision Systems Enhanced by Automatic Warnings*, UMTRI-2005-23, Ann Arbor, MI: University of Michigan Transportation Research Institute.

Tuaycharoen, N., and Tregenza, P.R. (2005) Discomfort glare from interesting images, *Lighting Research and Technology*, 37, 329–341.

Turner, D., Nitzburg, M., and Knoblauch, R. (1998) Ultraviolet headlamp technology for nighttime enhancement of roadway markings, *Transportation Research Record*, Paper 98-1187, Washington, DC: Transportation Research Board.

Ueki, K., Ohkubo, N., and Fujimura, H. (1992) Motorway accidents in tunnels in relation to lighting, *Road Lighting as an Accident Counter-Measure*, CIE Publication No. 93, Vienna: CIE.

Van Bommel, W.J.M., and de Boer, J.B. (1980) *Road Lighting*, London: The MacMillan Press.

Van Bommel, W.J.M., and Tekelenburg, J. (1986) Visibility research for road lighting based on a dynamic situation, *Lighting Research and Technology*, 18, 37–39.

Van den Berg, T.J.T.P. (1993) Quantal and visual efficiency of fluorescence in the lens of the human eye. *Investigative Ophthalmology and Vision Science,* 34, 3566–3573.

Van den Berg, T.J.T.P., IJspeert, J.K., and de Waard, P.W.T. (1991) Dependence of intraocular straylight on pigmentation and transmission through the ocular wall, *Vision Research*, 31, 1361–1367.

Van Derlofske, J., and Bullough, J.B. (2003) *Spectral effects of high-intensity discharge automotive forward lighting on visual performance*, SAE Technical Paper Series 2003-01-0559, Warrendale, PA: Society of Automotive Engineers.

Van Derlofske, J., Bullough, J.B., and Hunter, C.M. (2001) *Evaluation of high-intensity discharge automotive forward lighting*, SAE Technical Paper Series 2001-01-0298, Warrensburg, PA: Society of Automotive Engineers.

Van Derlofske, J., Boyce, P.R., and Hunter C.M. (2003) Evaluation of in-pavement warning lights on pedestrian crosswalk safety, *IMSA Journal*, 20 and 54.

Van Derlofske, J., Bullough, J.B., Dee, P., Chen, J., and Akashi, A. (2004) *Headlamp parameters and glare*, SAE Technical Paper Series 2004-01-1280, Warrendale, PA: Society of Automotive Engineers.

Van Derlofske, J., Chen, J., Bullough, J.D., and Akashi, Y. (2005) *Headlight glare exposure and recovery*, SAE Paper 05B-269, Warrendale, PA: Society of Automotive Engineers.

Van Derlofske, J., Bullough, J.B., and Gribbin, C. (2007) Comfort and visibility characteristics of spectrally-tuned high intensity discharge forward lighting systems, *European Journal of Scientific Research*, 17, 73–84.

Van Houten, R., Healy, K., Malenfant, J.E.L., and Retting, R. (1998) The use of signs and signals to increase the efficacy of pedestrian-activated flashing beacons at crosswalks, *Proceedings of the 77th Transportation Research Board*, Washington, DC: Transportation Research Board.

Van Lierop, F.H., Rojas, C.A., Nelson, G.J., Dielis, H., and Suijker, J.L.G. (2000) 4,000K low wattage metal halide lamps with ceramic envelopes: A breakthrough in color quality, *Journal of the Illuminating Engineering Society*, 29, 83–88.

Van Nes, F.L., and Bouman, M.A. (1967) Spatial modulation transfer in the human eye, *Journal of the Optical Society of America*, 47, 401–406.

Varady, G., Freiding, A., Eloholma, M., Halonen, L., Walkey, H., Goodman, T., and Alferdinck, J. (2007) Mesopic visual efficiency III: Discrimination threshold measurements, *Lighting Research and Technology*, 39, 355–364.

Viikari, M., Eloholma, M., and Halonen, L. (2005) 80 years of V (λ) use: A review, *Light and Engineering*, 13, 24–36.

Voevodsky, J. (1974) Evaluation of a deceleration warning light in reducing rear end collisions, *Journal of Applied Psychology*, 59, 270–273.

Vos, J.J. (1984) Disability glare — a state of the art report, *CIE Journal*, 3, 39–53.

Vos, J.J. (1999) Glare today in historical perspective: Towards a new CIE glare observer and a new glare nomenclature, *Proceedings of the CIE 24th Session, Warsaw*, Vienna: CIE.

Vos, J.J., and Boogaard, J. (1963) Contribution of the cornea to entoptic scatter, *Journal of the Optical Society of America*, 53, 869–873.

Vos, J.J., and Padmos, P. (1983) Straylight, contrast sensitivity and the critical object in relation to tunnel lighting, *Proceedings of the CIE 20th Session, Amsterdam*, Vienna: CIE.

Wada, T., Kurokawa, K., Itozawa, K., and Saito, H. (2006) *Development of a New Instrument Cluster with Electrochromic Device (ECD)*, SAE Paper 2006-01-0945, Warrendale, PA: Society of Automotive Engineers.

Wald, G. (1945) Human vision and the spectrum, *Science*, 101, 653–658.

Waldram, J.M. (1938) The revealing power of street lighting installations, *Transactions of the Illuminating Engineering Society (London)*, 3, 173–186.

Waldram, J.M. (1972) The calculation of sky haze luminance from street lighting, *Lighting Research and Technology*, 4, 21–26.

Walker, M.F. (1977) The effects of urban lighting on the brightness of the night sky, *Publications of the Astronomical Society of the Pacific*, 89, 405–409.

Walker, T. (2007) Remote monitoring systems assessed, *Lighting Journal*, 72, 49–53.

Wang, J.S., and Knipling, R.R. (1994) *Lane Change/Merge Crashes: Problem Size Assessment and Statistical Description*, Report DOT-HS-808075, Washington, DC: US Department of Transportation.

Watkinson, J.C. (2005) *Additivity of Discomfort Glare*, MS Thesis, Rensselaer Polytechnic Institute, Troy, NY.

Weale, R.A. (1992) *The Senescence of Human Vision*, Oxford, UK: Oxford University Press.

Wells, S., Mullin, B., Norton, R., Langley, J., Connor, J., Lay-Yee, R., and Jackson, R. (2004) Motorcycle rider conspicuity and crash related injury: Case-control study, *British Medical Journal*, 328, 857–862.

Werner, J.S., Peterzell, D.H., and Scheetz, A. J. (1990) Light, vision and aging. *Optometry and Visual Science*, 67, 214–229.

Wertheim, A.H. (1981) On the relativity of perceived motion, *Acta Psychologica*, 48, 97–110.

Whillans, M.G. (1983) Colour-blind driver's perception of traffic signals, *Canadian Medical Association Journal*, 128, 1187–1189.

White M.E., and Jeffrey D.J. (1980) *Some Aspects of Motorway Traffic Behavior in Fog*, Report LR 958, Crowthorne, UK: Transport Research Laboratory.

Whitlock and Weinberger Transportation (1998) *An Evaluation of a Crosswalk Warning System Utilizing In-Pavement Flashing Lights*, Santa Rosa, CA: Whitlock and Weinberger Transportation.

Whittaker, J. (1996) An investigation into the effects of British summer time on road traffic accident casualties in Cheshire, *Journal of Accident and Emergency Medicine*, 13, 189–192.

Wierda, M. (1996) Beyond the eye: Cognitive factors in drivers' visual perception, in A.G. Gale, I.D. Brown, C.M. Haslegrave, and S.P. Taylor (eds.) *Vision in Vehicles-V*, Amsterdam: North-Holland.

Williams, T.D. (1983) Aging and central visual field area, *American Journal of Optometry and Physiological Optics*, 60, 888–891.

Wolf, E., and Gardiner, J.S. (1965) Studies on the scatter of light in the dioptric media of the eye as a basis of visual glare, *Archives of Ophthalmology*, 74, 338–345.

Wolfe, J.M., Kluender, K.R., Levi, D.M., Bartoshuk, L.M., Herz, R.S., Klatzky, R.L., and Lederman, S.J. (2006) *Sensation and Perception*, Sunderland, MA: Sinauer Associates Inc.

Wood, J.M. (2002) Age and visual impairment decrease driving performance as measured on a closed-road circuit, *Human Factors*, 44, 482–494.

Wordenweber, B., Wallaschek, J., Boyce, P., and Hoffman, D.D. (2007) *Automotive Lighting and Human Vision*, Berlin: Springer.

Wulf, G., Hancock, P.A., and Rahirni, M. (1989) Motorcycle conspicuity: An evaluation and synthesis of influential factors, *Journal of Safety Research*, 26, 153–176.

Wyszecki, G. (1981) Uniform color spaces, *Golden Jubilee of Colour in the CIE*, Bradford, UK: The Society of Dyers and Colourists.

Yannis, G., Papadimitrou, E., Lejeune, P., Treny, V., Hemdorff, S., Bergel, R., Haddak, M., Hollo, P., Cardoso, J., Bijleveld, F., Houwing, S., and Bjornskau, T. (2005) *State of the Art Report on Risk and Exposure Data*, Deliverable 2.1 Project 506723 Building the European Road Safety Observatory, Brussels: European Union Directorate-General Transport and Energy.

Yerrell, J.S. (1971) *Headlamp Intensities in Europe and Britain, LR 383*, Crowthorne, U.K: Road Research Laboratory.

Yerrell, J.S. (1976) Vehicle headlamps, *Lighting Research and Technology*, 8, 69–79.

Index

Q

W

X

Y

For Product Safety Concerns and Information please contact our EU representative GPSR@taylorandfrancis.com / Taylor & Francis Verlag GmbH, Kaufingerstraße 24, 80331 München, Germany